Electric Aircraft Dynamics

Electric Aircraft Dynamics

A Systems Engineering Approach

Ranjan Vepa

CRC Press
Taylor & Francis Group
Boca Raton London New York

CRC Press is an imprint of the
Taylor & Francis Group, an **informa** business

First edition published 2020
by CRC Press
6000 Broken Sound Parkway NW, Suite 300, Boca Raton, FL 33487-2742

and by CRC Press
2 Park Square, Milton Park, Abingdon, Oxon, OX14 4RN

Library of Congress Cataloging-in-Publication Data

Names: Vepa, Ranjan, author.
Title: Electric aircraft dynamics : a systems engineering approach / Ranjan
Vepa.
Description: First edition. | Boca Raton, FL : CRC Press, 2020. | Includes
bibliographical references and index. | Identifiers: LCCN 2020009505 (print) |
LCCN 2020009506 (ebook) | ISBN
9780367194246 (hardback) | ISBN 9780429202315 (ebook)
Subjects: LCSH: Electric airplanes--Design and construction. |
Airplanes--Motors. | Airplanes--Electric equipment.
Classification: LCC TL683.3 .V47 2020 (print) | LCC TL683.3 (ebook) | DDC
629.134/35--dc23
LC record available at https://lccn.loc.gov/2020009505
LC ebook record available at https://lccn.loc.gov/2020009506

ISBN: 978-0-367-19424-6 (hbk)
ISBN: 978-0-367-51358-0 (pbk)
ISBN: 978-0-429-20231-5 (ebk)

Typeset in Times
by Deanta Global Publishing Services, Chennai, India

Instructors, to access downloadable PowerPoint Lecture Slides please visit www.routledgetextbooks.com/
textbooks/instructor_downloads/.

To my father, Narasimha Row, and mother, Annapurna

Contents

Preface..xv
Acronyms...xvii

Chapter 1 Introduction 1

    1.1   Introduction to Electric Aircraft...1
    1.2   The Systems Engineering Method 1
    1.3   Hybrid and All-Electric Aircraft: Examples...........................3
    1.4   Battery Power 4
    1.5   Range and Endurance of Electric Aircraft..............................4
    1.6   Propulsion Motors 5
    1.7   Propellers, Aeroacoustics and Low Noise Design..................6
    1.8   Electric Propulsion Issues 6
    1.9   Key Technology Limitations 7
    1.10   Future Work..7
    Chapter Summary..8
    References 8

Chapter 2 Electric Motors..11

    2.1   Introduction to DC Motors...11
          2.1.1   DC Motor Principles 11
          2.1.2   DC Motor Characteristics 13
          2.1.3   Classification of DC Motors....................................14
          2.1.4   Dynamic Modeling of DC Motors 14
          2.1.5   Control of DC Motors 16
    2.2   Introduction to AC Motors 18
          2.2.1   Synchronous Motors..18
          2.2.2   Three-Phase Motors..19
          2.2.3   Loading and Back-EMF in Synchronous Motors 20
          2.2.4   Characteristics of AC Motors..................................20
          2.2.5   Induction Motors 21
          2.2.6   Squirrel-Cage Rotor 21
          2.2.7   Controlling AC Motors 22
    2.3   Reluctance Motors: Reluctance Principle..............................24
          2.3.1   Types of Construction 24
          2.3.2   Reluctance Torque...25
          2.3.3   Switched Reluctance Motor 26
          2.3.4   Operation of a Switched Reluctance Motor 26
    2.4   Brushless DC Motors 26
          2.4.1   Brushless or Electronic Commutation 27

2.4.2   Dynamic Modeling 27
2.4.3   Switching and Commutation......................................27
Chapter Summary...28
References 28

Chapter 3   Batteries...29

3.1   Introduction to Batteries..29
3.1.1   Battery Structure and Specifications.........................30
3.1.2   Rechargeable Batteries 31
3.1.3   Charge, Capacity and Discharge Features 32
3.1.4   Temperature Effects and Capacity Fading 32
3.2   Battery Dynamic Modeling: Physical, Empirical, Circuit
        and Hybrid Models 33
3.2.1   Battery SOC Estimation..35
3.3   Types and Characteristics of Batteries 37
3.3.1   Lithium-Ion (Li-Ion) Batteries 40
3.3.2   Gel Polymer Electrolytes..42
3.3.3   Lithium–Sulfur (Li–S) Batteries...............................42
3.3.4   Metal-Air and Li-Air Batteries 43
3.4   Applications...46
3.4.1   Batteries for Electric Aircraft 46
Chapter Summary...46
References 46

Chapter 4   Permanent Magnet Motors and Halbach Arrays.....................49

4.1   Motors for All-Electric Propulsion....................................49
4.2   High Torque Permanent Magnet Motors.............................49
4.2.1   Rare Earth Elements 49
4.2.2   Neodymium Magnets and Samarium–Cobalt
         Magnets 49
4.3   Magnetic and Electromagnetic Effects 49
4.3.1   Magnetic Materials on a Microscopic Scale..............49
4.3.2   Diamagnetism 50
4.3.3   Paramagnetism...50
4.3.4   Remnant Magnetic Moment.......................................51
4.3.5   Ferromagnetism..51
4.3.6   Curie Temperature..51
4.3.7   Magneto-Striction 52
4.3.8   Ferrimagnetism 52
4.3.9   The Maximum Energy Product...................................52
4.3.10   Coercivity 52
4.3.11   High Temperature Coercivity....................................53
4.3.12   Curie Temperature of NdFeB....................................53
4.3.13   Intrinsic Coercivity..53

	4.3.14	Intrinsic and Normal Coercivity Compared............... 53
	4.3.15	Permanent Magnets with Reduced Rare Earth Elements 53
4.4	Halbach Array Motors.. 54
	4.4.1	Complex Halbach Arrays 55
	4.4.2	Ring Type Structures... 57
4.5	Modeling the Magnetic Field Due to a Halbach Array 57
Chapter Summary... 59
References 60

Chapter 5   Introduction to Boundary Layer Theory and Drag Reduction........... 61

5.1	Principles of Airfoil and Airframe Design 61
5.2	Flow Over an Aerofoil.. 61
5.3	Aerodynamic Drag 62
5.4	Boundary Layer Flow.. 66
	5.4.1	The Navier–Stokes (NS) Equations 67
	5.4.2	Viscous Energy Dissipation 69
	5.4.3	Non-Dimensionalizing and Linearizing the NS Equations.. 70
	5.4.4	Analysis in the Boundary Layers 71
	5.4.5	Boundary Layer Equations.. 74
	5.4.6	Vorticity and Stress in a Boundary Layer 75
	5.4.7	Two-Dimensional Boundary Layer Equations........... 76
	5.4.8	The Blasius Solution.. 77
	5.4.9	The Displacement, Momentum and Energy Thicknesses 80
5.5	Computation of Boundary Layer Velocity Profiles 82
	5.5.1	The von Karman Method: The Integral Momentum Equation... 82
	5.5.2	Wall Shear Stress, Momentum Thickness, Displacement Thickness and Boundary Layer Thickness for the Blasius Solution 86
	5.5.3	The Methods of Pohlhausen and Holstein and Bohlen 88
	5.5.4	Refined Velocity Profiles within the Boundary Layer.. 91
	5.5.5	Laminar Boundary Layers: Integral Methods Using Two Equations... 97
	5.5.6	Effect of Suction, Blowing or Porosity....................... 99
	5.5.7	Reduction of the Equations 100
	5.5.8	Special Cases... 102
	5.5.9	Thwaites Correlation Technique 102
5.6	Transition and Separation... 104
	5.6.1	Walz–Thwaites' Criterion for Transition/Separation......106
	5.6.2	The Transitional Boundary Layer 108

5.7   Turbulent Boundary Layers.. 109
         5.7.1   Predicting the Turbulent Boundary Layer................ 111
         5.7.2   The Entrainment Equation Due to Head.................. 111
         5.7.3   Drela's Method for a Turbulent Boundary Layer 113
5.8   Strategy for Aircraft Drag Reduction.................................... 114
Chapter Summary... 115
References 115

Chapter 6   Electric Aircraft Propeller Design 117

6.1   Introduction 117
6.2   Aerofoil Sections: Lift and Drag.. 118
6.3   Momentum Theory... 120
6.4   Actuator Disk... 121
6.5   Blade Element Theory 123
6.6   Dynamics and Modeling of the Inflow................................ 126
6.7   Integrating the Thrust and Torque....................................... 127
6.8   Blade Element Momentum Theory 128
         6.8.1   Application to Ducted Propellers 137
6.9   Lifting Line Theory.. 140
6.10   Blade Circulation Distribution: Potential Flow-Based
         Solutions 148
6.11   Standard Propeller Features and Design Considerations 151
6.12   Propellers for Distributed Propulsion.................................. 152
Chapter Summary... 153
References 153

Chapter 7   High Temperature Superconducting Motors.................................. 155

7.1   High Temperature Superconductors (HTS)........................... 155
         7.1.1   The Meissner State and the Meissner Effect............ 156
         7.1.2   Features of Superconducting Materials.................... 156
7.2   HTS Motors.. 157
         7.2.1   HTS DC Motors 159
         7.2.2   HTS Synchronous and Induction Motors................ 160
         7.2.3   Cryostats for HTS Motors 161
         7.2.4   Control of 3-Phase HTS PMSM............................. 162
7.3   Homopolar Motors 163
         7.3.1   Superconducting Homopolar Motors 165
7.4   Design of HTS Motors for Aircraft Propulsion 165
Chapter Summary... 165
References 166

Chapter 8   Aeroacoustics and Low Noise Design... 169

8.1   Aeroacoustic Analogies.. 169
         8.1.1   Sound Pressure Level.. 173

8.2   Integral Methods of Lighthill, Ffowcs Williams and Hawkings, and Kirchhoff 173

8.3   Monopoles, Dipoles and Quadrupoles 177

       8.3.1   Tonal Characterization of Aeroacoustically Generated Noise 179

8.4   Application to Propellers and Motors 180

       8.4.1   Sources of Airfoil and Propeller Noise 181

       8.4.2   Hamilton-Standard Procedure for Estimating the Noise Due to Propeller Aerodynamic Loading 184

8.5   Theoretical Modeling of the Noise Fields 186

       8.5.1   Theoretical Modeling of the Propeller Noise Fields 186

       8.5.2   Farassat's Formulation of the FW–H Equation 190

       8.5.3   Formulation of the Far-Field Noise Based on a Rotating Source 191

       8.5.4   Lilley's Analogy and Its Application to Ducts 195

Chapter Summary 203

References 203

Chapter 9   Principles and Applications of Plasma Actuators 207

9.1   Flow Control and Plasma Actuation 207

9.2   Passive Methods of Flow Control 208

       9.2.1   Riblets 209

       9.2.2   Dimples 210

       9.2.3   Fences 210

       9.2.4   Vortex Generators (VGs) and Micro-VGs 210

       9.2.5   Vortilons 211

       9.2.6   Winglets 212

       9.2.7   Cavities 213

       9.2.8   Gurney Flaps 213

9.3   Passive Methods Coupled with Plasma Actuation 214

9.4   Reduction of Skin-Friction Drag by Feedback 215

       9.4.1   Feedback Control of Transition 216

       9.4.2   Modeling the Flow Due to DBD Plasma Actuators 219

       9.4.3   Decomposition of Simulated Flow Features 224

       9.4.4   Application of Wavelet Decomposition and De-Noising 225

       9.4.5   A Review of Wavelet Decomposition Based on the Wavelet Transform 226

       9.4.6   Application to the Regulation of Laminar Flow over an Airfoil 229

9.5   Control Laws for Active Flow Control 234

       9.5.1   Integral Equations for the Boundary Layer 236

       9.5.2   The Inverse Boundary Layer Method: Uniform Solutions 238

       9.5.3   Uniform and Prescribed Shape Factor 239

9.5.4   The Vorticity–Velocity Formulation with Control
        Flow Inputs.. 241
9.5.5   Active Control of Velocity Profiles 243
9.5.6   Hybrid Active Laminar Flow Control with
        Plasma Actuation..244
9.5.7   Application of the Control Laws to a Typical
        Airfoil... 245
Chapter Summary... 253
References 254

Chapter 10 Photovoltaic Cells.. 259

10.1   History of the Photoelectric Effect................................... 259
10.2   Semiconductors: Silicon Photo Diodes 259
10.3   Photoconductive Cells 261
10.4   The Photovoltaic Effect 262
       10.4.1   The Photovoltaic Cell: The Solar Cell 263
       10.4.2   Solar Cell Characteristics.................................... 265
       10.4.3   Modeling the Power Output of a Solar Cell 266
       10.4.4   Maximum Power Point Tracking 268
       10.4.5   The Shockley–Queisser Limit............................. 269
10.5   Multi-Junction Silicon PV Cells 270
       10.5.1   Modeling the Power Output of Multi-Junction
                Cells... 271
Chapter Summary... 273
References 273

Chapter 11 Semiconductors and Power Electronics 275

11.1   Semiconductors and Transistors....................................... 275
       11.1.1   Semiconductors and Semiconductor Diodes............ 275
       11.1.2   Transistors 276
11.2   Power Electronic Devices... 277
       11.2.1   Power Diodes: A Three-Layered Semiconductor
                Device... 280
       11.2.2   Thyristors and Silicon Controlled Rectifier (SCR) 281
       11.2.3   Controlled Devices: GTO and GTR....................... 286
       11.2.4   The MOSFET.. 286
       11.2.5   The IGBT 287
       11.2.6   Applications.. 288
Chapter Summary... 291
References 291

Chapter 12 Flight Control and Autonomous Operations 293

12.1   Introduction to Flight Control 293
        12.1.1   Range and Endurance of an Electric Aircraft.......... 293
        12.1.2   Equivalent Air Speed, Gliding Speed and
                    Minimum Power to Climb 296
12.2   Flight Path Optimization.. 299
        12.2.1   The Optimal Control Method................................... 302
        12.2.2   Cruise Optimization: Optimal Control
                    Formulation 302
        12.2.3   Optimization Procedure: Optimum Cruise
                    Velocity, Optimum Trajectory Synthesis 304
        12.2.4   Modeling with the Peukert Effect 307
12.3   Integrated Flight and Propulsion Control............................... 309
        12.3.1   Model-Based Design of Control Laws for
                    Distributed Propulsion-Based Flight Control........... 311
12.4   Flight Management for Autonomous Operation.................... 312
        12.4.1   Autonomous Control Systems 313
        12.4.2   Route Planning... 314
        12.4.3   Mission Planning for Autonomous Operations........ 314
        12.4.4   Systems and Control for Autonomy 315
12.5   Flight Path Planning.. 315
        12.5.1   Path Planning in Three Dimensions Using
                    a Particle Model 315
        12.5.2   Path Planning in the Horizontal Plane.................... 318
        12.5.3   Path-Following Control 320
Chapter Summary.. 320
References 321

Index.. 323

Preface

This book grew out of my interest in electric aircraft, which in turn was spurred by a concern for the rapidly deteriorating environment worldwide and the extreme impact this was having on weather conditions around the globe. Dealing with such a broad subject was a daunting task, and over the years I found that there were many different facets to the subject. However, after careful considerations of my interests, particularly in aiding the design of new aircraft for commercial airlines, I decided that I should adopt a systems engineering approach. I also felt that there was a need for greater emphasis on the expected developments which would facilitate an all-electric commercial flight. These interests and my personal perceptions on the need for new developments in electrical energy storage, high performance electric motors and new and optimized propeller designs for propelling the aircraft and reduction in the overall aerodynamic drag of the aircraft, as well as the noise pollution it could generate, led to the conception of this book. The book is conveniently organized into 12 chapters.

Several traditional areas of aeronautics, like aerodynamic design, structural design and aeroelastic analysis, as well as several others, have not been covered in the book. It does not mean, however, that these areas are any less important. There are several interesting and valuable books that have recently been published in these areas. Yet, it is essential that there are significant breakthroughs in certain areas, if electric aircraft are to be developed for large-scale civilian transportation. It is for this reason the book focuses on areas where fundamental breakthroughs are essential. It is expected that with these fundamental breakthroughs and with tools such as multi-disciplinary design optimization, a new generation of electric aircraft can be designed and built for future mass transportation of civilian passengers.

I thank my colleagues in the School of Engineering and Materials Science, at Queen Mary University of London and Prof. V. V. Toropov, in particular, for his support in this endeavor. I also express my special thanks to Jonathan Plant and Kyra Lindholm of CRC Press for their support in this endeavor.

I thank my family for their support, love, understanding and patience.

Ranjan Vepa
London, UK

Acronyms

AC	Alternating current
BCS	Bardeen, Cooper and Schrieffer
BEM	Blade element momentum
BJT	Bipolar junction transistor
BPF	Blade passing frequency
CG	Center of gravity
CWT	Continuous wavelet transform
DBD	Dielectric-barrier discharge
DC	Direct current
DEP	Distributed electric propulsion
DLR	Deutsches Zentrum für Luft- und Raumfahrt
DOD	Depth of discharge
DTC	Direct torque control
DWT	Discrete wavelet transform
EMF	Electromotive force
FET	Field effect transistor
FFT	Fast Fourier transform
FIR	Finite impulse response
FOC	Field oriented control
FW–H	Ffowcs Williams and Hawkings
GPE	Gel polymer electrolytes
HTS	High-temperature superconductors
IBM	International Business Machines
IGBT	Insulated gate bipolar transistor
JFET	Junction field effect transistor
kV	kilo Volt
kW	kilo Watt
LREE	Light rare earth elements
MD	McDonnell Douglas
MEMS	Microelectromechanical systems
MOSFET	Metal oxide semiconductor field effect transistor
MPP	Maximum power point
MPPT	Maximum power point tracking
MTOW	Maximum take-off weight
NACA	National Advisory Committee on Aeronautics
NS	Navier–Stokes
OC	Open circuit
PAN	Polyacrylonitrile
PEM	Proton exchange membrane
PEO	Polyethylene oxide
PIV	Particle image velocimetry
PMMA	Polymethyl methacrylate

PRM	Probabilistic road map
PV	Photovoltaic
PVdF	Polyvinylidene fluoride
PWM	Pulse width modulation
PZT	Lead (Pb) zirconate titanate
RRT	Rapidly exploring random tree
SCEPTOR	Scalable Convergent Electric Propulsion Technology Operations Research
SCR	Silicon controlled rectifier
SLAM	Simultaneous localization and mapping
SOC	State of charge
STFT	Short-time Fourier transform
TP-BVP	Two-point boundary value problem
TSP	Traveling salesman problem
UAV	Unmanned aerial vehicle
VG	Vortex generator
VSI	Voltage source inverter
WKB	Wentzel–Kramers–Brillouin
YBCO	Yttrium barium copper oxide

1 Introduction

1.1 INTRODUCTION TO ELECTRIC AIRCRAFT

Ever-increasing energy demands and rising fuel prices have motivated aircraft industries to develop alternative power sources for future aircraft. In the aviation sector, the requirements of flight reliability, low noise-emission levels, reduction in the dependence on fossil fuels, requirements of lower costs and lower weight, and longer life cycles, increase the complexity of the overall system and lead manufacturers to introduce major breakthroughs and innovations. Hybrid electric propulsion and all-electric propulsion for future aircraft are currently popular fields in the aircraft industry and are forming the basis for future commercial aircraft designs. Hybrid electric propulsion systems are composed of a networked set of gas turbines and batteries, while in all-electric propulsion systems, batteries are the only source of propulsive power on aircraft. Current battery technologies pose the most serious limitations to the development of all-electric and more-electric aircraft. However, rapid strides are being made in the evolution of battery technology and, for this reason, most aircraft industries are planning to introduce either more-electric or all-electric powered aircraft within the next two decades. Electric aircraft will pose new problems related to general aircraft architecture, geometry and shape, battery, motor, and propulsion system design, aerodynamics, drag reduction and boundary layer control, aircraft performance, stability and control, the design of flight controllers, optimum structural design, and a host of other issues. In this book, we hope to bring together a number of current aspects of electric aircraft that are being extensively researched within the aerospace community.

The main goal of this book is to bring together the theoretical and design issues relevant to the field of electric aircraft and present the key topics on the most pressing problems that designers are facing in making electric aircraft more popular and commercially viable. It is also intended to identify the current state-of-the-art and new developments in the research on all-electric propulsion, hybrid electric propulsion and more-electric propulsion, as well as on the impact of the associated systems on all aspects of electric aircraft design.

1.2 THE SYSTEMS ENGINEERING METHOD

A "system" is an interacting combination of elements, viewed in relation to a function. A typical example of a system is an aircraft and, in our context, an electric aircraft. The elements or subsystems that make up the electric aircraft are the propulsion subsystem, the power storage system, the power supply systems, the power converters, the propeller, the lifting bodies and wings and the control surfaces which constitute the major elements that must be synergistically combined to produce an electric aircraft. Another example is the unmanned aerial vehicle (UAV) which again is composed of a host of subsystems which when assembled together perform holistically the functions of the UAV. These systems are, vehicle airframe structure and aerodynamics, propulsion, avionics systems, flight controls and high lift, flight management, mission management, fuel or power source management, environment control for avionics bays, hydraulic and/or pneumatic systems, electrical and power distribution systems, auxiliary and emergency power, recovery, communication and navigation, flight management/flight control, electronic flight information systems, recording and telemetry, landing gear, safety systems, built-in test equipment, maintenance and computing and software.

Systems engineering (SE) is an interdisciplinary field focusing on design and management of complex systems over their life cycle from a holistic perspective. Systems-thinking principles and modeling methods are used to organize this process. An SE approach ensures that all aspects of a project or system are being considered and included for an integrated solution. A variety of competences such as: requirements engineering, cross-disciplinary team coordination, testing and evaluation (V & V), reliability and maintainability (dependability) and topics from other disciplines are important when dealing with complex technological systems. When designing a new aircraft, there is a need to define the basic system at the overall system level. An example is,

> The purpose of the aircraft is to carry 100 passengers and 1,000 kg of cargo a total distance of 400 miles at a Mach number of 0.2. The aircraft will operate primarily in sandy desert climates and at cruise altitudes of 10,000 to 15,000 ft.

Performance requirements establish how well a system or subsystem should perform, and constraints define the limits on the performance. They include broad specs: weight, maximum take-off weight (MTOW), center of gravity (CG) location, dimensions, reliability, safety and environment. At the subsystem level specification should contain some detail about its function and performance. For example, "The purpose of the subsystem is to protect the environmental control system from damage due to particulates in a sandy desert environment. The subsystem will remove at least 90% of the particulates from the air it receives." In addition to function and performance, generally the statement should include parameters such as: weight, cost, electrical loads and air distribution.

Clearly one is dealing with a large system and it is essential to approach the study of such a large system, such as an electric aircraft, in a structured and systematic way. First and foremost, systems engineering requires a great deal of coordination across disciplines as there is a need to deal with a large number of possibilities for design trade-offs across subsystems. Faults arising from poor design can propagate across the subsystems and are generally difficult to predict. Moreover, the mutual lack of understanding across engineering disciplines can lead to several problems and in particular, the resistance to changes in design principles and practices that one has to introduce to cover technological changes that are expected in the future. Any engineer acts as a systems engineer when responsible for the design and implementation of a total system. The difference with "traditional engineering" lies primarily in the greater emphasis on defining the goals and decomposing the system into subsystems, creating and generating a number of alternative designs, evaluating all available alternative designs and then coordinating and controlling the diverse tasks that are necessary to create a complex system. The role of systems engineer is one of a manager who utilizes a structured value delivery process. Thus, the role of the systems engineer "is to define the system goals and the conditions under which the system must operate so that the designer is free to create the best system possible." Systems engineering involves a process of systematic decomposition of the intended product or design into the subsystems and provides a balanced and disciplined approach to integration of the interacting component subsystems with the aim of delivering the synergistic goals of the system.

In the rest of this book, the systems engineering approach was adopted to decompose the overall electric aircraft into key subsystems so as to highlight and address some of the key technological challenges.

1.3 HYBRID AND ALL-ELECTRIC AIRCRAFT: EXAMPLES

Over a decade ago the Boeing company made history by showcasing a fuel-cell powered demonstrator aircraft [1]. A two-seat Dimona airplane, built by Diamond Aircraft Industries of Austria, was used as the airframe. With a 16.3 m wingspan, it was modified at Boeing's European research laboratory to include a proton exchange membrane fuel cell/lithium-ion battery hybrid system to power an electric motor coupled to a conventional propeller. During the flights, which took place at the airfield in Ocana, Spain, pilot Cecilio Barberan climbed to an altitude of 3,300 ft (1,000 m) above sea level using a combination of battery power and power generated by hydrogen fuel cells. Then, after reaching the cruise altitude and disconnecting the batteries, Barberan maintained level flight at a speed of about 60 mph (100 kph) for 20 minutes on fuel-cell generated power alone.

A year earlier, an unmanned jet powered by hydrogen fuel-cell technology, the *Hyfish* (as it was named) had taken flight near Bern in Switzerland. It was a cooperative project between the German Aerospace Center, called the Deutsches Zentrum für Luft- und Raumfahrt or DLR in Stuttgart, Germany, and its international partners, including Horizon Fuel Cell Technologies of Singapore, culminating in the maiden flight of the *Hyfish* in 2007. Scientists at the DLR Institute for Technical Thermodynamics in Stuttgart integrated Horizon Fuel Cell Technologies' ultra-light, compact fuel-cell system into this next-generation UAV, while keeping the total system weight to 13.2 lb. The *Hyfish* fuselage is about 4 ft long, and its wings are about 3 ft wide. During the flight, the UAV performed vertical climbs, loops and other aerial acrobatics at speeds reaching 124 mph, making the *Hyfish* the first fast plane with jet wings to fly with a hydrogen fuel cell as its only power source [2].

In November 2009, United Technologies Research Center (UTRC), the central research and innovation arm of United Technologies Corporation, achieved first flight of a hydrogen/air fuel cell powered rotorcraft. The successful technology demonstration was accomplished using a remote-controlled electric helicopter model modified to incorporate a custom proton exchange membrane (PEM) fuel cell power plant [3].

The power plant was a PEM fuel-cell prototype developed by the UTRC with a high-pressure hydrogen source and air were used. The self-sustained system with the power plant automatically started with the hydrogen supply and no additional batteries, and it was capable of carrying a 2.5 kg payload with a maximum output power of 1.75 kW and the system power density exceeded 500 W/kg.

In 2012, the use of efficient electric motors and batteries allowed pilots Bertrand Piccard and André Borschberg to keep the four-engine Solar Impulse aircraft aloft throughout the hours of darkness during a flight from Switzerland to Madrid that took 17 hours. After a change of pilot, the aircraft spent a further 19 hours in the air before landing in Morocco. The Solar Impulse has a wing span of 61 m, which is comparable with a commercial airliner, but at 1,500 kg the solar-powered plane weighs the same as a family car [4].

The first Airbus E-Fan prototype, which seats only one, was powered entirely by batteries. Powered by two lithium batteries, which provide 60 kW of power [5], the newer two-seater Airbus E-Fan prototype traveled from Lydd in Kent to Calais, France, a distance of 74 km, in 2015. With a wingspan of just under 10 m, the 600 kg aircraft was able to achieve speeds close to 200 km per hour at a cruising altitude of about 1,000 meters and managed to complete the journey in just 37 minutes [6].

The world's first four-seat passenger aircraft powered by a zero-emission hydrogen fuel-cell propulsion, accomplished a successful first public flight in 2016. It took off from the airport in Stuttgart and successfully performed a short 15-minute flight [7].

These developments have spurred several aircraft design teams to redouble their efforts to design an all-electric civil passenger aircraft for the future. The development of electric automobiles has certainly given a boost to the design of a fully battery-powered all-electric aircraft.

1.4 BATTERY POWER

The key to future all-electric flight is undoubtedly the limitations of current battery technologies. The use of batteries for electric power storage and to facilitate electric propulsion would signal a quantum leap toward the future and will gain commercial acceptance. Battery technology is already playing a key role in the development of electric cars. But to be useful in powering aircraft the energy density of current batteries must go up by a factor of ten. Innovative alternatives are being considered over segments of a typical flight which would require large amounts of power such as take-off and landing.

1.5 RANGE AND ENDURANCE OF ELECTRIC AIRCRAFT

Several studies have been conducted to maximize the range and endurance of electric aircraft. The problem is that electric aircraft cannot shed their weight as can a typical fossil fuel-powered aircraft and for this reason continues to maintain its weight right through the flight. Moreover, there is a need for fast-charging technologies, as well as technologies that would reduce the consumption of electric power during the take-off, cruise and landing phases of a typical flight by adopting optimizing strategies over all phases of a typical flight.

1.6 PROPULSION MOTORS

The US National Aeronautics and Space Administration (NASA) has launched by far the most ambitious programs to help the development of electric propulsion for aircraft. The Scalable Convergent Electric Propulsion Technology Operations Research, the SCEPTOR X-Plane Project, a three-year research project to achieve the first distributed electric propulsion-manned flight demonstrator in 2017, launched by NASA, was aimed at enhancing cruise efficiency at high flight speeds [8]. The goals of the project were to lower energy use by a factor of five, to raise the motor/controller/battery conversion efficiency from 28% to 92%, and to ensure that the benefits of integration were achievable. It was also desired to lower the operating costs substantially and achieve zero in-flight carbon emissions.

The NASA SCEPTOR flight demonstration project [9] aimed to retrofit an existing internal combustion engine-powered light aircraft with two types of distributed electric propulsion motors with several small "high-lift" propellers distributed along the leading edge of the wing. These propellers were designed to accelerate the flow over the wing at low speeds. Two much larger cruise propellers located at each of the wingtips were designed to provide the primary propulsive power. The aim of the high-lift system was to reduce the wing area by a factor of 2.5 compared to the original aircraft, thus facilitating the reduction of drag at the cruise speed. It was also aimed to increase the velocity at the maximum lift-to-drag ratio, while maintaining low-speed performance. The wingtip-mounted cruise propellers were designed to interact with the wingtip vortices, enabling a further increase in the efficiency and consequently a reduction of 10% in the maximum propulsive power required. The design approach was meant to create an aircraft that consumes an estimated 4.8× less energy at the selected cruise point when compared to the original aircraft. Stoll et al. [10] have also researched the applications of distributed electric propulsion to personal air vehicles. Kim, Perry and Phillip [11] have discussed one of the advantages of distributed electric propulsion systems utilizing boundary layer ingestion beneficially for improved propulsive efficiency and reduced turbulent kinetic energy losses in the wake of the vehicle. They expect additional aero-propulsive benefits to include the use of blown surfaces to locally increase dynamic pressure across aerodynamic surfaces, as well as the reduction of aircraft-induced drag through interactions between wingtip propulsors and the wing trailing vortex system.

The Maxwell X-57's Mod II vehicle [12], a derivative of the SCEPTOR program, features the replacement of traditional combustion engines on a baseline Tecnam P2006T aircraft, with electric cruise motors. The key features of the Maxell X-57 are its high aspect ratio wing and its 14 propellers (12 high-lift propellers along the leading edge of the wing for distributed electric propulsion and two large wingtip cruise propellers). The high-lift motors are only used to develop the additional lift required during take-off and landing while the wingtip cruise motors are the main propulsive motors during cruise. Both its original wing and the two gasoline-fueled piston engines were replaced with a large aspect ratio wing embedded with the 14 electric motors—12 on the leading edge driving the high lift propellers for take-offs and landings, and one larger motor on each wing tip for use while at the cruise altitude. A goal of the X-57, which was delivered to NASA in 2019, is to help develop certification standards for emerging electric aircraft markets, including urban air mobility vehicles, which also rely on complex distributed electric propulsion systems.

1.7 PROPELLERS, AEROACOUSTICS AND LOW NOISE DESIGN

A secondary objective of the SCEPTOR project was to lower the community noise by 15 dB, to provide a certification basis for distributed electric propulsion and validate the flight control system redundancy, its robustness and reliability, with improved ride quality. It is expected that NASA's all-electric X-57 plane could reduce noise, improve efficiency and reduce the environmental impact of air travel in the long run. Bonni et al. [13] have proposed a novel design of a conceptual low noise hybrid passenger aircraft. In their paper, Graham, Hall and Morales [14] discussed the potential contribution of future technology to reducing the levels of three key aircraft emissions: carbon dioxide, oxides of nitrogen and noise. They have reviewed several current and ongoing aircraft developmental projects and have presented their prognosis for the future. Rizzi et al. [15] have been considering the noise generated by aircraft such as the X-57. They believe that the prediction of noise generated by the high-lift system is not trivial. In addition to both broadband and tonal propeller source noise, there are numerous installation effects including propeller–propeller, propeller–nacelle and propeller–wing interactions, and other noise sources including electric motor and airframe noise, and wingtip cruise propeller noise, that must be assessed.

1.8 ELECTRIC PROPULSION ISSUES

The use of electric propulsion and the design of all-electric aircraft necessitates the redesign of the flight control laws, the redesign of the controllers for landing gears, doors and other such heavy-duty actuation tasks, the use of regenerative braking of the electric motors for controlling the thrust and speed of the aircraft, the application of electrically driven pumps and heaters for fuel pumping, propulsion ancillaries including motor starting, cabin pressurization, air conditioning, de-icing and other similar secondary tasks. Nearly all aircraft loads will require power converters for enhanced and continuous control. One is expected to take into account some of the characteristics of electrical actuation, such as better controllability and the availability of service on demand, faster response times, reconfigurability, maintaining functionality during faults, advanced diagnostics and prognostics techniques, intelligent maintenance and increased availability. While these features are expected to reduce overall operating costs and the environmental impact, they will increase the demands on the batteries. De-centralization of the power supplies and the use of multiple hybrid sources of power, including a limited use of fossil fuels over certain phases of the flight, are expected to relieve the load on the electrical power storage, power conditioning and supply units.

Another important consequence of the use of propulsion-based flight control is expected to be coupling between the longitudinal and lateral flight control systems, due to the fact that propeller thrust, where it is used for longitudinal or lateral control, is always dependent on air density, which in turn depends on altitude, a state of the longitudinal dynamic model. Thus, it is important that all flight controllers are carefully redesigned so as to account for this coupling when distributed electric propulsion with propulsion-based flight control is employed.

1.9 KEY TECHNOLOGY LIMITATIONS

Given that aviation fuel has a specific energy of 12 kWh per kilogram of fuel, the key limitation for electric aircraft is the energy density of its batteries. At the current time, some of the best lithium-ion and lithium-sulfur batteries have a specific energy of only 0.45 kWh per kilogram of the fuel. Although this is adequate for automobiles, and has already proved viable in cars, even for short flights up to 400 nautical miles, one would need batteries with energy densities up to 2 kWh per kilogram of the fuel, otherwise the weight of the batteries would be prohibitive. This would mean a complete rethink of the aerodynamic design of aircraft, as there is now a significant increase in the lift demanded from the wings. For this reason, designers have proposed rather radical solutions, such as the box wing by Siqueira et al. [16]. The closed box wing electric commuter aircraft concept was designed by Siqueira et al. [16]. The primary feature of the closed box wing is that it essentially functions like a high aspect ratio wing, but has a much lower wing span. Thus, it has the advantage of the structural design being simpler and with better stability characteristics in addition to the overall lower span of the vehicle.

1.10 FUTURE WORK

The study of all-electric aircraft design undertaken in this work has facilitated new directions of research and development which must be pursued actively. Apart from the areas already highlighted in relation to battery technology, power electronics, flight control system design, actuators for electric aircraft and optimum low noise propeller designs, there is a substantial need for the complete aerodynamic redesign of the aircraft, with the aim of drastically reducing the overall drag force generated by the aircraft, as this would be the major factor in determining the propulsive power required to propel the aircraft forward. At the same time there is also a need for configurations that would substantially increase the overall lift generated by the aircraft. Further research into these aspects can dramatically reduce the need for batteries with high energy densities beyond what is expected to be available a few decades from now.

Innovative methods of take-off and landing, including the use of ground-based power supplies for take-off, the use of permanently stationed air-winches powered by ground power supplies, which could winch the aircraft vertically into the sky prior to its setting off in cruise mode and the use of power-free gliding to land the aircraft may have to be considered to reduce the aircraft dependence on power supplies. Yet one could easily observe that the day is not far off when electric aircraft as well as all-electric personal air vehicles will be in regular usage for short- and medium-range travel while highly energy-efficient hybrid electric aircraft would be used for long-distance and inter-continental travel.

CHAPTER SUMMARY

In this chapter the notion of an all-electric aircraft was introduced and the key issues that may have to be addressed in their design are briefly identified. A systems approach is adopted and the aircraft is assumed to be decomposed into a number of key subsystems which will be discussed in the forthcoming chapters.

REFERENCES

1. Anon, Boeing fuel cell plane in manned aviation first, *Fuel Cells Bull.*, 2008(4):1, April 2008, doi:10.1016/S1464-2859(08)70146-9, Accessed December 29, 2019.
2. C. E. Howard, Hydrogen fuel-cell technology takes off, powering Hyfish UAV, *Mil. Aerosp. Electron.*, June 2007, https://www.militaryaerospace.com/, Accessed December 29, 2019.
3. Anon, UTRC pioneers fuel cell-powered rotorcraft flight, United Technologies Research Center, East Hartford, CT, http://www.utrc.utc.com/, Accessed December 29, 2019.
4. B. Woods, Plane completes 17-hour flight without fuel, https://www.zdnet.com/, Accessed December 29, 2019.
5. K. Tweed, Airbus's e-fan electric plane takes flight, *IEEE Spectrum*, https://spectrum.ieee.org/, Accessed December 29, 2019.
6. Anon, *Airbus Group's All-Electric E-Fan Aircraft Completes Historic Channel Crossing*, AIRBUS, https://www.airbus.com/newsroom, Accessed December 29, 2019.
7. Anon, First 4-seat aircraft powered by hydrogen fuel cells takes off today, http://hy4.org/zero-emission-air-transport-first-flight-of-four-seat-passenger-aircraft-hy4/, Accessed December 29, 2019.
8. M. D. Moore, *Distributed Electric Propulsion (Dep) Aircraft*, NASA Langley Research Center, 2015, https://aero.larc.nasa.gov/files/2012/11/Distributed-Electric-Propulsion-Aircraft.pdf, Accessed December 29, 2019.
9. N. K. Borer, M. D. Patterson, J. K. Viken, M. D. Moore, S. Clarke, M. E. Redifer, R. J. Christie, A. M. Stoll, A. Dubois, J. Bevirt, Design and performance of the NASA SCEPTOR distributed electric propulsion flight demonstrator, 16th AIAA Aviation Technology, Integration, and Operations Conference, June 13–17, 2016, Washington, DC. https://ntrs.nasa.gov/search.jsp?R=20160010157, Accessed December 29, 2019.
10. A. M. Stoll, J. Bevirt, P. P. Pei, E. V. Stilson, Conceptual design of the Joby S2 electric VTOL PAV, Aviation Technology, Integration, And Operations Conference, June 16–20, 2014, Atlanta, Georgia. https://www.jobyaviation.com/S2ConceptualDesign(AIAA).pdf, Accessed December 29, 2019.
11. H. D. Kim, A. T. Perry, A. Phillip, *A Review of Distributed Electric Propulsion Concepts for Air Vehicle Technology*, NASA Armstrong Flight Research Center, Edwards, CA, https://ntrs.nasa.gov/archive/nasa/casi.ntrs.nasa.gov/20180004729.pdf, Accessed December 29, 2019.
12. Anon, *NASA Takes Delivery of First All-Electric Experimental Aircraft*, Armstrong Flight Research Center, Edwards, CA, http://www.nasa.gov/x57, Accessed December 29, 2019.
13. J. Bonni, B. Burghoff, E. Dincer, J. F. L. Eichler, T. Ferguson, H. Mirza, M. Nuno, M. Y. Pereda, M. Valley, *Conceptual Design of a Low Noise Hybrid Passenger Aircraft*, Institute of Aeronautics and Astronautics, RWTH Aachen University, Aachen, Germany, https://www.dlr.de/content/de/downloads/2017/designentwurf-der-rwth-aachen-low-noise-hybrid-passenger-aircraft_2846.pdf?__blob=publicationFile&v=10, Accessed December 29, 2019.

14. W. R. Graham, C. A. Hall, M. V. Morales, The potential of future aircraft technology for noise and pollutant emissions reduction, Transport Policy Special Issue Aviation and the Environment, https://core.ac.uk/download/pdf/42337430.pdf, Accessed December 29, 2019.

15. S. A. Rizzi, D. L. Palumbo, J. Rathsam, A. Christian, M. Rafaelof, Annoyance to noise produced by a distributed electric propulsion high-lift system, AIAA Paper, NASA Langley Research Center, Hampton, VA, https://ntrs.nasa.gov/archive/nasa/casi.ntrs. nasa.gov/20170005762.pdf, Accessed December 29, 2019.

16. S. A. Siqueira, P. Skinner, A. M. Ibrahem, M. D. Liu, S. A. Barits, E. K. Bontoft, R. Vepa, V. V. Toropov, Design optimization of a closed box wing all-electric commuter aircraft concept, 2019 AIAA SciTech Forum, Session: MDO-05, Multidisciplinary Design Optimization for Vehicle Design II, Presentation: (Control ID 3041911), San Diego, CA, January 7–11, 2019.

2 Electric Motors

2.1 INTRODUCTION TO DC MOTORS

In this chapter an overview of electric motors, in particular those that could be used for the traction or propulsion of electric vehicles, their operational features and performance characteristics, will be briefly presented. It must be stated at the outset that there are a number of books on the basic principles and features of electric motors, such as Emadi [1], Filizadeh [2] and Piotr [3], on the modeling and control of electric motors such as Vepa [4], on the principles and operational features of reluctance motors such as Bilgin, Jiang and Emadi [5] and Miller [6], on the principle, modeling and control of brushless direct current (DC) motors, such as Krishnan [7] and the modeling and control of stepper motors, such as Athani [8]. It must also be said that there are indeed several other comprehensive books and papers on these subjects as well, which can be accessed from the references stated in the books cited above.

The aim of this chapter is to introduce the broad principles of motor dynamics and control, so as to facilitate the introduction of the dynamics and control of high-performance motors, in later chapters, which are particularly useful for the propulsion of all-electric aircraft.

2.1.1 DC MOTOR PRINCIPLES

The basic principle of torque generation, as a result of the principle of motor action, in a DC motor is well known. When a current carrying loop is placed in a magnetic field with its direction transverse to the plane of the loop, the loop experiences a net torque. Thus, if the loop is wound around a rotor, known as the armature, the net torque on the motor can be used to drive the rotor about its axis, provided the rotor is supported on bearings and housed in a fixed casing relative to the applied magnetic field. It is important that the direction of the current in the loop is continuously switched so as ensure that the torque acting on the armature is always unidirectional. An assembly of brushes and circumferentially split contacts known as a commutator are used for distributing the current to the conductors wound along the armature so the torque on each conductor about the rotor axis is always in the same direction, no matter what the azimuthal position of the conductor is on the surface of the rotor. As the conductors on one half of the rotor are under the influence of the one pole of a permanent magnet providing the magnetic field, while those on the other half are under the influence of the opposite pole, the commutator supplies the current such that all of the conductors are acted upon by a movement in the same direction, thus producing a net torque on the rotor shaft. The armature is constructed from an assemblage of laminated discs of alloy steel with slots on the surface parallel to the axis of rotation. Conductors forming coils are housed in the slots to form a complete winding. The torque induced on a conductor passing under the north pole of the permanent magnet supplying the magnetic field is in the opposite direction from that induced on a conductor situated under the south pole of the same permanent magnet. The whole purpose of the armature winding is to connect all the conductors so that the various induced torques shall always add up to a combined whole. No two conductors must be connected together in a way which would cause their induced torques to oppose each other, as such a connection would lead to an overall reduction in the induced torque. The resulting torque generated by a single coil is then given by

$$T = DF = BAI_a, \tag{2.1}$$

where D is the coil diameter and A is the coil area. The magnetic flux density may be related to the magnetic intensity due to the current flow as

$$B = \mu_0 \mu_r H, \tag{2.2}$$

where μ_0 is the permeability of free space, μ_r is the relative permeability and H is the magnetic field strength.

Alternatively, the product of the magnetic flux density and the area is the total external flux. For a large number of such rectangular coils, N, the number of coils is directly proportional to the circumference or diameter of the armature. Hence for a general motor armature the torque generated is

$$T = k_t DBAI_a = k_t D\Phi I_a. \tag{2.3}$$

One may define a *motor area coefficient* a_m as $a_m \approx A/D^2$, and write the torque as

$$T = k_t a_m D^3 BI_a. \tag{2.4}$$

Thus, the torque generated is directly proportional to the cube of the linear dimension. Hence it may be observed that size matters. A small motor with a high torque capability requires permanent magnets with a high magnetic field intensity. The discovery of magnetic alloy materials has greatly aided the development of small high torque permanent magnet motors. These are discussed in Chapter 4, in Section 4.2.

When the armature rotates, the conductors in the slots cut the lines of force of the magnetic field in which they revolve, so that an EMF is induced in the armature winding as in a generator. The induced EMF acts in opposition to the current in the machine and, therefore, to the applied voltage. Thus, it is customary to refer to this voltage as the back-EMF. The existence of the back-EMF may be deduced by Lenz's law, defining the direction of an induced EMF, which is such as to oppose the change causing it, that is, ultimately the applied voltage. Lenz's law states that: "An induced current is always in such a direction as to oppose the motion or change causing it."

The magnitude of the back-EMF may be calculated by using the formula for the induced EMF in a generator. It is important, in the case of a motor, to appreciate that this is proportional to the product of the flux and the speed. It is always less than the applied voltage (the difference is usually small) and it is the difference between these two quantities that actually drives the current in the armature winding. Since the back-EMF generated in a motor is directly proportional to the angular speed of the motor it may be expressed as

$$E_b = K_b\omega. \tag{2.5}$$

Given that the total resistance of the armature circuit is R_a and that the armature is driven by a constant external voltage V_a, the current in the armature may be expressed as

$$I_a = \frac{V_a - E_b}{R_a} = \frac{V_a - K_b\omega}{R_a}. \tag{2.6}$$

Thus, the torque acting on the armature is:

$$T = k_t a_m D^3 B\left(\frac{V_a - K_b\omega}{R_a}\right). \tag{2.7}$$

Theoretically, in view of the duality of electromagnetic induction and Lorentz's law, it may be shown that K_b is also given by:

$$K_b = K_t \equiv k_t a_m D^3 B. \tag{2.8}$$

Thus, we may express K_b in practice as

$$K_b = k_b k_t a_m D^3 B = k_b K_t. \tag{2.9}$$

The torque generated by the armature is:

$$T = k_t a_m D^3 B\left(\frac{V_a - K_b\omega}{R_a}\right) \equiv K_t\left(\frac{V_a - K_b\omega}{R_a}\right) = \frac{K_t}{R_a}\left(V_a - K_b\omega\right). \tag{2.10}$$

2.1.2 DC Motor Characteristics

Generally, the torque-speed characteristics are almost linear. A generic set of torque-speed, torque-current, torque-power and torque-efficiency characteristics are illustrated in Figure 2.1.

It is particularly important to note that when the torque exceeds a certain critical value known as the stall torque, the speed, power delivered and efficiency are equal to zero, although the current drawn is maximum. Thus, the motor is characterized by a maximum torque output, although this torque does not correspond to either the torque at the maximum power output or the torque when the efficiency is at maximum.

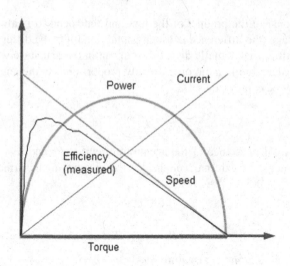

FIGURE 2.1   Typical set of DC motor characteristics.

2.1.3   CLASSIFICATION OF DC MOTORS

There are indeed several types of DC motors. They may be classified in different ways depending on the number of poles of the permanent magnets used in providing the magnet field, whether or not the motor uses a commutator, whether not the magnet used is a permanent magnet or a wound field electromagnet or the manner in which the armature conductors or coils and the wound field electromagnet are connected to a common DC supply, in series or in parallel, or to independent DC sources. Broadly, however, the different types of DC motors are illustrated in Figure 2.2.

Homopolar motors are discussed in Chapter 7. Universal motors have the capability to run from alternating current (AC) or DC.

2.1.4   DYNAMIC MODELING OF DC MOTORS

Considering a motor with a rotary inertia and viscous-friction load, the conditions for dynamic equilibrium are

$$T = \frac{K_t}{R_a}\left(V_a - K_b\omega\right) = J\frac{d\omega}{dt} + B_f\omega. \tag{2.11}$$

Assuming the motor is in steady state, and that the viscous friction is negligible in comparison with the back-EMF, the motor speed satisfies the relation, $\omega = V_a/K_b$. Thus, when the voltage across the armature is constant, one can expect the speed of the motor to also be constant. It also follows that the speed of the motor may be controlled by appropriately controlling the voltage across the armature. For the purpose of designing a controller for the motor, it is often essential to construct a realistic dynamic model of the motor and other components in the system, identify the key design parameters represented in the model and select them in some

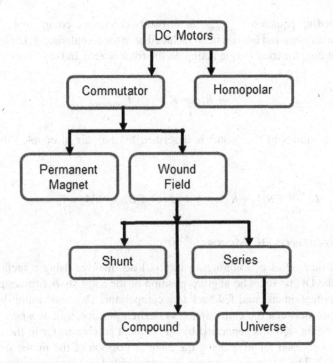

FIGURE 2.2 Types of DC motors.

optimal fashion so as to best realize the design objective. Thus, a starting point is the input-output description of the electrical and mechanical entities that represent the electro-mechanical energy conversion processes within the motor. Considering the case of transient armature currents, the mathematical model of the motor may be set up from the earlier equations, involving the applied voltage, armature current, motor speed and output torque. The total resistance of the armature circuit is given as R_a, the total inductance of the armature circuit is L_a and the armature is driven by a constant external voltage, V_a.

The current–voltage relationship in the armature may be expressed as

$$L_a \frac{dI_a}{dt} + R_a I_a = V_a - K_b \omega. \tag{2.12}$$

The torque acting on the armature is

$$T = K_t \times I_a + T_{\text{load}}, \tag{2.13}$$

where T_{load} is the external load torque acting on the rotor. Assuming the load torque is entirely due to the rotating inertial load on the rotor and the friction torque, the load torque, T_{load} is given by,

$$T_{\text{load}} = -J \frac{d\omega}{dt} - B_f \omega + T_{\text{disturbance}}. \tag{2.14}$$

In the preceding equation $T_{disturbance}$ is a disturbance torque component of the load which is usually ignored but has been included here for completeness. Under equilibrium conditions, the total torque acting on the rotor is zero and it follows that

$$J\frac{d\omega}{dt} + B_f\omega = K_t \times I_a + T_{disturbance}. \tag{2.15}$$

Thus, the dynamics of the motor is described by the pair of coupled differential equations,

$$L_a\frac{dI_a}{dt} + R_a I_a + K_B\omega = V_a, \quad J\frac{d\omega}{dt} + B_f\omega - K_t I_a = T_{disturbance}. \tag{2.16}$$

2.1.5 CONTROL OF DC MOTORS

A classical method of positioning an inertia load involves using a feedback loop enclosing the DC motor. The angular position of the rotor shaft is measured using an analog potentiometer and fed back to a comparator. The comparator determines the difference between the commanded or desired position signals where the commanded position signal is generated by an identical potentiometer in the command path as the one used for measuring the angular position of the motor shaft in the feedback path. The difference between the commanded or desired position signal, as well as a fraction of the velocity measurement, is fed back to the summing junction synthesizing the total input to the DC motor. The control input is then amplified by a pre-amplifier, followed by amplification by a power amplifier, and then applied to the voltage terminals of the DC motor. Thus, the difference between the output of the command potentiometer and the feedback potentiometer is essentially multiplied by the combined gain of the amplifiers and the error signal multiplied by the combined gain is then employed as the command input to the motor. The design of a feedback control system to position an inertia load employing an electric motor is probably one of the simplest and most common problems encountered. An important aspect of the control system design, apart from the selection of a suitable DC motor capable of achieving the dynamic response and matched to the objectives in cost, size, weight, torque, power and efficiency, is the selection of appropriate potentiometers, amplifiers and compensator circuits so as to generate the optimal control input.

Considering the velocity and position feedback loops, the applied voltage, V_a, may be expressed as

$$V_a = -K_p\left(\theta - \theta_d + K_d\omega\right), \tag{2.17}$$

where $d\theta/dt = \omega$. Using the D operator to represent the time derivative, $D \triangleq d/dt$, the motor equations may be expressed as

$$I_a = \frac{1}{L_a D + R_a}e = \frac{1}{L_a D + R_a}\left(V_a - K_b\omega\right), \quad \omega = \frac{1}{JD + B_f}\left(K_t I_a + T_D\right), \tag{2.18}$$

where T_D represents the disturbance torque. Eliminating the current I_a from the two equations for ω and I_a,

$$\omega = \frac{1}{JD + B_f}\left(K_t I_a + T_D\right) = \frac{1}{JD + B_f}\left(\frac{K_t}{L_a D + R_a}\left(V_a - K_b\omega\right) + T_D\right). \quad (2.19)$$

Re-arranging the terms,

$$\left(1 + \frac{1}{JD + B_f}\frac{K_t K_b}{L_a D + R_a}\right)\omega = \frac{1}{JD + B_f}\left(\frac{K_t V_a}{L_a D + R_a} + T_D\right). \quad (2.20)$$

Thus,

$$\omega = \frac{K_t V_a + T_D\left(L_a D + R_a\right)}{\left(JD + B_f\right)\left(L_a D + R_a\right) + K_t K_b}. \quad (2.21)$$

Assuming that the disturbance torque T_D is negligible,

$$\frac{\omega}{V_a} = \frac{K_t}{\left(JD + B_f\right)\left(L_a D + R_a\right) + K_t K_b}. \quad (2.22)$$

The preceding expression represents the transfer function relating the output angular velocity to the input armature voltage. The transfer function relating the angular position to the armature voltage is then given by

$$\frac{\theta}{V_a} = \frac{1}{D} \times \frac{K_T}{\left(JD + B_f\right)\left(L_a D + R_a\right) + K_T K_B}. \quad (2.23)$$

Introducing the feedbacks, the applied voltage, V_a, may be expressed in terms of the angular position as

$$V_a = -K_p\left(\left(1 + K_d D\right)\theta - \theta_d\right). \quad (2.24)$$

It follows that the angular position of the armature shaft is

$$\theta = -\frac{1}{D} \times \frac{K_T}{\left(JD + B_f\right)\left(L_a D + R_a\right) + K_T K_B} \times K_p\left(\left(1 + K_d D\right)\theta - \theta_d\right). \quad (2.25)$$

Hence, we obtain

$$\theta + \frac{1}{D}\frac{K_T K_p\left(1 + K_d D\right)}{\left(JD + B_f\right)\left(L_a D + R_a\right) + K_T K_B}\theta$$

$$= \frac{1}{D}\frac{K_T K_p}{\left(JD + B_f\right)\left(L_a D + R_a\right) + K_T K_B}\theta_d. \quad (2.26)$$

The transfer function of the closed loop system relating the angular position of the armature rotor to the commanded position:

$$\frac{\theta(t)}{\theta_d(t)} = \frac{K_T K_p}{D(JD + B_f)(L_a D + R_a) + DK_T K_B + K_T K_p(1 + K_d D)}. \tag{2.27}$$

The transfer function has the form of a ratio of two polynomials in the D operator. The stability of the closed loop system is guaranteed if the roots of the denominator, the characteristic polynomial, are all in the left half of the complex plane. Under steady state conditions we may set the time derivatives of $\theta(t)$ and $\theta_d(t)$ to zero. This is equivalent to setting $D = 0$, and in this case one has

$$\underset{t \to \infty}{\text{Limit}}\, \theta(t) = \underset{t \to \infty}{\text{Limit}}\, \theta_d(t), \tag{2.28}$$

provided the response of the closed loop system is stable and tends to a limit.

It should be observed that it is possible, in principle, to construct a reduced order model of the DC motor making assumptions such as $L_a \approx 0$.

2.2   INTRODUCTION TO AC MOTORS

Alternating-current (AC) motors may be divided broadly into two types, namely i) synchronous motors that run at a constant speed dependent only on the supply frequency, and ii) asynchronous motors that run at speeds not only dependent on the supply frequency, but also the load and, in some cases, the position of the brushes. The latter may be broadly further sub-divided into a) induction motors and b) commutator motors. The principal AC motors that are commonly used are the synchronous and the induction motors. Commutator motors are similar to DC motors as they use a commutator.

2.2.1   SYNCHRONOUS MOTORS

A synchronous motor has the same relationship to an AC generator or alternator as a DC motor has to a DC generator. In other words, if an AC generator is supplied with alternating current, it is capable of rotating as a motor and doing mechanical work. However, there is one important difference between the DC motor and the synchronous motor, apart from the fact the roles of the stator and rotor are interchanged. In a DC motor the applied DC voltage supplies both the field and the armature windings; in the synchronous motor two independent voltage sources, one DC and one AC, are required to excite the field or rotor windings and armature or stator windings, respectively. On one hand this feature is a disadvantage of a synchronous motor. But the advantages in using these fixed-speed motors in applications outweigh this disadvantage.

One method of providing the DC supply in large machines is to use an exciter driven directly by the shaft of the synchronous motor and coupled to it at one end, remote from the main motor. Thus, the exciter is connected through a commutator

system to the field or rotor windings. The main AC supply is connected to the armature or stator winding.

In order to maintain the driving torque on the rotor, the latter must rotate at a
speed such that its poles change positions at the same rate that the current in the
stator alternates with the supply frequency. Thus, a synchronous motor rotates at a
speed which is determined by the number of pole pairs in the rotor and the frequency
of the supply. This speed is called the synchronous speed. The motor runs only and
exclusively at this speed. It is natural to expect that the action of the motor is not self-
starting and certain distinctive methods have to be adopted to raise the speed from
zero to synchronism. The synchronous speed is given by

$$\omega_s = 2\pi f_s \left(2/P\right), \tag{2.29}$$

where f_s is the supply frequency and P is the number of stator pole pairs. In a synchronous motor a rotating field produced by means of polyphase alternating currents
is used to force a permanent magnet in the rotor to move in synchrony with the
applied field. The rotating field is analogous to that which would be produced by a
permanent magnet rotating uniformly about its midpoint, or by a revolving system
of poles excited by direct current, as in the case of the rotor of a synchronous motor,
with an electromagnet. Broadly, the different types of synchronous motors are illustrated in Figure 2.3.

2.2.2 THREE-PHASE MOTORS

In the case of a three-phase synchronous motor, the stator has a set of three pole
pairs, with each pole located at the vertex of a regular hexagon. The stator of such a

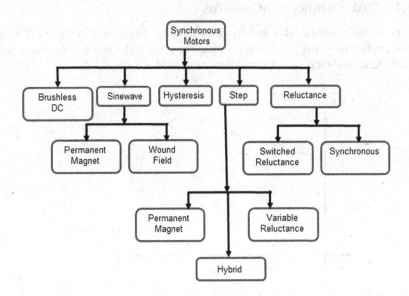

FIGURE 2.3 Types of synchronous motors.

three-phase motor produces a rotating field. The synchronous speed of rotation is the same for both the stator and the rotor, provided the supply frequency to the number of pole pairs is the same. It is clear that if the rotor in such a motor is brought up to the synchronous speed of the rotating field, it will remain locked to it and rotate synchronously with it.

2.2.3 Loading and Back-EMF in Synchronous Motors

When dealing with DC motors it may be recalled that the back-EMF was always approximately of the same order of magnitude as the applied voltage. Also, the small difference between the two was essentially the voltage driving the current in the motor. In the case of the synchronous motor as well, the back-EMF, which is now a rotating vector, is approximately equal in magnitude to the applied voltage vector. It is the vector difference between the applied voltage vector and the back-EMF vector that provides the driving voltage, which in turn drives the current in the stator windings. Once again, as in the DC motor, the back-EMF is practically a constant over a wide range of operating conditions so that a change in the field current produces a change in the opposite sense in the speed. In a no-load situation the input power represents the motor losses. The addition of mechanical load tends to increase the input power without altering the magnitude of the back-EMF vector. However, when the load on the synchronous motor is excessive, the motor seeks a decrease in the input power. But the situation is untenable, so when this condition is approached the rotor will fall out of synchronism and the machine comes to rest. The torque which causes the rotor to fall out of synchronism is called the pull-out torque.

2.2.4 Characteristics of AC Motors

A typical characteristic of a synchronous motor is illustrated in Figure 2.4. An important feature is the non-linearity when the rotor locks into synchronism, with the magnetic field rotating at the synchronous speed.

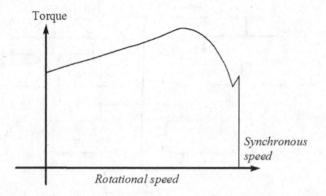

FIGURE 2.4   Torque-speed characteristic of a synchronous motor.

2.2.5 INDUCTION MOTORS

Induction motors may be designed to operate as single-phase, two-phase or as three-phase machines; the latter machine will be considered a typical example. While single-phase machines may be capacitor or shaded pole types, polyphase machines are wound rotor or squirrel-cage types. The stator of an induction motor is constructed as an assemblage of slotted stampings that house the stator windings. The ends of the windings are connected to a terminal box attached externally to the stator casing. The rotor is supported by a shaft that is fitted to a pair of bearings on either end. The bearings are supported by bearing housings attached to end shields that are bolted on both sides of the stator casing.

There are two types of rotor windings used in the construction of induction motors. These are known as the squirrel-cage type and the phase wound or wound type windings. The wound type rotor is provided with a distributed winding, usually consisting of three windings, one corresponding to each of three phases, the ends of which are typically connected to three insulated slip rings mounted on the rotor shaft. The object of this type of winding is to be able to connect to three-phase star connected resistance at the time of starting. The starter resistance is eventually disconnected and the slip rings are short-circuited during the normal operation of the induction motor.

2.2.6 SQUIRREL-CAGE ROTOR

The squirrel-cage type of rotor winding is usually confined to small induction motors due to the large current this type of motor requires during starting. An associated feature of this type of motor is the low starting torque provided by it. The main advantage of this type of rotor winding is that the motor is simple to construct, mechanically robust and efficient in operation. The action of the squirrel-cage induction motor, with a three-phase stator winding connected to a three-phase supply, also relies on the rotating field set up by the supply voltage. The rotating field induces an EMF in the conductors housed in the rotor. However, since the rotor winding is a closed circuit in normal operation, a relatively large current flows in it, the direction of which, by Lenz's law, is such as to oppose the change causing it.

As the change causing the current is the rotation of the stator field, the rotor attempts to run in the same direction as the field thereby reducing the relative angular velocity of the stator field with respect to the rotor. It is essential that in order to maintain the rotation a torque must be produced to overcome the load and cause the rotation. The torque is as a result of the currents flowing in the rotor conductors which are situated in and at right angles to the magnetic field. Now, if the rotor succeeds in completely catching up with the rotating field, as in the case of a synchronous motor, there would be no relative motion, no induced EMF, no current in the rotor windings and therefore no torque to maintain the rotation. The rotor of an induction never really runs as fast as the rotating field. There is a difference between the rotor speed and the synchronous speed of the rotating field. This difference is known as the slip speed and is usually expressed as a percentage of the synchronous speed of the rotating field. Thus, if for example, the synchronous speed of the rotating field is 2,000 rpm, and the actual rotor speed is 1,960 rpm, the difference is 40 rpm which is 2% of the synchronous speed. The induced EMF in the conductors housed in the rotors is directly proportional to the slip.

It follows that the torque required to maintain the rotation is also proportional to slip. Under no-load conditions, when the rotor only needs to overcome the friction torque to maintain a constant speed of rotation, the slip is relatively small. When the load on the motor is increased the slip also increases proportionately. The result is a reduced speed of operation.

Thus, the torque speed characteristics of an induction motor are, therefore, similar to those of DC shunt motor. If the torque is increased beyond a certain value, called the pull-out torque, the motor ceases to run. This pull-out torque is approximated to about two to three times the rated full load torque.

2.2.7 CONTROLLING AC MOTORS

To design a controller for an AC motor, the three-phase quantities must first be transformed to an equivalent two-dimensional stationary reference axis system by the Clarke transformation. The variable in the two-dimensional stationary frame are transformed into a synchronously rotating reference frame by the Park transformation. Both the Clarke and the Park transformations are discussed by Krishnan [7] and Vas [9]. After the control law synthesis, the three-phase AC control voltages that must be applied to the motor terminals are generated, so the values of the control voltages in the synchronously rotating reference frame should be transformed by the inverse Park and Clarke transformations. Achieving speed control of an electrical motor with or without the use of an external shaft position or velocity sensors is possible. "Sensorless" control (Vas [9]), is based on using the measurements from the stationary inputs and an accurate dynamic model to construct the control inputs, using a suite of estimators and filters to compensate for the absence of any measurements made on the rotating shaft.

The control input to the induction motor is the three-phase stator voltage, which is generated through the voltage source inverter in Figure 2.5. A three-phase voltage source inverter (VSI) is shown in Figure 2.6. Figure 2.6 shows a typical application of a three-phase inverter using six insulated gate bipolar transistors (IGBTs). Note that each phase uses a high-side and a low-side IGBT switch to apply positive and

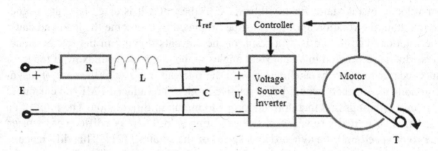

FIGURE 2.5   Controlling an induction motor.

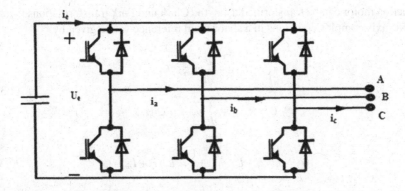

FIGURE 2.6   A typical voltage source inverter.

negative high-voltage DC pulses to the motor coils in an alternating mode. The VSI consists of three parallel output phases, each of which contains two IGBTs in parallel with diodes. The IGBT is a power transistor, which is used as a switch to conduct or block current in each of the output phases. In each of the phases, only one of the IGBTs is turned on at a time, while the other is turned off. If, for example, the upper IGBT in a phase is turned on and the lower is turned off, a positive current passes through the upper IGBT to the output. The electronic switch can block only one polarity of voltage, the one that keeps the diode reverse biased. (The operation of the IGBT is discussed in Chapter 11.)

In Figure 2.6, the output voltage to the motor is controlled by pulse width modulation (PWM). The inverter generates a waveform composed of many narrow pulses in each cycle and converts DC into AC. The width of switching pulses (i.e. the duty cycle) is varied in order to vary the average output voltage supplied to a motor. The IGBT blocks the flow of reverse current. In bridge circuits, where reverse current flow is needed, the additional diode in "anti-parallel" with the IGBT conducts current in the opposite direction. Each diode in the VSI provides a path through the diode of the bottom switch when the upper switch turns off, and vice versa. Thus, by applying positive base-to-emitter voltage of suitable magnitude to the transistor in the upper IGBT, the switch is turned on. Thus, by using PWM, the magnitudes of the outputs of the VSI are controlled and fed to the three phases, with the appropriate phase difference.

Typically, the controller is synthesized by a methodology known as field oriented control (FOC). The dynamics of the motor can be expressed in rotor or stator fixed coordinates by applying appropriate transformations. In rotor fixed coordinates it can be expressed in terms of two orthogonal components: the direct (d) and the quadrature (q) components. In rotor flux fixed coordinates, the induction motor resembles a DC motor, in that the torque is given as the product of the rotor flux magnitude and the quadrature current component. The rotor flux is only influenced by the direct current component. The control strategy of FOC is therefore to use the direct current component to regulate the rotor flux to a constant value and then to use the quadrature component of the stator current to control the torque as the two components can be modified independently. The control law synthesis is based on first transforming the three system into a stator-fixed reference system. The dynamics of a squirrel-cage induction motor can be transformed using the Clark and Park transformations to a set of two-axis complex equations in a stator-fixed reference frame given by

$$v_s = R_s \cdot i_s + \frac{d}{dt}\psi_s,$$

$$0 = R_r \cdot i_r + \frac{d}{dt}\psi_r - j\omega_r \cdot \psi_r,$$

(2.30)

$$\psi_s = L_s \cdot i_s + L_m \cdot i_r, \quad \psi_r = L_r \cdot i_r + L_m \cdot i_s,$$ (2.31)

$$T = \frac{3}{2} \cdot P \cdot \mathrm{Im}\left(\psi_s^* \cdot i_s\right).$$ (2.32)

In the above equations, where v_s denotes the complex stator voltage, ψ_s and ψ_r represent the complex stator and rotor flux, respectively. The complex quantities i_s and i_r are the stator and rotor currents. The quantities R_s and R_r are the stator and rotor resistances. L_s, L_r and L_m are stator, rotor and mutual inductances. Furthermore, ω_r is the electrical speed of the rotor, P is the number of pole pairs and T denotes the electromagnetic torque. In a rotor flux oriented synchronously rotating reference frame, where the "d-q" nomenclature represents the direct and quadrature components, the rotor flux can be commanded by the real part of the stator current or the "d" component i_{sd}. By applying the corresponding coordinate transformation, and forcing the quadrature component of the rotor flux ψ_{rq} to be null, rotor flux orientation is achieved. Then the direct component of the rotor flux, ψ_{rd} satisfies a first order differential equation and together with the quadrature component of the stator current i_{sq}, determines the control torque (see Vas [9]).

2.3   RELUCTANCE MOTORS: RELUCTANCE PRINCIPLE

The generation of reluctance torque is based on the fundamental tendency for electromagnetic flux to seek the path of least resistance. Consider the way a permanent magnet is attracted to a ferrous material, such as steel. The magnetic flux in the permanent magnet seeks out the steel as this provides a "low reluctance" path along which it can flow. This effect can also be used in an electric motor using electromagnets in the stator to attract low reluctance ferrous teeth in the rotor.

2.3.1   TYPES OF CONSTRUCTION

Two types of reluctance motor are possible. Synchronous reluctance motors use a stator very similar to that used in an induction motor. The stator is excited by an inverter drive applying sinusoidal currents to one motor phase after another, generally forming a three-phase system. The motor's rotor is salient, offering a low reluctance in one axis and a high reluctance in the other axis. The construction of the rotor allows it to lock synchronously onto the rotating stator field. These motors can be designed to have a high efficiency. However, they suffer from relatively low torque densities which may be a limiting factor in electric vehicle applications.

Essentially the reluctance motor looks like an induction motor with a modified squirrel-cage rotor. It can operate as a single-phase or three-phase motor. The rotating field generated by the stator results in a reluctance torque on the rotor, which pulls the rotor along into synchronous rotation. The rotor turns in synchrony with the rotating magnetic flux. The salient poles which offer low reluctance catch up with the poles of the stator to maintain synchronism with stator's rotating flux, with no load (and no slip). An increase in the load forces the rotor to lag behind the stator and the slip is non-zero. The lag cannot exceed a certain maximum value. There are essentially three types of rotors for reluctance motors: i) notch type, ii) flat type, and iii) barrier-slot type. These are illustrated in Figure 2.7. In the notched type, the notched regions are high reluctance pathways for the flux lines and the salient poles are the "pole" areas. The number of salient poles in this type of rotor is equal to the number of poles in the stator. In the barrier-slot rotor, the barriers are around the central region, while the slots and poles are in the periphery.

2.3.2　Reluctance Torque

The reluctance torque or alignment torque is the torque generated as a result of reluctance between the stator field and the ferromagnetic stator core. It is the main driving torque acting on the rotor in a reluctance motor. It cannot exceed a certain maximum value. The stator of the reluctance motor is similar to a single-phase induction motor and the rotor is generally a squirrel cage.

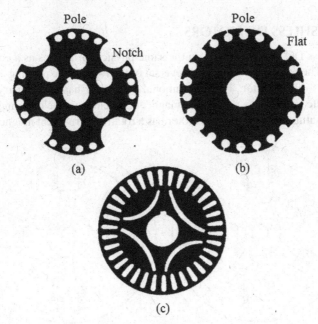

FIGURE 2.7 Types of rotors for reluctance motors: (a) notch type, (b) flat type, and (c) barrier-slot type.

2.3.3 SWITCHED RELUCTANCE MOTOR

The switched reluctance motor produces torque using a different mechanism from the motors previously described. The rotor has saliency and interspersed regions of high and low reluctance. In this case, control is provided by an uncommon form of power converter, known as an asymmetric half-bridge converter. The switched reluctance motor is a particular type of synchronous machine that has the wound field coils of a DC motor for its stator windings and has no coils or magnets on its rotor. In this machine, both the stator and rotor have salient poles. Hence, the machine is a doubly salient machine. One of the applications of a switched reluctance motor is in the construction of a stepper motor. Further details about stepper motors may be found in Athani [8].

The switched reluctance motor can be inexpensive, robust and produce high torque densities.

2.3.4 OPERATION OF A SWITCHED RELUCTANCE MOTOR

Referring to Figure 2.8, initially the stator poles cc' are aligned with the rotor pole pairs $s's$. Then the stator coils on aa' are switched on. The stator poles aa' are then aligned with rotor pole pairs rr' and a flux is established through stator poles a and a' and rotor poles r and r'. The flux tends to pull the rotor poles r and r' and align them with the stator poles a and a', respectively. When they are aligned, the stator current of the coil aa' is turned off and the stator coil b is excited, pulling s' and s toward b and b', respectively, in a clockwise direction. Thus, by switching the stator currents in a sequence, the rotor is continuously in rotation.

2.4 BRUSHLESS DC MOTORS

The brushless DC motor operates on the same physical principles as a conventional DC motor. The only difference is the reversal of the roles of the rotating and stationary elements of the motor. The conventional DC motor has a stationary magnetic field, generated either by a permanent magnet or a field wound electromagnet, and a rotating armature. A brushless DC motor has a rotating permanent magnet while the

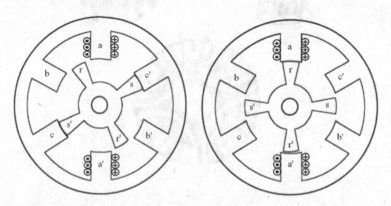

FIGURE 2.8   Operation of a switched reluctance motor.

armature windings are stationary. The term "brushless" is employed to indicate that the rotor is electronically commutated by sensing the rotor position. Commutation of the electronic current in the stationary armature windings by sequentially switching the current on and off in the windings, using solid state devices, so as to set up a rotating field. The switching sequence depends on the rotor's position which is sensed electronically.

A great advantage of brushless DC motors is the fact that they dispense with the need for brushes and commutator contacts or slip rings. Thus, the problems associated with sliding contacts and arcing are completely eliminated. The current in the armature circuit may be expressed in the usual manner, as in a DC motor with an additional term to include the inductive impedance of the armature.

2.4.1 Brushless or Electronic Commutation

The brushless DC motor is therefore said to be an electronically commutated DC motor. Although a variety of devices such as inductive and capacitive transducers, magneto resistors and optical encoders are potentially available for sensing the angular position of the rotor, Hall-effect sensors have emerged as the primary systems for this task. These sensors develop a polarized voltage depending on the control current and the magnetic field applied. They are highly sensitive and reliable micro-scale devices that are eminently suitable for measuring the angular position of the rotating magnetic field of the permanent magnet rotor.

2.4.2 Dynamic Modeling

The dynamics of a brushless DC motor are very similar to that of a conventional DC motor. What is different about the brushless DC motor is the fact that it is electronically commutated and controlled by PWM. Generally, brushless DC motors are current-controlled and so the motor dynamics are usually expressed in terms of the armature current. The primary control input to the motor stator is the armature current which is measured and fed back by PWM of the input current. The resultant DC voltage is typically switched by a six-switch inverter to a three-phase motor stator winding. A typical block diagram of a three-phase brushless DC motor showing the sensing and controller architecture is shown in Figure 2.9.

2.4.3 Switching and Commutation

The inverter switching sequence results in the commutation of the current and is determined by sensing the rotor position. The sensing of the rotor position is done by a distribution of Hall-effect sensors. The commutation is achieved by a set of switching transistors that are operated sequentially to switch on and switch off the current in each phase, in such a way as to keep the current flow in the appropriate direction necessary to maintain the rotor motion. The current feedback in each phase is a trapezoidal pulse with the linear portion of the pulse overlapping in two of the three phases. This permits a smooth transition from one phase to the next. A single cycle of the trapezoidal periodic pulse spans 360°: it starts at zero, is flat with maximum and

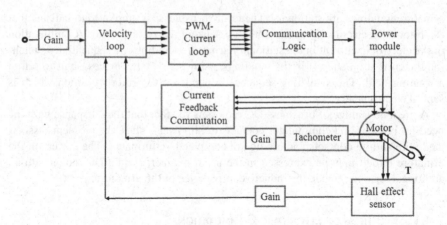

FIGURE 2.9 Typical block diagram of a three-phase brushless DC motor.

minimum magnitude over the 60°–120° and 240°–300° regions respectively, ends at zero and is linearly varying everywhere else. Although the principal component of the torque is sinusoidal, there is a higher harmonic torque ripple. Ignoring the torque ripple, one can show that the rotor is generally maintained at constant speed.

CHAPTER SUMMARY

In this chapter an overview of electric motors, in particular those that could be used for the traction or propulsion of electric vehicles, their operational features and performance characteristics, are presented. The motors include DC motors, AC synchronous and induction motors, reluctance motors and brushless DC motors.

REFERENCES

1. A. Emadi, *Energy-Efficient Electric Motors*, 3rd Edition, CRC Press, Boca Raton, FL, 2004.
2. S. Filizadeh, *Electric Machines and Drives: Principles, Control, Modeling, and Simulation*, CRC Press, Boca Raton, FL, 2017.
3. W. Piotr (Ed.). *Dynamics and Control of Electrical Drives*, Springer-Verlag, Berlin, Heidelberg, 2011.
4. R. Vepa, Principles of energy conversion, Chapter 2, pp 61–112. In *Dynamic Modelling, Simulation and Control of Energy Generation*, Lecture Notes on Energy Series, Vol. 20, Springer, 2013, ISBN 978-1-4471-5399-3.
5. B. Bilgin, J. W. Jiang, A. Emadi, *Switched Reluctance Motor Drives: Fundamentals to Applications*, CRC Press, Boca Raton, FL, 2018.
6. T. J. E. Miller. *Switched Reluctance Motors and Their Control*, Monographs in Electrical and Electronic Engineering, Oxford University Press, New York, 1993.
7. R. Krishnan. *Permanent Magnet Synchronous and Brushless DC Motor Drives*, CRC Press, Boca Raton, FL, 2009.
8. V. V. Athani. *Stepper Motors: Fundamentals, Applications and Design*, New Age International Publishers, New Delhi, India, 1997.
9. P. Vas, *Sensorless Vector and Direct Torque Control*, Oxford University Press, New York, 1998.

3 Batteries

3.1 INTRODUCTION TO BATTERIES

A battery is a device used to convert energy stored within molecules to useful energy by an electrochemical process. It is assembled from one or more electrochemical cells, which are arranged in a specific configuration depending on the output requirements. Energy is released from internal chemical reactions within an electrolyte. The reactions that take place within the electrolyte are generally the reduction and the oxidation types. These reactions take place within an electrolyte between the electrolyte and two electrodes, an anode and a cathode. The cathode is the external positive terminal, while the anode is the external negative terminal. Conventional current (holes) flows in an external circuit from the positive terminal of a power source to the negative terminal. Electrons move in the opposite direction to the conventional current. The cathode, which is the positive external terminal, supplies current (holes) and attracts electrons. Internally a chemical reduction occurs and this requires electrons from the external circuit. The anode which is the negative terminal, collects current (holes) and repels electrons. Internally a chemical oxidation occurs and this requires that it give up electrons from an external circuit. The electrolytes allow for the transport of ions and the resulting net flow of electrons in the opposite direction. The electron-deficient ions move within the electrolyte from the anode or negative terminal to the cathode or positive terminal, forcing electrons to move from the anode or negative terminal to the cathode or positive terminal, in an external circuit. Thus, the current (holes) flows out of the battery to perform work.

Batteries have a fascinating history. In the year 1800, the Italian Alessandro Volta invented the so-called Voltaic pile, using silver and zinc plates separated by leather moistened with brine or vinegar, which functioned as a single electric cell providing a current to an external circuit. In 1836, John Frederic Daniell, a British chemist and meteorologist, invented the Daniell cell, which was a copper pot filled with copper sulfate solution, into which was immersed an unglazed earthenware container filled with sulfuric acid and a zinc electrode. The copper electrode acted as the cathode while the zinc electrode was the anode and the unglazed earthenware provided a salt bridge between the copper sulfate solution and the sulfuric acid. The purpose of a salt bridge, connecting the two "half cells" was to prevent electron movement (not ions) in the electrolyte and to maintain charge balance. The principle of the Daniell cell is illustrated in Figure 3.1. In 1859, the Frenchman Gaston Planté constructed the first rechargeable lead-acid cell. In 1868, another Frenchman Georges Leclanché assembled the first carbon-zinc wet cell, which was followed in 1888 by Carl Gassner's invention of the carbon-zinc dry cell. It was in 1898, that the first commercial flashlight cell, now commonly known as the "D cell" was invented, and the very next year, Waldemar Junger invented the nickel-cadmium cell.

Almost half a century later, in 1946 Georg Neumann produced the sealed Ni–Cd cell, while in the 1960s, the alkaline rechargeable Ni–Cd cell became commercially

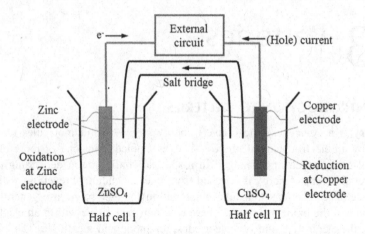

FIGURE 3.1   The principle of the Daniell cell.

available. However, it was only in the 1970s that a lithium-based sealed lead-acid cell was commercially marketed, while the first nickel metal hydride (NiMH) cell was made available some 20 years later. The lithium-ion cell was first built in 1991, while the very next year a rechargeable alkaline cell was first shown to be feasible. In 1999, the first lithium-ion polymer cell was shown to be practically feasible.

Typical electrodes employed to facilitate the reduction reaction or positive electrodes are gold, platinum, mercury, silver, copper, lead, nickel and cadmium, while the typical electrodes employed to facilitate the oxidizing reaction, or negative electrodes, are iron, zinc, aluminum, magnesium, sodium, potassium and lithium (most negative). The standard reduction potentials of all electrodes have been measured relative to the hydrogen electrode, which is the reference electrode. These various electrodes can be arranged in increasing order of their reduction potentials, which is called the electrochemical series.

3.1.1 BATTERY STRUCTURE AND SPECIFICATIONS

A typical dry battery cell, illustrated in Figure 3.2, is encased in a metal case, housing the electrodes and the two half cells. An external circuit is connected to the two electrodes, the cathode, which is the positive terminal, and the anode, which is the negative terminal. The voltage between the two electrodes depends on the difference between the two standard reduction potentials of the two electrodes used.

When the cell is connected to a load, a reduction-oxidation reaction transfers electrons from the anode to the cathode. As a result of this transfer of electrons, current flows in the external circuit from the positive terminal (cathode) to the negative terminal (anode). Consequently, the battery discharges, and the voltage across its two terminals drops. When this voltage falls below a cut-off, the battery is disconnected from the load.

The minimal battery specifications involve basic specifications, specifications related to its condition and other technical specifications. The basic specifications are: the cost and size (AAA, AA, C, D, Button,...), the specific energy (Watts/Kg, Watts/cm^3) at a current rate or C-rate, the battery capacity (at the room temperature)

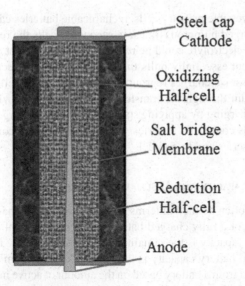

Steel cap
Cathode

Oxidizing
Half-cell

Salt bridge
Membrane

Reduction
Half-cell

Anode

FIGURE 3.2   A typical dry battery cell.

in Ampere-Hours or Ah (Ah for a specific C-rate), the battery life in terms of the number of charge/discharge cycles and the discharge characteristics in terms of C- and E-rates which may be defined as follows:

 i) A 1C-rate means that the discharge current will discharge the entire battery in 1 hour.
 ii) A 1E-rate means that the discharge power will discharge the entire battery in 1 hour.

The battery condition specifications are: the state of charge (SOC) (percentage (%) of charge remaining in the battery), the state of health, the depth of discharge (DOD) (as a percentage, %), the terminal voltage (V), the open-circuit voltage (V), the battery internal resistance, the battery's behavioral factors—such as the temperature range (storage, operation), the self-discharge rate and memory effects—and finally environmental factors, such as battery leakage, gassing, toxicity and resistance to shocks. The overall battery technical specifications include the nominal voltage (V), the cut-off voltage, the capacity (as discussed above), the energy or nominal energy (kWh, for a specific C-rate), the cycle life (number for a specific DOD), the specific energy (kWh/kg) (as discussed above), the specific power (kW/kg), the energy density (kWh/Vol), the power density (kW/Vol), the maximum continuous discharge current, the maximum 30-second discharge pulse current, the overall charge, the float voltage, the (recommended) charge current and the (maximum) internal resistance.

3.1.2   Rechargeable Batteries

Rechargeable batteries (also known as secondary cells) involve electrolytes with reversible cell reactions that allow them to be recharged to regain their full cell potential. This is achieved through the work done by passing currents of electricity.

As opposed to irreversible primary cells, rechargeable batteries can, in principle, be charged and discharged numerous times. In secondary cells the reduction-oxidation reaction within the electrolyte could be reversed with a sufficient amount of energy. The charging current essentially pulls the negative charges (electrons) back to the anode, while the positive charges regroup at the cathode. The cell would eventually approach equilibrium through the transferring of the electrons via an external circuit. To facilitate charging by applying standard voltages, rechargeable batteries are organized with cells connected in series, or in parallel, or in a combination of cells connected both in series and in parallel.

3.1.3 CHARGE, CAPACITY AND DISCHARGE FEATURES

Battery capacity is often defined in terms of charge units. Full charge capacity is the remaining capacity of a fully charged battery at the beginning of a discharge cycle, and the full design capacity is the remaining capacity of a newly manufactured battery. The theoretical battery capacity is the theoretical maximum amount of charge that can be extracted from a battery based on the amount of active material it contains. The standard battery capacity is the amount of charge that can be extracted from a battery when discharged under standard load and temperature conditions. The actual battery capacity is the amount of charge a battery delivers under operating conditions. The capacity of a battery is generally inversely proportional to the discharge rate. In a fully charged cell, the electrode surfaces contain the maximum concentration of the active species. When the cell is connected to a load, a current flows through it; the active species are consumed at the electrode surfaces and replenished by diffusion from the electrolyte. When the diffusion process cannot keep up with the reaction process, a concentration gradient builds up across the electrolyte. A higher load current generally results in a higher concentration gradient across the electrolyte, while a lower concentration of active species develops at the electrode surface. When the concentration falls below a certain critical value, corresponding to a voltage cut-off, the electrochemical reaction can no longer be sustained at the electrode surface. At this point, the charge that was unavailable at the electrode surface due to the lower gradient remains unusable and is responsible for the reduction in capacity. The unused charge is not physically "lost", but is simply unavailable due to the lag between reaction and diffusion rates. Decreasing the rate of discharge of the battery effectively reduces this lag as well as the concentration gradient. If the battery's load goes to zero, the concentration gradient reaches a plateau after a time, attaining equilibrium again. The concentration of active species near the electrode surface following the rest period makes some charge available for extraction. This is known as the charge recovery effect. Battery designers exploit this charge recovery effect to control the discharge rate and maximize battery lifetime under performance constraints. At sufficiently low discharge rates, the battery will behave like an ideal source.

3.1.4 TEMPERATURE EFFECTS AND CAPACITY FADING

The operating temperature significantly affects the battery discharge behavior. Below room temperature, chemical activity in the cell decreases and the internal resistance increases. The increase in the internal resistance reduces the full charge capacity and increases the slope of the discharge curve. At higher temperatures, the internal resistance decreases, increasing both the full charge capacity and voltage.

Some batteries, like the lithium-ion cells, lose a portion of their capacity with each discharge-charge cycle. This capacity fading results from spurious side reactions such as decomposition of the electrolyte, dissolution of the active material and the formation of a passive film. These are irreversible reactions that increase the cell's internal resistance, causing the battery to eventually fail.

3.2   BATTERY DYNAMIC MODELING: PHYSICAL, EMPIRICAL, CIRCUIT AND HYBRID MODELS

The battery and battery packs have emerged as the key components for energy storage and management in transport vehicles. Given a load applied to a battery over a certain period, information about when the battery fails, as well as its state of charge (SOC), or remaining capacity, at any instant of time, are indicators of the battery's dynamic performance. Modeling of the battery dynamics is essential for SOC estimation and prediction.

Battery dynamic models may be broadly classified as: physical models that are based on the dynamics of physical processes in the battery, empirical models that consist of ad hoc equations based on experimental data, circuit models that represent a battery as electrical circuits and hybrid models that offer a holistic view of the physical processes while also incorporating empirical relations.

There are various criteria for evaluating battery models, including accuracy in performance prediction, computational complexity, effort in configuring the model and the physical and analytical insight provided by the model. Physical models are the most accurate as well as most appropriate. While they are used by battery designers to optimize a battery's physical parameters, and although they provide a limited physical and analytical insight to system designers, they are generally slow to simulate and configure.

A popular empirical model is defined by Peukert's law. An ideal battery with a capacity C, discharged at a constant current rate would be expected to have a lifetime L given by $C = Li_{ideal}$. Peukert's law expresses battery capacity as a power law relationship, $C = L_{nom}i_{dch} = Li^{\chi}$ where $\chi > 1$. The exponent provides a simple way to account for the discharge rate dependence. An alternate way of stating the effective capacity of the battery decreases with increasing current taken from the battery. Therefore, the effective battery capacity discharge current i_{dch} increases with rising battery current $i(V)$ in accordance with the law:

$$i_{dch} = i\left(V\right)\left(i/i_{nom}\right)^{\chi-1}.$$ (3.1)

In Equation (3.1), i_{nom} is a constant.

There are two other empirical battery models which are also used. The first is a battery efficiency model based on the definition of battery efficiency which is the ratio of actual capacity to theoretical capacity. It is modeled as a function of the load current. The second is the Weibull fit model which is a statistical model of the discharge behavior of cells based on Weibull probability density function. Circuit models represent a battery as an electrical circuit that could be described by a set of continuous time or discrete-time equations, or as stochastic process models, and other dynamical models. Hybrid models combine a high-level representation of the battery parameters determined by experimental data with analytical expressions based on physical laws. Typically, the physical laws which form the basis for the model are: i) Faraday's law for electrochemical reaction and ii) Fick's laws for concentration changes during one-dimensional diffusion in an electrochemical cell

Modeling a battery pack that is assembled from single cells that are all exclusively connected in series is simple and straightforward, as all of the voltage outputs must be simply added up. On the contrary, when two or more cells are connected in parallel, there is a need to split the current between the parallel lines. In simulation, quite unlike in a real circuit, the current is usually externally supplied. If all the cells are identical, then it is reasonable to assume that the current splits equally between the parallel lines. The problem arises when the cells are not identical with respect to their impedance properties. So, one must necessarily adopt an approach which is not fully representative of the real situation. One approach is to assume that that the current splits equally even when the cells are dissimilar and make changes to the connections so as to ensure that the overall input-output relations and the impedances are in fact correct. An alternate approach is to use internal circuitry that actually equalizes the currents in the parallel lines. A third approach is to assume the use of DC/DC converters so the voltage in the parallel line is the same. However, the latter two approaches involve additional hardware and this is tantamount to changing the problem to fit the solution!

A more practical approach is to construct a reduced order model that addresses the particular issue under consideration and use this reduced order model to develop the representation for the full battery pack. Such an approach has been adopted by many battery pack manufacturers in relation to the temperature control or thermal management of the battery and for determining the state of charge (SOC) of the battery at any instant of time. The reduced order models are particularly suitable for use in extended or unscented Kalman filters for the estimation of the SOC, as shown by Wijewardana, Vepa and Shaheed [1].

Utilizing a battery cell level model, to capture the dynamics of a battery pack is a major concern for modeling in order to extend them towards hybrid electric vehicle and electric vehicle applications. To model a battery pack, one approach has been to represent the pack by the same equivalent circuit as the cell in Yurkovich [2] and derive the model parameters by a suitable method such as real-time identification or parameter estimation from the simulation of a full multi-cell model (Marco et al. [3], Barreras et al. [4], and Bruen, Marco and Gama [5]). Another approach adopted by Kim and Qiao [6] has been to develop the model based on the battery capacity as outlined in Section 7.4 of Vepa [7]. An alternate approach is to construct a relevant reduced order model for each cell and use this reduced order representation to derive the multi-cell model. Such an approach is adopted for thermal management studies or for state of charge estimation [8], as mentioned in the Introduction. Yet, another approach is to develop an equivalent linear impedance model and construct a multi-cell representation for parallel or series connected cells so that the overall input-output relations and the impedances are in fact correct.

Many cell level equivalent electrical battery circuit models were found in literature. Examples of such cell level models are found in Wijewardana, Vepa and Shaheed [1], Kroeze and Krein [9], Erdinc, Vural and Uzunoglu [10], Chen and Rincon-Mora [11], Vepa [7], Thanagasundram et al. [12], Benger et al. [13] and in Leng, Tan and Pecht [14]. The third-order cell level battery model which was presented in Antonuccia et al. [15] was tested for hybrid electric vehicles for Li-ion and NiMH batteries. Battery open circuit voltage as a function of SOC was given by Erdinc, Vural and Uzunoglu [10] and their parameter identification method was explicitly described and the experimental data was a very good source for the other researchers. A description of the Randles-Warburg model and a summary of latest developments in battery modeling are given in Vepa [7]. Zoroofi [16] presented a linear dynamic battery model and a non-linear battery model which has two ideal diodes to accommodate charging and discharging separately. Ross [17] presented a battery open circuit voltage as a function of specific gravity of the electrolyte and many new empirical formulations related to gas emissions in the electrodes of lead-acid batteries.

3.2.1 BATTERY SOC ESTIMATION

There are five categories of methods for estimating SOC. These methods are based on i) electrolyte specific gravity measurement, ii) stabilized float current measurement, iii) Coulombic measurements, iv) open circuit battery voltage method, and v) fully loaded battery voltage method. The specific gravity method is based on the electrolyte specific gravity (SG). The SG is a function of the amount of acid in the electrolyte. It is linearly related to the remaining battery capacity. Consequently, measuring SG and calibrating the measurement gives the SOC. In stationary applications, specific gravity is automatically measured via a hydrometer. A long rest period is required after discharge/charge before specific gravity can be accurately measured due to time constants of the electrolyte diffusion dynamics. For this reason, the sampling period for the measurement of SG is generally large. Manual measurement of SG is not feasible for mobile vehicle applications. The stabilized float current method relies on the fact that at the conclusion of the full charging cycle, after the battery is held at a constant float voltage, the charging current will stabilize. The float current is a more accurate and rapid indicator of battery SOC than the SG. However, because of the need for stability and as it can only be done after a full charging cycle, it is not suitable for real-time estimation of SOC. In the Coulombic measurements method, under the assumption that the battery terminal voltage is constant, the traditional definition of SOC is:

$$SOC(t) = 100 \left(\left(Q_c - \int_0^t i_{dch}(\tau) d\tau \right) \bigg/ Q_c \right). \tag{3.2}$$

The quantity, Q_c is the initial charge and i_{dch} is the discharge current. There are several practical difficulties in the application of this method to real-time estimation of SOC. The open circuit voltage method is based on the fact that the SOC is related to the open circuit voltage by:

$$SOC(t) = \left(\frac{V_{OC}(t) - V_{OC}(\infty)}{V_{OC}(0) - V_{OC}(\infty)} \right) \times 100\%. \tag{3.3}$$

If $V_{OC}(t)$ is estimated and, $V_{OC}(0)$ and $V_{OC}(\infty)$ are obtained from the battery dynamics or otherwise, the SOC can be easily estimated in real time using the methodology of Kalman filtering. A dynamic model for the open circuit voltage is used to construct an extended Kalman filter or an unscented Kalman filter.

To illustrate the Kalman filter approach consider a battery dynamic model during discharge given by

$$\dot{V}_p(t) = -\left(V_p - V_{OC} + R_d I_b\right)/R_d C_p, \quad V_{bm}(t) = V_p - I_b R_b. \tag{3.4}$$

In Equation (3.4), $V_{bm}(t)$ is the measured output voltage, V_p is the open circuit voltage state at battery terminals, I_b is the current flowing out of battery terminals, R_b is the battery external resistance, V_{OC} is the battery open-circuit voltage at initial time, R_d is the battery internal resistance during discharge and C_p is the effective polarizing capacitance. It is assumed that I_b is known exactly. Using Equation (3.3), it follows that the SOC can then be expressed in terms of the variables of Equation (3.4), as

$$SOC(t) = \left(\frac{V_p(t) - V_p(\infty)}{V_{OC} - V_p(\infty)} \right) \times 100\%. \tag{3.5}$$

If one defines the states,

$$x_1 = V_p, \quad x_2 = \frac{1}{R_d C_p}, \quad x_3 = \frac{V_{OC}}{R_d C_p}, \quad x_4 = \frac{1}{C_p} \quad \text{and} \quad x_5 = R_b, \tag{3.6}$$

the following discrete time model may be established:

$$x_1(k+1)/T = \left(x_1(k)/T\right) - x_1(k) x_2(k)$$

$$+ x_3(k) - I_b(k) x_4(k) + w_1(k),$$

$$x_2(k+1) = x_2(k) + w_2(k), \quad x_3(k+1) = x_3(k) + w_3(k),$$

$$x_4(k+1) = x_4(k) + w_4(k), \quad x_5(k+1) = x_5(k) + w_5(k). \tag{3.7}$$

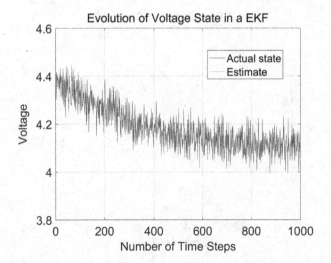

FIGURE 3.3 The Kalman filter estimate of the primary voltage state V_p.

The complete linearized model and the output equation may be expressed as,

$$\mathbf{x}_{k+1} = \mathbf{\Phi}_k \mathbf{x}_k + \mathbf{B}_{uk} \mathbf{u}_k + \mathbf{w}_k,$$ (3.8)

$$V_{bm}(k) = x_1 - I_b x_5 + v.$$ (3.9)

In Equations (3.7), (3.8) and (3.9) $w_i(k)$, $\mathbf{w}_k$ and v are respectively the scalar process noise, the vector process noise and the measurement noise sources. A typical estimate of the primary voltage state V_p, obtained by using a Kalman filter, programmed in MATLAB, is shown in Figure 3.3. The estimate of SOC is then calculated from the state estimates obtained from the Kalman filter. The evolution of the state of charge of a typical individual cell is shown in Figure 3.4. An improved estimate of the SOC may be obtained by retaining the non-linear dynamic model and using the extended or the unscented Kalman filtering approach as in Wijewardana, Vepa and Shaheed [1].

3.3 TYPES AND CHARACTERISTICS OF BATTERIES

As noted, batteries could be classified as primary, which are those that cannot be recharged and reused, and secondary batteries which are those that can be recharged and reused. Batteries belonging to the first category are zinc-carbon cells, heavy duty zinc-chloride cells, alkaline cells, lithium and silver batteries, mercuric oxide cells and zinc-air cells. Batteries belonging to the second category are lead-acid batteries, nickel-cadmium batteries, nickel metal-hydride batteries, rechargeable alkaline cells, lithium-ion batteries, lithium-ion polymer batteries and lithium-air batteries.

Considering a zinc-carbon battery, which has a zinc metal body (Zn) as its anode, a cathode made of manganese dioxide (MnO_2) and an acidic electrolyte made of a paste

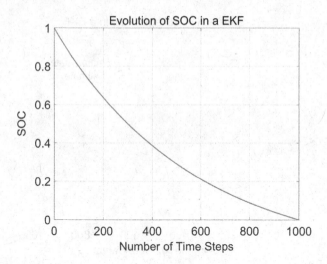

FIGURE 3.4 The evolution of the state of charge of a typical individual cell.

of zinc chloride and ammonium chloride dissolved in water. The half cell and overall reactions and the corresponding cell potentials may be summarized as follows:

$$Zn(s) \rightarrow Zn^{2+}(aq) + 2e^- \left[e° = -0.763 \text{ V} \right], \tag{3.10}$$

$$2NH_4^+(aq) + 2MnO_2(s) + 2e^-$$
$$\rightarrow Mn_2O_3(s) + H_2O(l) + 2NH_3(aq) + 2Cl^- \left[e° = 0.50 \text{ V} \right], \tag{3.11}$$

Overall reaction:

$$Zn(s) + 2MnO_2(s) + 2NH_4Cl(aq)$$
$$\rightarrow Mn_2O_3(s) + Zn(NH_3)_2 Cl_2(aq) + H_2O(l) \left[e° = 1.3 \text{ V} \right], \tag{3.12}$$

where "aq" represents the aqueous phase, "s" the solid phase, "l" the liquid phase.

As for heavy duty zinc-chloride cells, they have electrodes similar to zinc-carbon cells with an acidic zinc chloride aqueous electrolyte which is known to have a better resistance to leakage, known to be better at high current drain and has a better performance at low temperatures than zinc-carbon cells. Alkaline batteries typically use zinc in powder form to build the anode, a manganese dioxide cathode and an alkali as the electrolyte which is usually potassium hydroxide. The half cell and overall reactions and the corresponding cell potentials may be summarized as follows:

$$Zn_{(s)} + 2OH^-_{(aq)} \rightarrow ZnO_{(s)} + H_2O_{(l)} + 2e^- \left[e° = -1.28 \text{ V} \right], \tag{3.13}$$

$$2MnO_{2(s)} + H_2O_{(l)} + 2e^- \rightarrow Mn_2O_{3(s)} + 2OH^-_{(aq)} \left[e° = 0.15 \text{ V} \right], \tag{3.14}$$

Overall reaction:

$$Zn_{(s)} + 2MnO_{2(s)} \rightarrow ZnO_{(s)} + Mn_2O_{3(s)} \quad \left[e^\circ = 1.43\ \text{V}\right]. \qquad (3.15)$$

The silver oxide battery is an example of an alkaline cell, which has a zinc anode, silver dioxide cathode and sodium hydroxide (NaOH) or potassium hydroxide (KOH) as the electrolyte. The half cell and overall reactions and the corresponding cell potentials may be summarized as follows:

$$Zn_{(s)} + 2OH^-_{(aq)} \rightarrow ZnO_{(s)} + H_2O_{(l)} + 2e^- \quad \left[e^\circ = -1.28\ \text{V}\right], \qquad (3.16)$$

$$Ag_2O_{(s)} + H_2O_{(l)} + 2e^- \rightarrow 2Ag_{(s)} + 2OH^-_{(aq)} \left[e^\circ = 0.29\ \text{V}\right], \qquad (3.17)$$

Overall reaction:

$$Zn_{(s)} + Ag_2O_{(s)} \rightarrow ZnO_{(s)} + 2Ag_{(s)} \quad \left[e^\circ = 1.57\ \text{V}\right]. \qquad (3.18)$$

The zinc-air battery is an example of a metal air battery, which has a zinc anode usually constructed from zinc powder, a porous carbon cathode that facilitates reactions with O_2 and potassium hydroxide (KOH) as the electrolyte. The half cell and overall reactions and the corresponding cell potentials may be summarized as follows:

$$Zn_{(s)} + 4OH^-_{(aq)} \rightarrow Zn(OH)_4^{2-} + 2e^- \quad \left[e^\circ = -1.25\ \text{V}\right], \qquad (3.19)$$

$$Zn(OH)_4^{2-} \rightarrow ZnO_{(s)} + H_2O_{(l)} + 2OH^-_{(aq)}$$

$$(1/2)O_2 + H_2O_{(l)} + 2e^- \rightarrow 2OH^-_{(aq)} \quad \left[e^\circ = 0.34\ \text{V}\right], \qquad (3.20)$$

Overall reaction:

$$2Zn_{(s)} + O_{2(g)} \rightarrow 2ZnO_{(s)} \quad \left[e^\circ = 1.59\ \text{V}\right]. \qquad (3.21)$$

Considering the lead-acid batteries, which are typical examples of secondary or rechargeable cells, they use porous lead as the anode, lead-dioxide as the cathode and sulfuric acid, 6 molar H_2SO_4, as the electrolyte. The overall reactions during charge and discharge and the cell potentials may be summarized as follows:

During discharge, cathode (+) electrode: [$e^\circ = 2.1$ V],

$$PbO_{2(s)} + 4H^+_{(aq)} + SO_4^{2-}_{(aq)} + 2e^- \rightarrow PbSO_{4(s)} + 2H_2O_{(l)}, \qquad (3.22)$$

Anode, (–) electrode: $Pb(s) + SO_4^{2-}_{(aq)} \rightarrow PbSO_{4(s)} + 2e^-, \qquad (3.23)$

During charge, cathode (+) electrode:

$$PbSO_{4(s)} + 2H_2O_{(l)} \rightarrow PbO_{2(s)} + 4H^+_{(aq)} + SO_4^{2-}{}_{(aq)} + 2e^-, \qquad (3.24)$$

Anode, (−) electrode: $PbSO_{4(s)} + 2e^- \rightarrow Pb_{(s)} + SO_4^{2-}{}_{(aq)}. \qquad (3.25)$

Further details of lead-acid batteries may be found in Jung, Zhang and Zhang [18].

The more expensive nickel-cadmium batteries, which are smaller than lead-acid batteries and more expensive, are another example of rechargeable batteries and use a cadmium hydroxide, $Cd(OH)_2$ anode, a nickel hydroxide, $Ni(OH)_2$ cathode and potassium hydroxide, KOH as the electrolyte. The half cell and overall reactions during discharge may be summarized as follows:

$$Cd + 2OH^- \rightarrow Cd(OH)_2 + 2e^-, \qquad (3.26)$$

$$2NiO(OH) + Cd + 2e^- \rightarrow 2Ni(OH)_2 + 2OH^-, \qquad (3.27)$$

Overall reaction:

$$2NiO(OH) + Cd + 2H_2O \rightarrow 2Ni(OH)_2 + Cd(OH)_2. \qquad (3.28)$$

Secondary alkaline batteries are rechargeable cells with the lowest cost and a long shelf-life and are useful for moderate-power requirements, although their cycle life is less than most other secondary batteries. Yet they have little or no memory effect. They are popular with consumers as they combine the benefits of the alkaline cells with the added benefit of recharging and re-use. There are also no toxic ingredients and they can be disposed in regular, local waste disposal sites.

3.3.1 Lithium-Ion (Li-Ion) Batteries

Lithium-ion batteries use carbon (C), graphite (C_6) for the anode and a form of a transition metal-oxide (lithium cobalt oxide-$LiCoO_2$ or lithium manganese oxide $LiMn_2O_4$) impregnated with lithium for the cathode. For the electrolyte, they use a mixture of solid lithium, lithium salts ($LiPF_6$, $LiBF_4$, or $LiClO_4$) or organic solvents. The half cell and overall reactions during discharge and the open-circuit voltage may be summarized as follows:

$$CoO_2 + Li^+ + e^- \leftrightarrow LiCoO_2, \quad E^\circ = 1\,V, \qquad (3.29)$$

$$Li^+ + C_6 + e^- \leftrightarrow LiC_6, \quad E^\circ \sim -3.01\,V, \qquad (3.30)$$

Overall reaction during discharge:

$$CoO_2 + LiC_6 \leftrightarrow LiCoO_2 + C_6, \qquad (3.31)$$

Overall open circuit cell voltage:

$$E_{oc} = E^+ - E^- = 1 - (-3.01) = 4 \text{ V}. \qquad (3.32)$$

The lithium ions are inserted and exerted into the lattice structure of anode and cathode during charging and discharging respectively. During discharge, current flows through the external circuit. During charging, the electron flow is reversed and the current is in the opposite direction. The energy density is almost twice that of the standard Ni–Cads. It is essential that the charging is completely controlled for Li-ion cells. It is very important to equalize the cells to a minimum of 2.85 V initially before fully charging the cells. Thus lithium-ion cells require a specialized charger, which initially delivers a constant current for charging. The controlled charging has two essential purposes: to limit the current and to equalize the cell voltages. Moreover, the temperature sensitivity of the cells is an issue that must considered in the design of the charger.

Generally, different electrolyte materials are used in the vicinity of the anode and the cathode. The electrolytes play a significant role of transmitting electrons and lithium ions during charging and discharging. The desirable qualities of electrolytes are:

1. High ionic conductivity at wide range of temperatures to increase the lithium ions diffusion and resist polarization during charging/discharging;
2. Desirable thermal stability to ensure the battery's operation under appropriate temperature;
3. A wide electrochemical window, to prevent side reaction between electrodes and electrolyte;
4. Essential mechanical properties, safety (high flash point), low cost, ease of manufacturing and enhanced safety and non-toxic (environmentally friendly).

Typical liquid electrolytes, containing lithium salts are $LiPF_6$, $LiBF_6$, $LiClO_4$, $LiBC4O_8$ or Li $[PF_3(C_2CF_5)_3]$ dissolved in organic alkyl carbonate solvent. Generally, these electrolytes are not very stable and break down in contact with O_2 ions, and are thermally unstable, making them hazardous and requiring careful packaging.

There are several challenges, in the development of lithium batteries, which have been outlined by Kerman et al. [19]. One problem seems to be the inability to use lithium-metal electrodes, although there is substantially greater energy density within the lithium-metal electrodes, compared with lithium-ion electrodes. Unfortunately, lithium-metal electrodes can develop whisker or filament-like fast-growing and inflammable structures known as dendrites (Zhang, Xu and Henderson [20]). These filament-like lithium growths or dendrites can cause the battery to short-circuit eventually. One solution is to replace the lithium-metal electrode with a carbon electrode with an internal lattice structure that enables lithium ions to stay connected to the electrode. Thus, one is still retaining the essential principle for the electrode design for the lithium-ion battery. The battery with the carbon electrode essentially has a lower energy storage capacity than the same battery with a solid lithium-metal electrode. To reduce the hazard of the battery catching fire, the liquid electrolyte in the battery is replaced with a non-inflammable solid-state electrolyte, such as a polymer or even a plastic. The use of a lithium electrode is now also feasible, as long as the lithium ions can travel through the solid-state electrolyte as fast as they do in a liquid electrolyte. The other advantage of using a solid-state electrolyte is that it is expected to inhibit the formation of the fast-growing dendrites on the lithium-metal electrode. The inhibition of the growth of dendrites could double the battery's energy capacity. Controlling the formation of dendrite growth provides an alternate method of enhancing the battery life, as proposed by Liu et al. [21]. An important method of increasing the battery energy capacity is to influence the electrode–electrolyte interface kinetics which was discussed in the review by Yu and Manthiram [22]. With a complete understanding of the electrode–electrolyte interface kinetics, it is possible that it can be manipulated to provide strategies for enhancing battery capacity.

3.3.2 Gel Polymer Electrolytes

Gel polymer electrolytes are synthesized by incorporating a liquid electrolyte into polymer base membranes by soaking membranes in lithium-based organic electrolytes. There are four polymer host materials for GPEs, which are polyethylene oxide (PEO), polyacrylonitrile (PAN), polymethyl methacrylate (PMMA) and polyvinylidene fluoride (PVdF).

3.3.3 Lithium–Sulfur (Li–S) Batteries

Lithium–sulfur (Li–S) batteries use Li-metal or Li-metal compound for the anode, a porous form of carbon (graphite) + sulfur for the cathode and GPEs for the electrolyte. The half cell and overall reactions during discharge and the open-circuit voltage may be summarized as follows:

$$2Li \rightarrow 2Li^+ + 2e^-, \quad S + 2e^- \rightarrow S^{2-}, \tag{3.33}$$

Overall reactions and overall cell open-circuit voltage:

$$2Li + S \rightarrow Li_2S + 2e^-, \quad E_{oc} = 2.23 \text{ V}. \tag{3.34}$$

There are several challenges in the development of high capacity Li–S batteries. Some of these challenges are, low sulfur utilization, low Coulombic efficiency, fast capacity fade and short cycle life and Li dendrite growth at Li-metal anode, which was discussed in the preceding section. The growth of the dendrites can lead to short circuit failures. The insulating nature of sulfur can inhibit the lithium ions from traveling across the electrodes, as well as resulting in low utilization of sulfur (low capacity). The low mass-loading of sulfur (≤ 2 mg/cm^2) also results in capacity reduction. Moreover, the dissolution of the polysulfide can result in capacity fading. There some possible solutions to the Li dendrite growth at a Li-metal anode, such as pre-lithiation of the anode (or passivation of Li-metal electrodes). To compensate for the insulating nature of sulfur, one could use a conducting agent. To alleviate the effect of the low mass-loading of sulfur, a composite coating could be used to produce a uniform and crack-free cathode. To mitigate the effects of the dissolution of polysulfide, a stabilizing agent could be used.

3.3.4 METAL-AIR AND LI-AIR BATTERIES

A metal-oxygen battery (also referred as a "metal-air" battery) is a cell in which one of the reactants is gaseous oxygen, O_2. Oxygen enters the cell typically at the positive electrode—after being separated from an inflow of air—and dissolves in the electrolyte. The negative electrode is typically a metal monolith or foil. Electrolytes can be aqueous or non-aqueous, solid or composite in carbonate solvents, like propylene carbonate, ethylene carbonate and dimethyl carbonate. Metal-air batteries, such as lithium-air (Li-air) batteries, utilize alkaline earth metal anodes and porous carbon electrodes where the active oxygen in the pores serves as the cathode. O_2 does not have to be stored in the battery and can be obtained from the atmosphere, so the battery can be considered to be air breathing. Li-air batteries tend to be very popular because of their high specific charge density. Amongst the metal-air batteries, Li-air batteries can carry 30 times the charge of Li-ion batteries, which is 12 times more than Li–S batteries, ten times as much charge as zinc-air batteries, double the charge of magnesium-air batteries and 60% more than aluminum-air batteries and are therefore the batteries with the highest energy capacity. The high specific energy density of metal-air batteries is due in part to the coupling of a reactive metal anode to an air electrode (cathode), providing a battery with a continuous supply of the cathode reactant. It is the reaction at the cathode that delivers most of the energy as most of the cell's voltage drop occurs at the air cathode. The disadvantages are: drying out due to the air, electrolyte flooding, self-discharge due to corrosion of the metal electrode and limitations to both the power density and the operating temperature range. It is essential that the highly porous cathode must be able to sustain an oxygen reduction reaction and so catalysts are impregnated into it. The metal-air batteries are typically divided into two groups based on the electrolyte type of (i) aqueous (ii) non-aqueous, as well as (iii) hybrid and (iv) solid-state. Further details about all metal-air and metal-sulfur batteries may be found in Neburchilov and Zhang [23].

As lithium travels from anode to cathode, lithium (Li) is easily ionized to form Li^+ plus one electron. The anodic reaction and the cell voltage are:

$$Li \rightarrow Li^+ + e^-, \quad E_0 = -3.04 \text{ V}. \tag{3.35}$$

Li-air batteries can use a variety of electrolytes and a porous carbon electrode. Depending on the nature of the electrolyte several overall reactions are possible. Oxygen reduction in a non-aqueous electrolyte can be substantially different than that in an aqueous electrolyte.

To consider the cathodic reactions, it is essential to state the Gibbs energy potential associated with the electrolyte. Gibbs energy is the capacity of a system to do non-mechanical work. $\Delta G°$ is a measure of the non-mechanical work done on the system (negative sign for work done by the system) and corresponds to the maximum possible useful work that can be extracted from a reaction. When an amount of charge, Q moves through a potential difference, ΔE, the work equals:

$$w = -Q\Delta E = -nFE_0 \equiv \Delta G°. \tag{3.36}$$

One cathodic reaction and oxygen reduction in a non-aqueous electrolyte and the overall cell voltage are:

$$2Li + O_2 \rightarrow Li_2O_2, \quad E_0 = 0.06 - (-3.04) \text{ V} = 3.1 \text{ V}. \tag{3.37}$$

The corresponding Gibbs energy is $\Delta G° = -145$ kCals .

Another cathodic reaction and oxygen reduction in an alternate non-aqueous electrolyte and the overall cell voltage are:

$$4Li + O_2 \rightarrow 2Li_2O, \quad E_0 = -0.13 - (-3.04) \text{ V} = 2.91 \text{ V}. \tag{3.38}$$

The corresponding Gibbs energy is $\Delta G° = -268$ kCals .

In the case of an aqueous electrolyte

$$2Li + 2H_2O \rightarrow 2LiOH + H_2, \quad E_0 = -0.96 - (-3.04) \text{ V} = 2.08 \text{ V}. \tag{3.39}$$

The corresponding Gibbs energy is $\Delta G° = -145$ kCals .

Upon discharge, metal cations present in the electrolyte react with dissolved oxygen and electrons from the electrode to form a metal-oxide or metal-hydroxide discharge product. In some chemistries, the discharge product remains dissolved in the electrolyte. In other systems, it precipitates out of solution, forming a solid phase that grows in size as the discharge proceeds. The main discharge reaction in the absence of a catalyst in the porous carbon cathode is the reduction of oxygen to form insoluble Li_2O_2 and Li_2O. Thus, Li_2O_2 and Li_2O will deposit at the location where the charge transfer occurs, inside the pores, thus reducing the effective porosity of the carbon electrode. The capacity of a gas diffusion electrode in a non-aqueous electrolyte is limited by the surface available for the deposition and the pore volume available for the storage of the solid reduction products.

The reactions in secondary Li-air cells are rechargeable and the reactions are reversible. The overall reversible reactions secondary Li-air cells are:

$$\text{Cathode: } 2Li + O_2 \leftrightarrow Li_2O_2, \quad \text{Anode: } Li \leftrightarrow Li^+ + e^-. \tag{3.40}$$

In secondary metal-oxygen batteries, the recharge process proceeds via the decomposition of the discharge phase back to O2 and dissolved metal cations. In light of the processes associated with discharge and charging, reversible metal-oxygen batteries with solid discharge products are often referred to as precipitation-dissolution systems.

Most developmental efforts have been focused on four different chemical designs of Li-air batteries classified by their inner structures, namely: (a) aqueous electrolyte-based Li-air battery, (b) non-aqueous electrolyte-based Li-air battery where the non-aqueous electrolytes generally contain Li+ salts and organic solvents such as alkene-ester, ether, dimethyl sulfoxide (DMSO) and tetra(ethylene) glycol dimethyl ether, (c) hybrid electrolyte-based Li-air battery where the hybrid electrolyte contain an aprotic (aprotic solvents cannot donate hydrogen) organic electrolyte at the anode end (Li electrode, covered by an anode protective layer) and an aqueous electrolyte at the cathode side separated by a ceramic conducting separator film and (d) a solid-state Li-air battery with polymer electrolytes. The challenges in the development of Li-air batteries are the overall lack of stability, the degradation of the solvent, the effect of small amounts of impurities on cell performance, the need to improve and further develop the negative electrode, the incorporation of advanced positive electrodes into the cell, the necessity of using red-ox mediators for bypassing charge transport limitations and the design for solid-state applications with hybrid insertion electrodes.

Li-air cells can be considered to be a battery or a fuel cell and so can replace a fuel cell in a fuel-cell powered aircraft or can serve as a battery. It can be said that both the fuel cell and battery technologies essentially merge and come together in Li-air batteries! There are, however, still some important differences between Li-air batteries and fuel cells. The features of several popular secondary cells are compared in the Table 3.1.

TABLE 3.1
Features of Several Popular Secondary Cells

Feature	Lead-acid	Ni–Cad	Ni–MH	Li-ion
Cell Volt.	2	1.2	1.2	3.6
Sp. Energy (Wh/Kg)	1–60	20–0	1–100	5–120
Sp. Power (W/Kg)	<300	150–300	<220	120–1,200
Energy Density (kWh/m^3)	30–60	25	80–100	100–250
Power Density (MW/m^3)	<0.8	0.15	1.5–5	0.5–2
Cycle life	250–1,000	600–1,200	600–1,200	3,600
Discharge time	>1 min	1 min–8 h	>1 min	30 s–2 h
Energy cost (£/kWh)	80	400	380	400
Power cost (£/kW)	150	400	720	800
Efficiency	80–90	85	88	98–99

3.4 APPLICATIONS

With almost half the power weight density as fossil fuels, Li-air batteries can be used to power an electric and hybrid aircraft, thus reducing the carbon footprint of the aircraft. However, other improvements must be made to mitigate that their power weight density is less than half that of fossil fuels, as this would imply a large battery weight which an aircraft must carry on board. They could be considered the greenest of batteries! Li-air batteries can be used to improve grid stability by acting as a buffer to compensate for the intermittent nature of renewable energy sources.

3.4.1 BATTERIES FOR ELECTRIC AIRCRAFT

Assuming 200–300 Wh/kg and a battery pack with a total weight of up to 500 kg, we can assume that between 400–600 kWh can be stored in four battery packs locally for aircraft applications. A typical Tesla Model S automobile battery pack weighs approximately 320 kg and provides 85 kWh, so our assumption is not unreasonable. On a comparative scale aviation fuel has a specific energy of nearly 12 kWh/kg. There is indeed a need to reduce pack weight by a factor of > 20 and Li-air batteries can more than half it! It should be noted that a typical Tesla Model S battery pack uses 7,104 lithium-ion battery cells, with 16 modules wired in series, where each module has six groups of 74 cells wired in parallel and the six groups are then wired in series within the module, which in total are equivalent to 35,417 AA cells! With two 65 kg lithium battery packs to drive the two 30 Kw motors, the two-seater Airbus E-Fan cruises at 185 kph and flies for one hour. A 50-seater commuter aircraft would require in excess of 1,500 kWhs of battery power or ten of the battery packs mentioned earlier, weighing a minimum of 5,000 Kg! If we can increase the specific energy of a battery pack by a factor of 20, a four-hour flight of a 50-seater aircraft may just be feasible, and we may have to opt for hybrid engines in the medium term. This would involve removing the turbine in "turbo-jet" and drive compressor with a battery while using an electric motor for taxiing. Houston, we certainly have a problem but the solution may be found sooner rather than later!

CHAPTER SUMMARY

In this chapter an overview of various types of batteries, particularly those that could be used for supplying power to electric vehicles for the purpose of propelling them or providing a traction force, their key features, performance characteristics and limitations, are presented.

REFERENCES

1. S. Wijewardana, R. Vepa, M. H. Shaheed, Dynamic battery cell model and state of charge estimation, *J. Power Sources*, 308:109–120, March 2016.
2. B. J. Yurkovich, Electro-thermal battery pack modelling and simulation, Thesis presented to the Graduate School of The Ohio State University, in partial fulfilment of the requirements for the degree of Master of Science in Engineering, pp 38–58, 2010.

3. J. Marco, N. Kumari, W. D. Widanage, P. Jones, A cell-in-the-loop approach to systems modelling and simulation of energy storage systems, *Energies*, 8:8244–8262, 2015, doi:10.3390/en8088244.

4. J. V. Barreras, M. J. Swierczynski, E. Schaltz, S. J. Andreasen, C. Fleischer, D. U. Sauer, A. E. Christensen, Functional analysis of battery management systems using multi-cell HIL simulator, Proceedings of the 2015 Tenth International Conference on Ecological Vehicles and Renewable Energies (EVER), Monte-Carlo: IEEE Press. doi:10.1109/EVER.2015.7112984.

5. T. Bruen, J. Marco, M. Gama, Current variation in parallelized energy storage systems, Vehicle Power and Propulsion Conference 2014 (VPPC 2014), Portugal, October 27–30, 2014. Published in Proceedings of the 10th Vehicle Power and Propulsion Conference 2014, pp. 1–6.

6. T. Kim, W. Qiao, A hybrid battery model capable of capturing dynamic circuit characteristics and nonlinear capacity effects, *IEEE Trans. Energy Convers.*, 26(4):1172–1180, December 2011.

7. R. Vepa (Ed.), Batteries: Modelling and state of charge estimation, Chapter 7, pp 323–347. In *Dynamic Modeling, Simulation and Control of Energy Generation*, Lecture Notes in Energy Series, Springer, 2013, ISBN 978-1-4471-5400-6.

8. S. Tong, M. P. Klein, J. W. Park, A comprehensive battery equivalent circuit based model for battery management application, Proceedings of the of the 6th Annual Dynamic Systems and Control Conference, March 2013, Palo Alto, CA.

9. R. C. Kroeze, P. T. Krein, Electrical battery model for use in dynamic electric vehicle simulations, IEEE 2008 Power Electronics Specialists Conference, pp 1336–1342, 2008, doi:10.1109/PESC.2008.4592119.

10. O. Erdinc, B. Vural, M. Uzunoglu, A dynamic lithium-ion battery model considering the effect of temperature and capacity fading, IEEE 2009 International Conference on Clean Electrical Power, June 9–11, 2009, doi:10.1109/ICCEP.2009.5212025.

11. M. Chen, G. A. Rincon-Mora, Accurate electrical battery model capable of predicting runtime and I–V performance, *IEEE Trans. Energy Convers.*, 21(2):505–511, 2006.

12. S. Thanangasundram, R. Arunachala, K. Makinejad, T. Teutsch. *A Cell Level Model for Battery Simulation*, European Electric Vehicle Congress, Brussels, Belgium, November 20–22, 2012.

13. R. Benger, H. Wenzl, H.-P. Beck, M. Jiang, D. Ohms, G. Schaedlich, Electrochemical and thermal modelling of lithium-ion cells for use in HEV or EV application, EVS24, Stavanger, Norway, May 13–16, 2009, *World Electr. Veh. J.*, 3(2),0342–0351, 2009.

14. F. Leng, C. M. Tan, M. Pecht, Effect of temperature on the aging rate of Li ion battery operating above room temperature, *Sci. Rep.*, 5:12967, 2015, doi:10.1038/srep12967.

15. V. Antonuccia, G. Brunaccinia, A. De Pascaleb, M. Ferraroa, F. Melinob, V. Orlandinib, F. Sergi, Integration of P-SOFC generator and ZEBRA batteries for domestic application and comparison with other P-CHP technologies, The 7th International Conference on Applied Energy – ICAE2015, *Energy Procedia*, 75:999–1004, 2015.

16. S. Zoroofi, Modeling & simulation of vehicular power systems, MSc Thesis, Submitted in Partial Fulfillment of Requirements for the Degree Master of Science, Department of Energy and Environment, Division of Power Engineering Chalmers University of Technology, Goteborg, Sweden, 2008.

17. M. M. D. Ross. A simple but comprehensive lead-acid battery model for hybrid system simulation, Workshop on Photovoltaic Hybrid Systems - PV Horizon, Montréal, Canada, September 10, 2001.

18. J. Jung, L. Zhang, J. Zhang (Eds.). *Lead-Acid Battery Technologies: Fundamentals, Materials and Applications*, Electrochemical Energy Storage and Conversion Series, CRC Press, Boca Raton, FL, 2015.

19. K. Kerman, A. Luntz, V. Viswanathan, Y.-M. Chiang, Z. Chena, Review—Practical challenges hindering the development of solid state Li ion batteries, *J. Electrochem. Soc.*, 164(7):A1731–A1744, 2017.

20. J.-G. Zhang, W. Xu, W. A. Henderson, Characterization and modeling of lithium dendrite growth, Chapter 2, pp 1–42. In *Lithium Metal Electrodes and Rechargeable Lithium Metal Batteries*, Springer, 2017, ISBN 978-3-319-44053-8.

21. Y. Liu, Q. Liu, L. Xin, Y. Liu, E. Yang, E. A. Stach, J. Xie, Making Li-metal electrodes rechargeable by controlling the dendrite growth direction, *Nat. Energy*, 2:17083, June 5, 2017, https://www.nature.com/articles/nenergy201783, Accessed September 25, 2018.

22. X. Yu, A. Manthiram, Electrode–electrolyte interfaces in lithium-based batteries, *Energy Environ. Sci.*, 3(11):527–543, 2018.

23. V. Neburchilov, J. Zhang (Eds.), *Metal-Air and Metal-Sulphur Batteries: Fundamentals and Applications*, Electrochemical Energy Storage and Conversion Series, CRC Press, Boca Raton, FL, 2016.

4 Permanent Magnet Motors and Halbach Arrays

4.1 MOTORS FOR ALL-ELECTRIC PROPULSION

The induction motor and permanent magnet motor are the common types of motor used for all-electric propulsion, due to their efficient operation. There are other types of motor that are proposed to be used for all-electric propulsion including switched reluctance motors, transverse flux motors and synchronous reluctance motors. There are significant differences in the construction of permanent magnet and induction motors.

4.2 HIGH TORQUE PERMANENT MAGNET MOTORS

4.2.1 RARE EARTH ELEMENTS

The discovery of magnetic alloy materials such as alnico and ferrite, rare-earth magnetic materials like cobalt and samarium, as well as the development of commercial products such as neodymium-iron-boron (NdFeB) has greatly aided and spurred the development of small high torque permanent magnet motors. Neodymium is a member of the family of materials known as light rare earth elements (LREE), along with lanthanum (used in optics) and samarium (also used in magnetic materials).

4.2.2 NEODYMIUM MAGNETS AND SAMARIUM–COBALT MAGNETS

Neodymium magnets, which are the most common type of rare-earth magnets, are made from an alloy of neodymium, iron and boron (Nd2Fe14B). Samarium–cobalt (SmCo5) magnets are typically two to three times stronger than ferrite or ceramic-type permanent magnets and could be used for the construction of electric motors. The sintered neodymium-iron-boron (NdFeB) hard magnetic material was patented by Sumitomo Special Metals in 1983 and it brought about a big change in the performance of electric motors.

4.3 MAGNETIC AND ELECTROMAGNETIC EFFECTS

4.3.1 MAGNETIC MATERIALS ON A MICROSCOPIC SCALE

Recall that the magnetic induction field (B) or the induction field induces a voltage, and necessarily an electric field on a test wire, coil or charge. Since the force on a charge is defined in terms of the field, it is the induction field that is associated with the Lorentz force law which is important to electric motors. The induction field is sometimes called the "magnetic flux density", as it measures the density of the magnetic field, that is, the concentration of magnetic flux tubes. The magnetic field intensity is introduced as

$$B = \mu_0 \mu_r H = \mu_0 \left(H + M \right),\tag{4.1}$$

where μ_r is the relative permeability of free space and is equal to 1 (for free space) and M is known as the magnetization. It is a vector field that expresses the density of the magnetic dipole moments induced in a magnetic material by an applied external field. The magnetic field intensity is the magnetic field of an electrically charged current. When an external field is applied to magnetic material, the magnetic induction field is enhanced by the magnetization properties of the material. The intrinsic magnetic moments within the material are responsible for magnetization and are related microscopic electric currents due to the orbital motion and spin of electrons and nuclei in the atoms.

The classical electromagnetic effect is the change in magnetization due to the application of an electric current. In this respect all magnetic materials generally fall into four broad categories: i) diamagnetic, ii) paramagnetic, iii) ferromagnetic and iv) ferrimagnetic.

The total contribution to the atomic magnetic moment is from three distinct sources, namely the magnetic orbital moment due to electrons orbiting the atomic nucleus, the magnetic spin moment due to the spin of the electrons about their axis (related to the so-called Bohr magneton) and the magnetic moment of the nucleus because of the spin of the nucleus about its own axis.

4.3.2 DIAMAGNETISM

In diamagnetic materials the total or net magnetic moment always adds up to zero. The distinctive feature of diamagnetism is that an applied electromagnetic field induces a magnetic field in opposition of an externally applied field, thus causing a repulsive effect. It is only exhibited by a diamagnetic material in the presence of an externally applied magnetic field. Thus, the force acting in a strong electromagnetic field is repulsive and proportional to the square of the current.

4.3.3 PARAMAGNETISM

Paramagnetic materials are generally attracted to magnetic fields. They have a relative magnetic permeability greater than one (or, equivalently, a positive magnetic susceptibility). For low levels of magnetization, Curie's law is a good approximation for the magnetization of paramagnets is given by $M = \chi H = \left(C/T \right) H$ where M is the resulting magnetization or magnetic moment, χ is the magnetic susceptibility, H is the auxiliary magnetic field, measured in amperes/meter, T is absolute temperature, measured in K and C is a material-specific Curie temperature.

Paramagnetic materials have a higher permeability than a vacuum. Diamagnetic materials have a lower permeability. The nature of paramagnetic materials causes them to be attracted to magnetic fields, whereas diamagnetic are actually repelled. In most cases diamagnetism is too weak to be observed, except in superconductors, which have perfect diamagnetism as a result of their zero resistance, a feature that is called the Meissner Effect. Consequently, the magnetic field is completely expelled from the interior of a superconducting material.

4.3.4 REMNANT MAGNETIC MOMENT

Paramagnetic materials are characterized by a net remnant magnetic moment per unit volume M, and may be described by the so-called Langevin function:

$$M = N_{dp}m_{dp}\left(\coth\left(\mu_0 m_{dp} H/kT\right) - \left(\mu_0 m_{dp} H/kT\right)^{-1}\right) \qquad (4.2)$$

where N_{dp} is total number of magnetic dipoles per unit volume, m_{dp} is the magnetic dipole moment (in Am^2) of the atom, μ_0 the permeability in a vacuum (1.256 637 10^{-6} H/m), k the Boltzmann constant (1.38×10^{-23} $J/°K$) and T the absolute temperature.

When this law is linearized for small arguments of the function $coth(.)$, one obtains a linear law, which is equivalent to Curie's law. (Note: The following electronic constants are standard: $q = 1.666 \times 10^{-19}$ C, $k = 1.38 \times 10^{-20}$ Joule/deg., $h = 6.62 \times 10^{-34}$ Joule sec and $c = 3 \times 10^8$ m/sec.)

4.3.5 FERROMAGNETISM

Iron, nickel, cobalt and some of the rare earths (gadolinium, dysprosium, alloys of samarium and neodymium with cobalt) exhibit a unique magnetic behavior which is called ferromagnetism when a small externally imposed magnetic field, say from an electric coil with a DC current flowing through it, can cause the magnetic domains to line up with each other and the material is said to be magnetized. In a ferromagnetic material the magnetic moment of the atoms on different sub-lattices are all in the same direction. The most important class of magnetic materials is the ferromagnets, which includes iron, nickel, cobalt, manganese and their compounds. These materials have a non-linear permeability which is best displayed by a B-H curve. The B-H curve of ferromagnetic materials show a rapid rise in the magnetic flux (B) which then approaches a steady value at an upper limit as the magnetic field intensity increases, a phenomenon known as "saturation".

A typical B-H curve is illustrated in Figure 4.1. However, most ferromagnetic materials used in motors are also electrically conductive and suffer from eddy current losses.

4.3.6 CURIE TEMPERATURE

All ferromagnets have a maximum temperature at which the ferromagnetic property disappears as a result of thermal agitation. This temperature is called the Curie temperature. It is a temperature above which certain magnetic materials experience a sudden change in their magnetic properties. Thus, the Curie temperature is a critical point beyond which a material's intrinsic magnetic moments change direction.

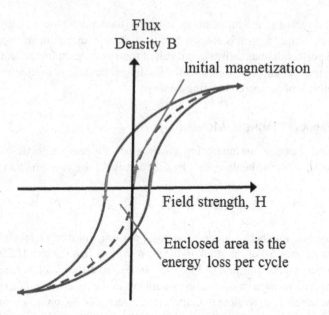

FIGURE 4.1 Typical B-H curve for ferromagnetic material.

4.3.7 Magneto-Striction

Magneto-striction is a property of all ferromagnetic materials that manifests itself as a relatively slight change of shape under the application of a magnetic field. Conversely, application of stresses with the subsequent change in shape or deformation can cause a magnetic field to be generated in these materials.

4.3.8 Ferrimagnetism

A ferrimagnetic material is one in which the magnetic moment of the atoms on different sub-lattices are opposed, as in anti-ferromagnetism. In ferrimagnetic materials, the opposing moments are unequal and a spontaneous magnetization remains.

4.3.9 The Maximum Energy Product

Ferromagnetic material magnets offer high levels of performance owing to their very high maximum energy product (MEP) compared to other magnetic materials. The maximum energy product is the measure of the magnetic energy which can be stored, per unit volume, by a magnetic material.

4.3.10 Coercivity

MEP is calculated as the maximum product of a material's residual magnetic flux density (degree of magnetization) and its *coercivity* is the ability to resist demagnetization once magnetized. In NdFeB, the remnant flux density and coercivity are both significantly higher than for other magnetic materials. Coercivity and remanence are both related to the tendency of some ferromagnetic materials to retain a magnetic field after an externally applied field is removed. Remanence is the flux density which remains, and coercivity is a measure of the field strength which must be applied to reduce the remnant flux to zero.

4.3.11 High Temperature Coercivity

A key ingredient in allowing NdFeB magnets to operate at high ambient temperatures is dysprosium, a heavy rare earth element (HRE) that is added to NdFeB in order to increase the high temperature coercivity (ability to withstand demagnetization) of the magnets above ~100°C. This has been essential in making it possible to use these magnets in high power density applications. Ames Labs of the US Department of Energy, in Ames, Iowa, have found that cerium can substitute for dysprosium when properly co-alloyed with cobalt.

4.3.12 Curie Temperature of NdFeB

The Curie temperature is the point above which magnetic properties for the alloy or metal are lost. Earlier attempts to use cerium in magnets were unsuccessful as it actually lowered the Curie temperature. When cobalt is added to the mix, you get an alloy that performs better than anything else above 150°C.

4.3.13 Intrinsic Coercivity

Several important variables come into play when creating a magnetic alloy for electric motor applications.

Intrinsic coercivity at elevated temperatures is key for many common applications. This is the intrinsic ability of the material to resist demagnetization. It is the strength of the magnetic field that is necessary to make the magnetic polarization (strength of magnetization) to zero.

4.3.14 Intrinsic and Normal Coercivity Compared

Normal coercivity is the applied field required to reduce the external field generated by the magnet to zero. Intrinsic coercivity is the applied field required to fully demagnetize the material. Thus, intrinsic coercivity is the magnitude of the demagnetizing field that would overcome the material's magnetization and reduce its net value to zero.

4.3.15 Permanent Magnets with Reduced Rare Earth Elements

Samarium–cobalt (SmCo) are attractive alternatives and can withstand higher temperatures than NdFeB without suffering demagnetization. They are more expensive than NdFeB magnets as they use the rare earth material samarium and expensive cobalt. (Lanthanum, another rare earth element, could also be used). Ferrite magnets are the classical materials for permanent magnets and are manufactured from iron oxide combined with the metals strontium, barium or cobalt.

4.4   HALBACH ARRAY MOTORS

A Halbach array, first introduced by Halbach [1], is an arrangement of magnets designed to enhance the magnetic flux in a certain domain. Figure 4.2 shows a linear array of permanent magnets designed to create a flux field on one side of the array and almost cancel the flux on the other side, except at the two ends of the arrangement. The magnetization pattern is indicated by the arrows in (a) and the flux field in (b).

A typical arrangement of magnets in Halbach array is shown in Figure 4.3; the magnets are oriented to enhance the field on the top of the disc and cancel it below the disc, so as to provide a net axial flux field.

Figure 4.4 shows a typical cross-sectional view of a dipole constructed from a Halbach array. Also shown in Figure 4.4 are the directions of magnetization of the permanent magnets that make up the array. The result of the arrangement is a net uniform field inside the array directed vertically as well as the almost complete cancellation of the field outside the array. The Halbach arrays were soon used to develop highly efficient, high performance electric motors for a range of applications [3]. A typical axial field brushless DC motor with a disc of permanent magnets has been developed by Gallo [4]. Analytical expressions for the flux field solutions for an axial Halbach array of permanent magnets have been developed by Thompson [5]. Figure 4.5 shows a typical Halbach disc with 32 segments, where each segment has a slot for a permanent magnet and the field of each magnet has changed its direction by 22.5° after each segment. The net field generated is therefore axial.

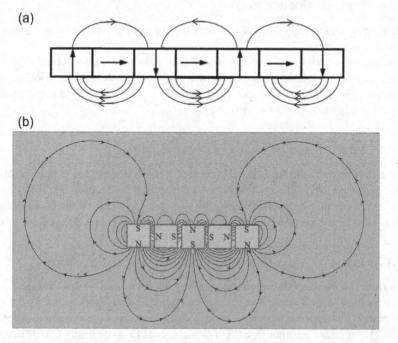

FIGURE 4.2   A linear Halbach array of permanent magnets, shown in (a), designed to enhance the flux on one side of the array and almost cancel the flux on the other side. Also shown in the figure, (b) are the typical field lines.

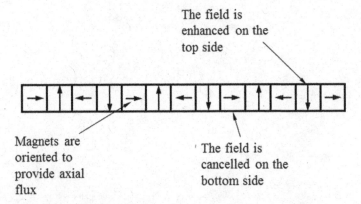

FIGURE 4.3   A Halbach array for generating an axial field.

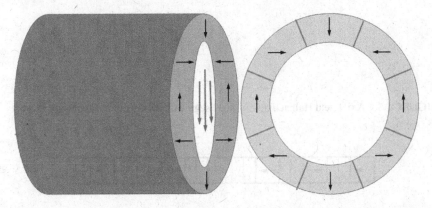

FIGURE 4.4   A Halbach array for emulating the field generated by a dipole magnet (redrawn from Merritt et al. [2]).

Halbach arrays have been successfully used to develop linear motors [6]. Siemens [7] have successfully developed an electric motor for aircraft propulsion applications.

4.4.1   Complex Halbach Arrays

In order to achieve a smooth distribution of the flux lines, which tends to increase the back-EMF, it is required to increase the number of segments. Figure 4.6 illustrates the arrangements of two Halbach linear arrays, one where the shift in the direction of the field in 90° after each segment, and another where the shift in the direction of the field in 45° after each segment.

An axial flux permanent magnet motor could use a typical pair of Halbach arrays [8], [9]. A representation of a double-sided Halbach magnet array as used in an axial flux permanent magnet motor is shown in Figure 4.7.

The peak magnetic flux density $\tilde{B}$ for segmented Halbach arrays with the magnet height h_m and the number of magnets n_m per spatial period λ is given by,

$$\tilde{B} = B_r \left(1 - \exp\left(-2\pi h_m/\lambda\right)\right)\sin\left(\pi/n_m\right). \tag{4.3}$$

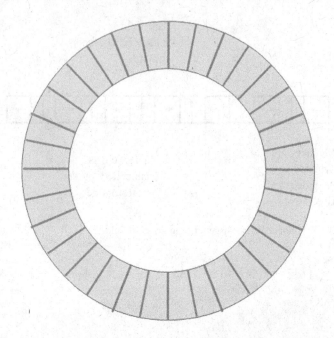

FIGURE 4.5   Axial field Halbach disc with 32 segments for slots with permanent magnets.

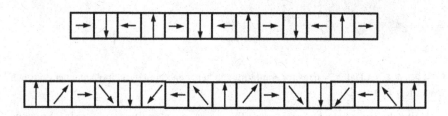

FIGURE 4.6   Figure shows a 90° (top) and a 45° (bottom) Halbach linear array of permanent magnets. The magnets in the 45° array would be half the size, and twice as many would be needed.

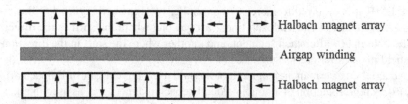

FIGURE 4.7   Double-sided Halbach magnet array arrangement for an axial flux permanent magnet motor without the back iron, and with a molded air gap winding in between (Winter et al. [10]).

In Equation (4.3) B_r denotes the remanent magnetic flux density of the chosen magnet material. The force on a charged particle in an electromagnetic field, where the electric field is denoted by $\mathbf{E}$, the magnetic field by $\mathbf{B}$ and $\mathbf{v}$ is the velocity of the charged particle, is the Lorentz force, and is given by the relation,

$$\mathbf{F} = \mathbf{q}\left(\mathbf{E} + \left(\mathbf{v} \times \mathbf{B}\right)\right). \tag{4.4}$$

Thus, the force on a current-carrying conductor of length $\mathbf{L}$ with a current $\mathbf{I}$ is given by,

$$\mathbf{F} = \mathbf{I}\left(\mathbf{L} \times \mathbf{B}\right). \tag{4.5}$$

For an axial motor with two equal and opposite force acting at distance from each other a torque is generated. Under the assumption of a sinusoidal current I_{rms} and magnetic flux ϕ derived by $\tilde{B}$, the electromagnetic torque magnitude T generated by axial flux permanent magnet motor is

$$T = \frac{m_1}{\sqrt{2}} pN\phi I_{\text{rms}}. \tag{4.6}$$

Here, m_1 is the number of phases, p is the number of pole pairs and N is the number of turns per phase.

4.4.2 Ring Type Structures

An alternate type of arrangement is a radial Halbach array shown in Figure 4.8. In this case, there are three Halbach segments per pole and six poles. The direction of the magnetic field of each permanent magnet has changed by 40° after each segment.

4.5 MODELING THE MAGNETIC FIELD DUE TO A HALBACH ARRAY

There is a need for a simple and accurate analytical model for prediction and optimization of the electromagnetic performance, of a brushless machine having a Halbach array of permanent magnets, together with a physical understanding of the relationship between parameters and performance. Thus, there is a need to develop an analytical representation for the field distributions produced by a Halbach array of permanent magnets, so it is possible to accurately predict the average output torque for a range of input currents. All motors generate a torque following the interaction of electrical currents with a magnetic field. The force on a charged particle in an electromagnetic field is given the name "Lorentz force". Since current flow in a conductor is due to the continuous motion of electrical charge, the application of a magnetic field on a current-carrying conductor of finite length results in a net force. By configuring two conductors such that the forces on both these conductors are in

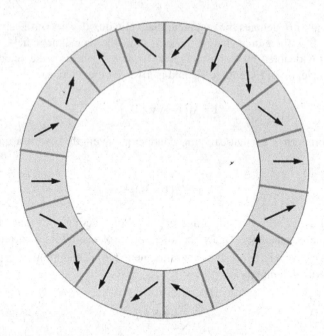

FIGURE 4.8 Radial Halbach array, with three segments per pole and six poles.

opposite directions, but offset by a distance, a net torque is generated. The torque produces rotation of armature. The other important law in relation to electric motors is Lenz's law which results in the induced EMF always countering the motion of the armature. Consequently, the source of back-EMF is the rotation of armature and the electromagnetic field. Thus, the strength and variation of the magnetic field is of primary importance in predicting the performance characteristics of a motor. The magnetic field is governed by Maxwell's laws of electromagnetism which are:

$$\nabla \times \mathbf{H} = \mathbf{J} + \frac{\partial \mathbf{D}}{\partial t}, \quad \nabla \times \mathbf{E} = -\frac{\partial \mathbf{B}}{\partial t}, \quad \nabla \cdot \mathbf{D} = \rho, \text{ and } \nabla \cdot \mathbf{B} = 0, \tag{4.7}$$

where the electric displacement field, $\mathbf{D}$, the electric current density $\mathbf{J}$, the magnetic flux density or magnetic flux per unit area $\mathbf{B}$ are related to the electric field $\mathbf{E}$ and the magnetic field strength $\mathbf{H}$, by the material relations, $\mathbf{D} = \varepsilon \mathbf{E}$, $\mathbf{J} = \sigma \mathbf{E}$ and $\mathbf{B} = \mu \mathbf{H}$ where ε, σ and μ are the permittivity, the conductivity and the permeability, respectively, of the medium. The last relation may also be expressed as

$$\mathbf{B} = \mu \mathbf{H} = \mu_0 \left(\mu_r \mathbf{H} + \mathbf{M} \right), \tag{4.8}$$

where μ_0 is the permeability of free space, μ_r is the relative permeability of the medium and $\mathbf{M}$ is the magnetization. The magnetic flux density $\mathbf{B}$ can also be derived from a vector potential function, $\mathbf{A}$. Thus,

$$\mathbf{B} = \nabla \times \mathbf{A}. \tag{4.9}$$

The vector potential function permits one to express the magnetic flux due the presence of the permanent magnets as well as the air gaps i.e. in the magnetic regions, the slots and the airgaps by a Poisson equation for the vector potential function $\mathbf{A}$, where the right-hand side of the Poisson equation is zero only in the air gaps but non-zero in the magnetic regions and the slots. Thus, from Equation (4.8) and the first of Equations (4.7), under steady-state conditions,

$$\mathbf{H} = \frac{1}{\mu_r}\left(\frac{\mathbf{B}}{\mu_0} - \mathbf{M}\right), \quad \frac{1}{\mu_r \mu_0}\nabla \times \mathbf{B} = \mathbf{J} + \nabla \times \frac{\mathbf{M}}{\mu_r}. \tag{4.10}$$

From Equation (4.9):

$$\nabla \times (\nabla \times \mathbf{A}) = \nabla(\nabla \cdot \mathbf{A}) - \nabla^2 \mathbf{A} = \mu_r \mu_0 \mathbf{J} + \mu_0 \nabla \times \mathbf{M}. \tag{4.11}$$

Thus, if the vector potential function, $\mathbf{A}$ is also solenoidal i.e. when $\nabla \cdot \mathbf{A} \equiv 0$,

$$\nabla^2 \mathbf{A} = -\mu_r \mu_0 \mathbf{J} - \mu_0 \nabla \times \mathbf{M}. \tag{4.12}$$

When, $\mathbf{J} \equiv 0$, as in the magnetic region,

$$\nabla^2 \mathbf{A} = -\mu_0 \nabla \times \mathbf{M}. \tag{4.13}$$

When the $\mathbf{M} \equiv 0$ and $\mu_r = 1$, as in the slots

$$\nabla^2 \mathbf{A} = -\mu_0 \mathbf{J}. \tag{4.14}$$

When both $\mathbf{J} \equiv 0$ and $\mathbf{M} \equiv 0$, as in the air gaps, one has the vector Laplacian equation,

$$\nabla^2 \mathbf{A} = \mathbf{0}. \tag{4.15}$$

The specific arrangements of the Halbach array in a particular motor determine the right-hand side of the Poisson equation in the magnetic regions, while the current density in the conductors determines the right-hand side in the slots. Consequently, the magnetic flux density $\mathbf{B}$ is expressed as a Fourier series in the circumferential direction which can be derived from the solution for the vector potential $\mathbf{A}$. The actual analysis may be carried out by a typical finite element analysis.

CHAPTER SUMMARY

In this chapter, the focus is on current high performance permanent magnet electric motors, including Halbach arrays which have been briefly considered. These motors are known to be particularly suitable for the traction or propulsion of electric vehicles and for this an overview of their features and construction is presented as they could be significantly different from conventional electric motors.

REFERENCES

1. K. Halbach, Design of permanent multipole magnets with oriented rare earth cobalt material, *Nucl. Instrum. Methods*, 169(1):1–10, 1980.
2. B. T. Merritt, R. F. Post, G. R. Dreifuerst, D. A. Bender, *Halbach Array Motor/ Generators – A Novel Generalized Electric Machine*, Lawrence Livermore National Laboratory, UCRL-JC-119050, October 28, 1994, Presented at the Halbach Festschrift Symposium, Berkeley, CA, February 3, 1995.
3. B. C. Mecrow, A. G. Jack, D. J. Atkinson, S. R. Green, G. J. Atkinson, A. King, B. Green, Design and testing of a four-phase fault-tolerant permanent-magnet machine for an engine fuel pump, *IEEE Trans. Energy Convers.*, 19(4):671–678, 2004.
4. C. A. Gallo, *Halbach Magnetic Rotor Development*, NASA/TM—2008-215056, National Aeronautics and Space Administration, Glenn Research Center, Cleveland, OH, February 2008.
5. W. K. Thompson, *Three-Dimensional Field Solutions for Multi-Pole Cylindrical Halbach Arrays in an Axial Orientation*, National Aeronautics and Space Administration (NASA)—2006-214359, September 2006.
6. A. K. M. Parvez-Iqbal, F. A. M. Mokhtar, K. S. M. Sahari, I. Aris, A review of permanent magnet linear motor with Halbach arrays, *J. Eng. Appl. Sci.*, 11(8):1752–1761, 2016.
7. Anon. *World-Record Electric Motor Makes First Flight*, Press Release, Siemens AG, Munich, July 4, 2016, https://press.siemens.com/global/en/pressrelease/world-record-electric-motor-makes-first-flight, Accessed December 29, 2019.
8. L. Yan, J. Hu, Z. Jiao, I.-M. Chen, C. K. Lim, Magnetic field modeling of linear machines with double-layered halbach arrays, International Conference on Fluid Power and Mechatronics (FPM), August 2011, Beijing, China. doi:10.1109/FPM.2011.6045891.
9. B. P. Ruddy, High force density linear permanent magnet motors, PhD Dissertation in Mechanical Engineering, Massachusetts Institute of Technology, September 2012.
10. O. Winter, S. Ucsnik, M. Rudolph, C. Kral, E. Schmidt, Ironless in-wheel hub motor design by using multi-domain finite element analyses, IEEE International Symposium on Power Electronics, Electrical Drives, Automation and Motion, Sorrento, Italy, pp 1474–1478, June 20–22, 2012.

5 Introduction to Boundary Layer Theory and Drag Reduction

5.1 PRINCIPLES OF AIRFOIL AND AIRFRAME DESIGN

The fundamental component of an airframe which provides the basis of the design of any lifting vehicle in the airfoil. A wing is simply an ordered collection of wing or airfoil sections. An essential component of an airframe is a wing, or indeed a pair of wings, which are primarily responsible in providing the lifting vehicle with its lift force distributions, which in turn define its lifting characteristics. Yet the whole of the airframe including the wings, the fuselage, control surfaces and pylons contribute to the drag forces acting on the vehicle. A key element in the design of a lifting vehicle is the estimation, reduction and subsequent optimization of the drag forces. The need for optimization arises as one is interested in minimizing the drag forces while ensuring the lifting characteristics of the vehicle are not compromised. The geometry, and in fact the entire topology of the airframe, must therefore be designed to ensure minimum aerodynamic drag without having to compromise any of the lifting and performance characteristics of the design in its entirety.

5.2 FLOW OVER AN AEROFOIL

In a real flow field, potential flow theories of flow around an airfoil are generally applicable, which generally implies that all viscous forces may be neglected, provided the Kutta–Joukowski condition for a smooth flow at the trailing edge are imposed. The lift of an airfoil is created by a pressure differential between the bottom, or pressure side, and top, or suction side. The drag force developed is assumed to be in the same direction as the relative wind velocity direction. The net lift force is assumed to be normal to the direction of the drag force. Under such circumstances, the lift, drag and pitching moment characteristics of an airfoil can be assumed to be functions of the angle of attack alone. Given the lift force L and the drag force D per unit span and the chord length c, the coefficients of lift and drag may be defined as

$$C_\ell = \frac{L}{\frac{1}{2}\rho V^2 c},$$ (5.1)

$$C_d = \frac{D}{\frac{1}{2}\rho V^2 c}.$$ (5.2)

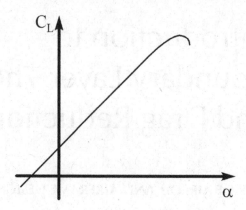

FIGURE 5.1   Lift coefficient characteristic (from Vepa [1]).

Over the useful range of the angle of attack, α, which usually means the that $\alpha < \alpha_{\text{stall}}$, where α_{stall} is the angle of attack at which there is loss of lift due to the flow field stalling, so the corresponding lift coefficient is maximum, the lift coefficient c_l behaves linearly with α and could be expressed as

$$C_\ell(\alpha) = a_0(\alpha - \alpha_0), \quad \alpha_0 < \alpha < \alpha_{\text{stall}}, \quad C_\ell(\alpha_{\text{stall}}) = C_{\ell,\max}. \tag{5.3}$$

The angle α_0 is known as the zero-lift angle of attack and defines the lower limit of the range of α values over which the Equation (5.3) is valid. The stall-region of the $C_\ell = C_\ell(\alpha)$ characteristic, illustrated in Figure 5.1, occurs immediately after the maximum coefficient of lift is achieved. This region shows a dramatic decrease in lift and a corresponding increase in drag that is a result of separation of the flow when the flow is no longer attached over portions of the suction side of the airfoil.

5.3   AERODYNAMIC DRAG

If one is interested in minimizing the total drag force acting on a vehicle, it is absolutely vital that one is able to estimate the total rag force acting on it. The estimation of aerodynamic drag which is the sum of three components: form, friction (i.e. profile) drag and induced drag, is normally best done from test data generated within the confines of a wind tunnel. A model of the vehicle, or one of its components, is suspended in a wind tunnel by special support and coupled to an *aerodynamic balance* which is used to measure the forces and moments acting on the component. The wind tunnel experiments are usually performed for dozens of different parametric values (different angles of attack or sideslip angle) and similar configurations. The data is then appropriately non-dimensionalized and re-interpreted to obtain the aerodynamic drag values corresponding to the real component. The most commonly adopted method of estimating the drag of the aircraft is to estimate the drag of the components separately and then sum up the individual contributions to the total drag.

A typical plot of the variations of the total drag coefficient and the profile and induced drag contributions to it versus the equivalent airspeed at sea level is shown in Figure 5.2(a). Unfortunately, the minimum value of the drag and the maximum lift

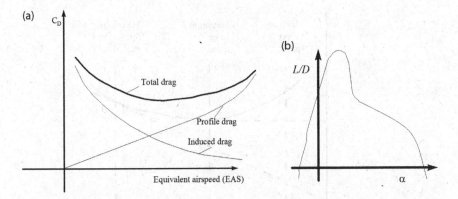

FIGURE 5.2 (a) Plot of wing drag components versus equivalent (sea level) air speed (from Vepa [1]). (b) Variation of the L/D ratio with the angle of attack (from Vepa [1]).

do not occur for the same angle of attack; the drag is a minimum at about 1°, while the lift is a maximum at about 15°. Thus, there is a need to compromise, and this is done by considering the variation of the ratio of the lift to the drag with the angle of attack. A plot of this ratio is shown in Figure 5.2(b) which is usually at about $\alpha \approx 4°$. The curve shows that there is a steep fall in the value of L/D when the angle of attack exceeds the stalling angle. This is due to the rate of increase of drag increasing sharply before stall.

The speed at which the maximum value of the L/D ratio can be attained corresponds to minimum thrust. As the angle of attack is increased, the coefficient of drag and equilibrium flight requires a lower flight speed. As the stalling angle is approached, the effect of the increasing drag coefficient outweighs that of the decreasing velocity and more thrust is required at the stalling speed than at other speeds. On the other hand, there is a limit to the amount of thrust that can be delivered, and as a consequence the maximum speed depends primarily on engine power.

For an aircraft in level flight, based on the total drag curve, one obtains analytical approximations to a number of related airspeeds. Three points on the drag curve are particularly important. These are: the minimum drag speed, the minimum power speed and the speed at the maximum speed/drag ratio. They are illustrated in Figure 5.3. The drag curve in the figure is estimated for the case of level flight when the lift is maintained a constant. Hence, it follows that when the drag is at minimum, the ratio of the lift to the drag (L/D) is a maximum. From Equations (5.1) and (5.2), the lift and drag forces on a wing section of area S could be expressed in terms of the wing section lift and drag coefficients as

$$L = (1/2)\rho V^2 \times S \times C_L \text{ and } D = (1/2)\rho V^2 \times S \times C_D. \tag{5.4}$$

Further, as the total drag is the sum of the profile drag and induced drag, the drag coefficient may be expressed as the sum of two components: the profile drag coefficient, C_{D0} and the induced drag coefficient, C_{Di}.

$$C_D = C_{D0} + C_{Di}, \tag{5.5}$$

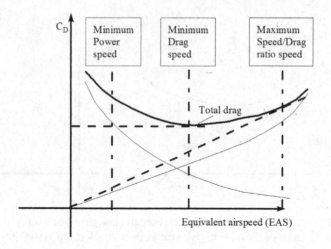

FIGURE 5.3 Definition of various optimum flight speeds (from Vepa [1]).

which may be expressed as,

$$C_D = C_{D0} + C_L^2/(\pi e AR) = C_{D0} + KC_L^2. \tag{5.6}$$

Equation (5.6) is referred to as the *drag polar*. In Equation (5.6), AR is the aspect ratio, C_L is the wing lift coefficient, $e = 1/k$, and k is a constant that is known as Oswald's efficiency factor and $K = 1/(\pi e AR)$.

The minimum drag speed is then obtained by maximizing the (L/D) ratio with respect to the airspeed. But the (L/D) ratio in steady level flight is expressed as

$$\frac{L}{D} = \frac{(1/2)\rho U_0^2 \times S \times C_L}{(1/2)\rho U_0^2 \times S \times C_D} = \frac{C_L}{C_D} = \frac{C_L}{C_{D0} + KC_L^2}. \tag{5.7}$$

Thus the (L/D) ratio is a maximum when C_D/C_L is minimum. However, the C_D/C_L ratio is given by

$$\frac{C_D}{C_L} = \frac{C_{D0} + KC_L^2}{C_L} = \frac{C_{D0}}{C_L} + KC_L. \tag{5.8}$$

Thus C_D/C_L is minimum when

$$K - C_{D0}/C_L^2 = 0. \tag{5.9}$$

The minimum of the above expression occurs when

$$C_L = \sqrt{C_{D0}/K} \equiv (C_L)_{\text{min drag}}. \tag{5.10}$$

Hence, the (L/D) ratio is

$$(L/D)_{max} = 1/2\sqrt{KC_{D_0}}. \tag{5.11}$$

But the lift force is given by

$$L = (1/2)\rho U_0^2 C_L S = (1/2)\rho_0 U_E^2 C_L S. \tag{5.12}$$

Moreover, the aircraft is flown at constant lift, and the equivalent airspeed corresponding to minimum drag is

$$U_{min\ drag} = \sqrt{\frac{2L}{\rho_0 \left(C_L\right)_{min\ drag} S}}. \tag{5.13}$$

A key performance parameter related to the minimum drag speed is the minimum power speed. To obtain the minimum power speed, one must consider the minimum power required to maintain a constant speed in level flight. The power, which is the product of the drag and airspeed, is minimum when the drag decreases as fast as the speed increases. If the drag decreases any faster or slower, the product of the drag and airspeed is not a minimum when plotted against the airspeed. Since the airspeed is proportional to $\sqrt{1/C_L}$, the power required is minimum when $C_D/\left(C_L\right)^{3/2}$ is a minimum. But

$$\frac{C_D}{\left(C_L\right)^{3/2}} = \frac{C_{D_0} + KC_L^2}{\left(C_L\right)^{3/2}} = \frac{C_{D_0}}{\left(C_L\right)^{3/2}} + K\sqrt{C_L}. \tag{5.14}$$

The minimum of the above expression occurs when

$$C_L = \sqrt{3C_{D_0}/K} \equiv \left(C_L\right)_{min\ power} = \sqrt{3}\left(C_L\right)_{min\ drag}. \tag{5.15}$$

The equivalent airspeed corresponding to minimum power is

$$U_{min\ power} = \sqrt{\frac{2L}{\rho_0 \left(C_L\right)_{min\ power} S}} = \frac{U_{min\ drag}}{3^{1/4}}. \tag{5.16}$$

To obtain the speed at which the speed to drag ratio is a maximum, one must consider the minimum drag to speed ratio at a constant lift in level flight. Since the airspeed is proportional to, $\sqrt{1/C_L}$, the drag to speed ratio is a minimum when $C_D/\left(C_L\right)^{1/2}$ is minimum. But

$$\frac{C_D}{\left(C_L\right)^{\frac{1}{2}}} = \frac{C_{D_0} + KC_L^2}{\left(C_L\right)^{\frac{1}{2}}} = \frac{C_{D_0}}{\left(C_L\right)^{\frac{1}{2}}} + K\left(\sqrt{C_L}\right)^3. \tag{5.17}$$

The minimum of the above expression occurs when

$$C_L = \sqrt{\frac{C_{D_0}}{3K}} \equiv \left(C_L\right)_{\text{max drag/speed}} = \frac{1}{\sqrt{3}} \left(C_L\right)_{\text{min drag}}. \tag{5.18}$$

The equivalent airspeed corresponding to minimum power is

$$U_{\text{min power}} = \sqrt{\frac{2L}{\rho_0 \left(C_L\right)_{\text{max drag/speed}} S}} = \frac{U_{\text{min drag}}}{\left(1/3\right)^{1/4}}. \tag{5.19}$$

The estimation of these speeds provides an idea of the range of speeds at which the aircraft is expected operate in steady and stable flight. Moreover, it is clear from the analysis that the two key coefficients that play a role in the performance of the vehicle are the profile drag coefficient C_{D_0} and the lift coefficient C_L. In this chapter the estimation of the profile drag, which is entirely due to the skin friction effects due to the viscosity of the fluid in the flow, and the profile drag coefficient is considered in some detail. Viscous effects are generally confined to the flow adjacent to the wing surface to region known as the boundary layer.

5.4   BOUNDARY LAYER FLOW

Fluid flow is governed by the Navier–Stokes equations which are essentially an application of Newton's second law of motion to a fluid within a control volume. In the absence of viscosity, they reduce to the well-known equations of Euler. In the early years of the development of fluid mechanics, two groups emerged, one based on experimental observations of real viscous fluids and another based on the inviscid equations of Euler. While the equations of Euler have several merits, and form the basis of the potential flow theory, they cannot explain why airfoils in real viscous flows experience drag, as potential flow predicts that airfoils cannot experience drag. This shortcoming of potential flow theory and the experience of viscous flows, known as d'Alembert's paradox, was resolved after Ludwig Prandtl [2] proposed his boundary layer theory and the two groups began to understand each other. Prandtl showed that when considering the flow of real fluid around a solid body, the domain can be partitioned into two regions as shown in Figure 5.4; one region is relatively far away from the body where the effects of viscosity could be completely ignored, so the flow is essentially inviscid and the other region, termed the *boundary layer*,

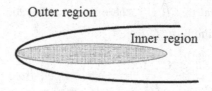

FIGURE 5.4   The boundary layer concept.

a thin layered region close to the solid body, where the effects of viscosity must be considered. The *boundary layer* can be defined as "that part of the flow that is directly influenced by the presence of the body's surface and responds to surface excitations of the flow within a characteristic time-scale". Considering the derivation of the Navier–Stokes equations, it must be observed that the derivation assumed that the length scales and time-scales were based on the dimensional units rather than meaning physical length and time-scales.

The key point in the derivation of the boundary layer theory is the physically meaningful assumption that there are different scales of distances in differing coordinate directions and that there is a physically meaningful time-scale associated with the flow. Thus, in two-dimensional flow the existence of two different length scales, in the two orthogonal directions of flow, is assumed. In the streamwise direction or in the direction of the main flow it is some characteristic length scale of the object in the flow in that direction, while in the normal direction it is the thickness of the boundary layer which may be broadly defined as the region where viscous forces play a role in the flow; they can be neglected, in principle, everywhere else. It is important to recognize that the actual thickness of the boundary layer will vary along the direction of the flow, often increasing monotonically.

An in-depth understanding of boundary layers and their influence on drag, is essential for drag reduction which is indeed a central issue in the design of electric aircraft.

5.4.1 THE NAVIER–STOKES (NS) EQUATIONS

Consider a small parallelepiped enclosing a volume of fluid and located at the Cartesian coordinates, $[x_1 \, x_2 \, x_3]$. The velocity vector at this location is assumed to be $\mathbf{u} = [u_1 \, u_2 \, u_3]$. The stress acting in the jth direction on a face of the parallelepiped that has its normal in the ith direction is denoted by σ_{ij} and the external body forces on the element by $\mathbf{F} = [F_1 \, F_2 \, F_3]$. Assuming the density ρ to be constant and applying Newton's second law of motion,

$$\rho \frac{Du_i}{Dt} = \frac{\partial \sigma_{ji}}{\partial x_j} + F_i. \tag{5.20}$$

In the above equation, the total or substantive derivative with respect to the time t following the motion D/Dt, sometimes referred to as "the Stokes derivative", consists of an unsteady term and a convective term and is

$$\frac{D}{Dt} = \frac{\partial}{\partial t} + \mathbf{u} \cdot \nabla, \quad \nabla = \left[\frac{\partial}{\partial x_1} \quad \frac{\partial}{\partial x_2} \quad \frac{\partial}{\partial x_3} \right]. \tag{5.21}$$

It is also convenient to separate the stress tensor, σ_{ij}, into an inviscid part due to the fluid pressure p and a viscous stress tensor, σ'_{ij}, given by

$$\sigma_{ij} = (1/3)(\sigma_{11} + \sigma_{22} + \sigma_{33})\delta_{ij} + \sigma'_{ij} \equiv -p\delta_{ij} + \sigma'_{ij}, \tag{5.22}$$

where δ_{ij} is defined by, $\delta_{ii} = \delta_{jj} = 1$, $\delta_{ij} = 0$, $i \neq j$. The viscous stress is generally assumed to be linearly proportional to rate of strain tensor e_{ij} and consequently,

$$\sigma'_{ij} = \mu e_{ij} = \mu \left(\frac{\partial u_i}{\partial x_j} + \frac{\partial u_j}{\partial x_i} \right). \tag{5.23}$$

The equations of motion of the parallelepiped of fluid reduce to,

$$\rho \frac{\partial u_i}{\partial t} + \rho \mathbf{u} \cdot \nabla u_i = -\frac{\partial p}{\partial x_i} + \mu \frac{\partial}{\partial x_j} \left(\frac{\partial u_i}{\partial x_j} + \frac{\partial u_j}{\partial x_i} \right) + F_i. \tag{5.24}$$

These are the celebrated Navier–Stokes (NS) equations for the motion of the fluid. The NS equations may be written in a convective vector form as:

$$\frac{\partial}{\partial t} \mathbf{u} + (\mathbf{u} \cdot \nabla) \mathbf{u} = -\rho^{-1} \nabla p + v \nabla^2 \mathbf{u} + \mathbf{F}. \tag{5.25}$$

However, this form is not invariant under a change of coordinate system. Hence it is preferable to write the NS equations in a rotational form as,

$$\frac{\partial}{\partial t} \mathbf{u} + (\nabla \times \mathbf{u}) \times \mathbf{u} + \frac{1}{2} \nabla (\mathbf{u} \cdot \mathbf{u}) = -\rho^{-1} \nabla p + v \nabla^2 \mathbf{u} + \mathbf{F}. \tag{5.26}$$

The NS equations may also be expressed in terms of the vorticity which is defined as, $\omega = \nabla \times \mathbf{u}$. This form of the NS equations is useful when it is essential to eliminate the pressure explicitly from the equations. Thus, taking the vector cross-product of the vector gradient operator ∇, and the NS equations and assuming a conservative external body force field,

$$\frac{\partial}{\partial t} \omega + \nabla \times (\omega \times \mathbf{u}) = v \nabla^2 \omega + \nabla \times \mathbf{F} = v \nabla^2 \omega. \tag{5.27}$$

This form of the NS equations is known as the vorticity transport equations and show that the local changes in vorticity are due to the stretching of vortex lines, which are lines whose tangents are parallel to the local vorticity vector, and due to convection and viscous conduction. The stretching of vortex lines is absent in two-dimensional flow. In two-dimensional flow, introducing the stream function $\bar{\psi} = \bar{\psi}(x, y)$ to satisfy the continuity equation, so that the velocity components and the vorticity component normal to plane containing the velocity components are respectively given by,

$$u = \partial \psi / \partial y, \quad v = -\partial \psi / \partial x \text{ and } \omega = -\nabla^2 \bar{\psi}. \tag{5.28}$$

The vorticity transport equations reduce to a single scalar equation for $\bar{\psi} = \bar{\psi}(x, y)$ given by,

$$\left(\frac{\partial}{\partial t} + \frac{\partial \bar{\psi}}{\partial y} \frac{\partial}{\partial x} - \frac{\partial \bar{\psi}}{\partial x} \frac{\partial}{\partial y} \right) \nabla^2 \bar{\psi} = v \nabla^4 \bar{\psi}. \tag{5.29}$$

While a number of important features of the flows of incompressible fluids may be derived from the vorticity transport equations, two of these relate to the generation of vorticity within a viscous fluid and the nature of vorticity creation by viscosity. Vorticity cannot be generated within an incompressible viscous fluid subject only to conservative external forces, but is only diffused inwards from the boundaries. The creation of vorticity by viscosity over a time, is a non-analytic process while the decay or dissipation of it is analytic.

5.4.2 VISCOUS ENERGY DISSIPATION

To understand the analytic nature of vorticity dissipation, it is essential to recognize that the energy dissipated is always positive definite in the presence of viscosity. Consider the work done by the stress field σ, per unit time and per unit mass, on a deforming element of fluid which may be expressed as

$$W = \frac{1}{\rho}\frac{\partial\left(\sigma_{ji}u_i\right)}{\partial x_j} = \frac{1}{\rho}\left\{u_i\frac{\partial\sigma_{ji}}{\partial x_j} + \sigma_{ji}\frac{\partial u_i}{\partial x_j}\right\}. \tag{5.30}$$

In the above expression, repeated indices are assumed to be summed from 1 to 3. Using the equations of motion with, $F_i = 0$

$$u_i\frac{\partial\sigma_{ji}}{\partial x_j} = \rho u_i\frac{Du_i}{Dt} = \frac{\rho}{2}\frac{D\left(u_ju_i\right)}{Dt}. \tag{5.31}$$

The right-hand side of the above equation is the rate of change of the kinetic energy. Thus, the work done by the stress field is

$$W = \frac{1}{2}\frac{D\left(u_ju_i\right)}{Dt} + \frac{\sigma_{ji}}{\rho}\frac{\partial u_i}{\partial x_j}$$

$$= \frac{1}{2}\frac{D\left(u_ju_i\right)}{Dt} - \left\{\frac{p}{\rho}\delta_{ij} - \frac{\mu}{\rho}\left(\frac{\partial u_i}{\partial x_j} + \frac{\partial u_j}{\partial x_i}\right)\right\}\frac{\partial u_i}{\partial x_j}, \tag{5.32}$$

which reduces to

$$W = \frac{1}{2}\frac{D\left(u_ju_i\right)}{Dt} + \frac{\mu}{\rho}\left(\frac{\partial u_i}{\partial x_j} + \frac{\partial u_j}{\partial x_i}\right)\frac{\partial u_i}{\partial x_j}$$

$$= \frac{1}{2}\frac{D\left(u_ju_i\right)}{Dt} + \frac{\mu}{2\rho}\left(\frac{\partial u_i}{\partial x_j} + \frac{\partial u_j}{\partial x_i}\right)^2. \tag{5.33}$$

The above expression reflects the rate change of kinetic and potential energies. The total rate of change of potential energy is given by the volume integral

$$E = -\frac{\mu}{2}\int_0^V\left(\frac{\partial u_i}{\partial x_j} + \frac{\partial u_j}{\partial x_i}\right)^2 dV. \tag{5.34}$$

In two-dimensional flow, it reduces to,

$$E = -\mu \int_0^V \left\{ \left(\frac{\partial u_1}{\partial x_2} + \frac{\partial u_2}{\partial x_1} \right)^2 + 2\left(\frac{\partial u_1}{\partial x_1} \right)^2 + 2\left(\frac{\partial u_2}{\partial x_2} \right)^2 \right\} dV. \tag{5.35}$$

Thus, in the presence of viscosity, the total rate of change of potential energy is a positive definite function, indicating that only energy dissipation is possible.

5.4.3   NON-DIMENSIONALIZING AND LINEARIZING THE NS EQUATIONS

To non-dimensionalize the field velocity vector and the distances, respectively, consider the conservation of mass or continuity equation (scalar) for an incompressible flow in vector notation which is,

$$\nabla \mathbf{u} = \mathbf{0}. \tag{5.36}$$

In the absence of body forces the NS equations in a convective vector form are,

$$\frac{\partial}{\partial t}\mathbf{u} + (\mathbf{u} \cdot \nabla)\mathbf{u} = -\rho^{-1}\nabla p + \nu \nabla^2 \mathbf{u}. \tag{5.37}$$

It is now assumed that the total flow field comprising of the velocity and pressure can be decomposed into a steady laminar flow field $\left(U(x,y), V(x,y), W(x,y), P(x,y)\right)$ and a small disturbance flow field with time dependent fluctuations, (u', v', w', p') superposed on the steady field. The x-direction is assumed to be the streamwise direction, the y-direction being normal to the surface and the z-direction is assumed to be the spanwise direction. If this is introduced into the continuity and Navier–Stokes equations, and the equations are made non-dimensional by the following transformations,

$$(x, y, z, t) = \mathsf{I}(x', y', z', t'), \tag{5.38a}$$

$$(U, V, W) = (\mathsf{U}U', \mathsf{U}V', \mathsf{U}W') \equiv \mathsf{U}\mathbf{U}, \tag{5.38b}$$

$$(u', v', w') = (\mathsf{U}u'', \mathsf{U}v'', \mathsf{U}w'') \equiv \mathsf{U}\Delta\mathbf{u}, \quad p' = \rho\mathsf{U}^2 p'' \equiv \rho\mathsf{U}^2\tilde{p}, \tag{5.38c}$$

where U is the far field or characteristic velocity of the free steam and l is a non-dimensionalizing distance. The vector NS equations may be written as:

$$\partial(\mathbf{U} + \Delta\mathbf{u})/\partial t' + \left((\mathbf{U} + \Delta\mathbf{u}) \cdot \nabla'\right)(\mathbf{U} + \Delta\mathbf{u})$$

$$= -\rho^{-1}\nabla'(p + \tilde{p}) + (1/\mathrm{Re})\nabla'^2(\mathbf{U} + \Delta\mathbf{u}), \tag{5.39}$$

where $\mathrm{Re} = \rho U l / \mu$ is the Reynolds number. After dropping the primes and double primes, and higher order terms in $\Delta \mathbf{u}$ one obtains the following set of equations (after the equations satisfied by the laminar base flow are subtracted),

$$\partial \Delta \mathbf{u} / \partial t + (\Delta \mathbf{u} \cdot \nabla) \mathbf{U} + (\mathbf{U} \cdot \nabla) \Delta \mathbf{u} = -\rho^{-1} \nabla \tilde{p} + (1/\mathrm{Re}) \nabla^2 \Delta \mathbf{u}. \qquad (5.40)$$

Assuming that,

$$\begin{aligned} & \big(U(x,y), V(x,y), W(y,z), P(x,y) \big) \\ & = \big(U(x,y), V(x,y), 0, P(x,y) \big), \end{aligned} \qquad (5.41)$$

and neglecting second order terms [3], the perturbation equations are

$$\frac{\partial u}{\partial x} + \frac{\partial v}{\partial y} + \frac{\partial w}{\partial z} = 0, \qquad (5.42\text{a})$$

$$\frac{\partial u}{\partial t} + U \frac{\partial u}{\partial x} + u \frac{\partial U}{\partial x} + v \frac{\partial U}{\partial y} + V \frac{\partial u}{\partial y} = -\frac{\partial p}{\partial x} + (1/\mathrm{Re}) \nabla^2 u, \qquad (5.42\text{b})$$

$$\frac{\partial v}{\partial t} + U \frac{\partial v}{\partial x} + u \frac{\partial V}{\partial x} + V \frac{\partial v}{\partial y} + v \frac{\partial V}{\partial y} = -\frac{\partial p}{\partial y} + (1/\mathrm{Re}) \nabla^2 v, \qquad (5.42\text{c})$$

$$\frac{\partial w}{\partial t} + U \frac{\partial w}{\partial x} + V \frac{\partial w}{\partial y} = -\frac{\partial p}{\partial z} + (1/\mathrm{Re}) \nabla^2 w. \qquad (5.42\text{d})$$

5.4.4   ANALYSIS IN THE BOUNDARY LAYERS

Ludwig Prandtl's analysis was defined by intuition and experimental observations. Prandtl divided the viscous flow over most surfaces into an external region where the appropriate length and velocity scales are U and L, and a boundary layer region where flow properties vary rapidly in the direction normal to the solid surface. In this outer region, our earlier order of magnitude of analysis is acceptable, and the flow equations reduce to inviscid flow equations. In the inner boundary layer region, across a small distance (the order of a few millimeters for an airfoil that is one meter long), the velocity varies rapidly from zero to freestream velocity. In the streamwise direction, the variations are considerably small. While in the streamwise direction the length scale was L, Prandtl proposed using a length scale δ while assessing the gradients in the normal y-direction, across the boundary layer. This length scale is a rough measure of the boundary layer thickness, at some point between the forward quarter-chord and mid-chord. As far as velocities, the u-component of velocity within the boundary layer is found from a large number of measurements to be of the same order as the freestream velocity U and consequently u is normalized by U. The v- component of velocity is considerably smaller than u or U. Rather than picking a velocity scale for v, Prandtl decided to let his order of magnitude tell him how v should be normalized. So, one chooses

$$\bar{x} = x/L, \quad \bar{y} = y/\delta, \quad \bar{u} = u/U \quad \text{and} \quad \bar{v} = v/V, \tag{5.43}$$

for the coordinates and velocities. The characteristic time-scale follows as:

$$t = x/u = (L/U)\bar{x}/\bar{u} = (L/U)\bar{t}. \tag{5.44}$$

It follows that, $\bar{t} = t/t_c = t/(L/U)$ and a characteristic time for the flow is proportional to (L/U). Returning to the continuity equation in dimensional coordinates, it is

$$\frac{\partial \rho}{\partial t} + \frac{\partial \rho u}{\partial x} + \frac{\partial \rho v}{\partial y} + \frac{\partial \rho w}{\partial z} = 0. \tag{5.45}$$

Assuming steady ($\partial/\partial t \equiv 0$) incompressible ($\rho = \rho_0 = $ a constant) flow in two dimensions, one has

$$\frac{\partial u}{\partial x} + \frac{\partial v}{\partial y} = 0. \tag{5.46}$$

In the transformed coordinates,

$$\frac{U}{L}\frac{\partial \bar{u}}{\partial \bar{x}} + \frac{V}{\delta}\frac{\partial \bar{v}}{\partial \bar{y}} = 0. \tag{5.47}$$

If one wishes to preserve the nature of the continuity equation in the transformed coordinates i.e. let

$$\frac{\partial \bar{u}}{\partial \bar{x}} + \frac{\partial \bar{v}}{\partial \bar{y}} = 0, \tag{5.48}$$

then it follows that

$$\frac{U}{L} = \frac{V}{\delta} \Rightarrow V = \frac{U\delta}{L}. \tag{5.49}$$

Thus, one observes that the velocities in a direction normal to the flow direction are of the same order of magnitude as δ. The pressure is scaled so as to make it non-dimensional and consequently one assumes that

$$\bar{p} = p/\rho U^2. \tag{5.50}$$

Now consider the momentum equation in the direction of the flow,

$$\rho\left(\frac{\partial u}{\partial t} + u\frac{\partial u}{\partial x} + v\frac{\partial u}{\partial y} + w\frac{\partial u}{\partial z}\right) = -\frac{\partial p}{\partial x} + \mu\left(\frac{\partial^2 u}{\partial x^2} + \frac{\partial^2 u}{\partial y^2} + \frac{\partial^2 u}{\partial z^2}\right). \qquad (5.51)$$

Assuming steady ($\partial/\partial t \equiv 0$) incompressible ($\rho = \rho_0 =$ a constant) flow in two dimensions, with $v = \mu/\rho$, one has,

$$u\frac{\partial u}{\partial x} + v\frac{\partial u}{\partial y} = -\frac{1}{\rho}\frac{\partial p}{\partial x} + v\left(\frac{\partial^2 u}{\partial x^2} + \frac{\partial^2 u}{\partial y^2}\right). \qquad (5.52)$$

In the transformed coordinates,

$$\frac{U^2}{L}\bar{u}\frac{\partial\bar{u}}{\partial\bar{x}} + \frac{U^2\delta}{L\delta}\bar{v}\frac{\partial\bar{u}}{\partial\bar{y}} = -\frac{U^2}{L}\frac{\partial\bar{p}}{\partial\bar{x}} + v\left(\frac{U}{L^2}\frac{\partial^2\bar{u}}{\partial\bar{x}^2} + \frac{U}{\delta^2}\frac{\partial^2\bar{u}}{\partial\bar{y}^2}\right). \qquad (5.53)$$

The transformed x-momentum equation may be expressed as

$$\frac{U^2}{L}\left(\bar{u}\frac{\partial\bar{u}}{\partial\bar{x}} + \bar{v}\frac{\partial\bar{u}}{\partial\bar{y}}\right) = -\frac{U^2}{L}\frac{\partial\bar{p}}{\partial\bar{x}} + \frac{U^2}{L}\frac{v}{UL}\left(\frac{\partial^2\bar{u}}{\partial\bar{x}^2} + \frac{L^2}{\delta^2}\frac{\partial^2\bar{u}}{\partial\bar{y}^2}\right). \qquad (5.54)$$

Writing $\mathrm{Re} = UL/v$, which is the Reynolds number based on the length scale L,

$$\bar{u}\frac{\partial\bar{u}}{\partial\bar{x}} + \bar{v}\frac{\partial\bar{u}}{\partial\bar{y}} = -\frac{\partial\bar{p}}{\partial\bar{x}} + \frac{1}{\mathrm{Re}}\left(\frac{\partial^2\bar{u}}{\partial\bar{x}^2} + \frac{L^2}{\delta^2}\frac{\partial^2\bar{u}}{\partial\bar{y}^2}\right); \qquad (5.55)$$

Considering the limit when $\mathrm{Re} \to \infty$, one may ignore the term $\partial^2\bar{u}/\partial\bar{x}^2 \to 0$. Moreover, one may choose,

$$(1/\mathrm{Re})(L/\delta)^2 \approx 1 \Rightarrow \delta \approx L/\sqrt{\mathrm{Re}} \text{ or } \mathrm{Re} \approx (L/\delta)^2. \qquad (5.56)$$

The transformed x-momentum equation reduces to,

$$\bar{u}\frac{\partial\bar{u}}{\partial\bar{x}} + \bar{v}\frac{\partial\bar{u}}{\partial\bar{y}} = -\frac{\partial\bar{p}}{\partial\bar{x}} + \frac{\partial^2\bar{u}}{\partial\bar{y}^2}. \qquad (5.57)$$

Similarly, in the y-direction, the momentum equation,

$$u\frac{\partial v}{\partial x} + v\frac{\partial v}{\partial y} = -\frac{1}{\rho}\frac{\partial p}{\partial y} + v\left(\frac{\partial^2 v}{\partial x^2} + \frac{\partial^2 v}{\partial y^2}\right), \qquad (5.58)$$

reduces to,

$$\frac{U^2\delta}{L^2}\bar{u}\frac{\partial\bar{v}}{\partial\bar{x}}+\frac{U^2\delta^2}{L^2\delta}\bar{v}\frac{\partial\bar{v}}{\partial\bar{y}}=-\frac{U^2}{L}\frac{\partial\bar{p}}{\partial\bar{y}}+v\left(\frac{U\delta}{L^3}\frac{\partial^2\bar{v}}{\partial\bar{x}^2}+\frac{U\delta}{L\delta^2}\frac{\partial^2\bar{v}}{\partial\bar{y}^2}\right).$$ (5.59)

Hence,

$$\frac{\delta}{L}\times\bar{u}\frac{\partial\bar{v}}{\partial\bar{x}}+\frac{\delta}{L}\times\bar{v}\frac{\partial\bar{v}}{\partial\bar{y}}=-\frac{\partial\bar{p}}{\partial\bar{y}}+\frac{v}{UL}\left(\frac{\delta}{L}\frac{\partial^2\bar{v}}{\partial\bar{x}^2}+\frac{L}{\delta}\frac{\partial^2\bar{v}}{\partial\bar{y}^2}\right).$$ (5.60)

Given that $(1/\mathrm{Re})(L/\delta)^2\approx1$ and neglecting terms of order δ/L and above, the y-momentum equation reduces to $0=-\partial\bar{p}/\partial\bar{y}$.

5.4.5 Boundary Layer Equations

The key points are that the term $\partial^2 u/\partial x^2\cong0$ in the original x-momentum equation has been ignored and the pressure does not vary across the boundary layer. The complete Navier–Stokes equations are made up respectively of the continuity, momentum and energy equations and are

$$\frac{\partial\rho}{\partial t}+\frac{\partial}{\partial x_i}\left(\rho u_i\right)=0,$$ (5.61a)

$$\frac{\partial}{\partial t}\left(\rho u_i\right)+\frac{\partial}{\partial x_i}\left(\rho u_j u_i+p\delta_{ij}-\tau_{ij}\right)=0,$$ (5.61b)

$$\frac{\partial}{\partial t}\left(\rho E\right)+\frac{\partial}{\partial x_i}\left(\rho u_j E+u_j p+q_j-u_i\tau_{ij}\right)=0.$$ (5.61c)

Thus, the original continuity remains unchanged

$$\frac{\partial\rho}{\partial t}+\frac{\partial\rho u}{\partial x}+\frac{\partial\rho v}{\partial y}+\frac{\partial\rho w}{\partial z}=0,$$ (5.62)

while the momentum equations which are

$$\rho\left(\frac{\partial u}{\partial t}+u\frac{\partial u}{\partial x}+v\frac{\partial u}{\partial y}+w\frac{\partial u}{\partial z}\right)=-\frac{\partial p}{\partial x}+\mu\left(\frac{\partial^2 u}{\partial x^2}+\frac{\partial^2 u}{\partial y^2}+\frac{\partial^2 u}{\partial z^2}\right).$$ (5.63a)

$$\rho\left(\frac{\partial v}{\partial t}+u\frac{\partial v}{\partial x}+v\frac{\partial v}{\partial y}+w\frac{\partial v}{\partial z}\right)=-\frac{\partial p}{\partial y}+\mu\left(\frac{\partial^2 v}{\partial x^2}+\frac{\partial^2 v}{\partial y^2}+\frac{\partial^2 v}{\partial z^2}\right),$$ (5.63b)

$$\rho\left(\frac{\partial w}{\partial t}+u\frac{\partial w}{\partial x}+v\frac{\partial w}{\partial y}+w\frac{\partial w}{\partial z}\right)=-\frac{\partial p}{\partial z}+\mu\left(\frac{\partial^2 w}{\partial x^2}+\frac{\partial^2 w}{\partial y^2}+\frac{\partial^2 w}{\partial z^2}\right),$$ (5.63c)

are respectively reduced to

$$\rho\left(\frac{\partial u}{\partial t}+u\frac{\partial u}{\partial x}+v\frac{\partial u}{\partial y}+w\frac{\partial u}{\partial z}\right)=-\frac{\partial p}{\partial x}+\mu\frac{\partial^2 u}{\partial y^2}. \tag{5.64a}$$

$$0=-\partial p/\partial y, \tag{5.64b}$$

$$\rho\left(\frac{\partial w}{\partial t}+u\frac{\partial w}{\partial x}+v\frac{\partial w}{\partial y}+w\frac{\partial w}{\partial z}\right)=-\frac{\partial p}{\partial z}+\mu\frac{\partial^2 w}{\partial y^2}. \tag{5.64c}$$

Boundary layer flows often have the property of self-similarity or near-self-similarity. Suppose the solution has the property that when $\bar{u}=u/U$ is plotted against y/δ a universal function is obtained, with no further dependence on x, although δ is not necessarily a constant or defined *a priori*. Such a solution is called a "similarity solution".

5.4.6 VORTICITY AND STRESS IN A BOUNDARY LAYER

The vorticity is defined in a two-dimensional flow as

$$\omega=\omega_z=\left(\partial u/\partial y\right)-\left(\partial v/\partial x\right). \tag{5.65}$$

In the transformed boundary layer coordinates,

$$\omega=\frac{U}{L}\frac{L}{\delta}\frac{\partial\bar{u}}{\partial\bar{y}}-\frac{U}{L}\frac{\delta}{L}\frac{\partial\bar{v}}{\partial\bar{x}}\equiv\frac{U}{L}\bar{\omega}. \tag{5.66}$$

Hence,

$$\bar{\omega}=\frac{L}{\delta}\frac{\partial\bar{u}}{\partial\bar{y}}-\frac{\delta}{L}\frac{\partial\bar{v}}{\partial\bar{x}}=\sqrt{\text{Re}}\frac{\partial\bar{u}}{\partial\bar{y}}-\frac{1}{\sqrt{\text{Re}}}\frac{\partial\bar{v}}{\partial\bar{x}}. \tag{5.67}$$

Hence as $\text{Re}\to\infty$,

$$\bar{\omega}\approx\sqrt{\text{Re}}\,\partial\bar{u}/\partial\bar{y}. \tag{5.68}$$

Hence vorticity may be approximated in the boundary layer as,

$$\omega=\omega_z=\partial u/\partial y. \tag{5.69}$$

The normal components of the stress perpendicular and parallel to the direction of expressed non-dimensionally respectively are,

$$\frac{\tau_{xx}}{\rho U^2}=-\frac{p}{\rho U^2}+\mu\frac{2}{\rho U^2}\frac{\partial u}{\partial x}=-\frac{p}{\rho U^2}+2\frac{\mu}{\rho UL}\frac{\partial\bar{u}}{\partial x}=-\frac{p}{\rho U^2}+\frac{2}{\text{Re}}\frac{\partial\bar{u}}{\partial x}, \tag{5.70a}$$

$$\frac{\tau_{yy}}{\rho U^2} = -\frac{p}{\rho U^2} + \mu \frac{2}{\rho U^2}\frac{\partial v}{\partial y} = -\frac{p}{\rho U^2} + 2\frac{\mu}{\rho UL}\frac{\partial \bar{v}}{\partial \bar{y}} = -\frac{p}{\rho U^2} + \frac{2}{Re}\frac{\partial \bar{v}}{\partial \bar{y}}. \quad (5.70b)$$

As $Re \to \infty$,

$$\frac{\tau_{xx}}{\rho U^2} \approx \frac{\tau_{yy}}{\rho U^2} \approx -\frac{p}{\rho U^2}. \quad (5.71)$$

The shearing stress over surfaces parallel to the flow is,

$$\tau_{xy} = \mu\left(\partial u/\partial y\right) + \mu\left(\partial v/\partial x\right). \quad (5.72)$$

Similar to the case of vorticity, it can be shown that in the boundary layer,

$$\tau_{xy} \approx \mu\, \partial u/\partial y. \quad (5.73)$$

5.4.7 TWO-DIMENSIONAL BOUNDARY LAYER EQUATIONS

In incompressible flow ($\rho = \rho_0 =$ a constant),

$$\frac{\partial u}{\partial x} + \frac{\partial v}{\partial y} + \frac{\partial w}{\partial z} = 0, \quad (5.74a)$$

$$\frac{\partial u}{\partial t} + u\frac{\partial u}{\partial x} + v\frac{\partial u}{\partial y} + w\frac{\partial u}{\partial z} = -\frac{1}{\rho}\frac{\partial p}{\partial x} + \frac{\mu}{\rho}\frac{\partial^2 u}{\partial y^2}, \quad (5.74b)$$

$$0 = -\partial p/\partial y. \quad (5.74c)$$

In two dimensions,

$$\frac{\partial}{\partial x}u + \frac{\partial}{\partial y}v = 0, \quad \frac{\partial u}{\partial t} + u\frac{\partial u}{\partial x} + v\frac{\partial u}{\partial y} = -\frac{1}{\rho}\frac{\partial p}{\partial x} + \frac{\mu}{\rho}\frac{\partial^2 u}{\partial y^2}, \quad 0 = -\frac{\partial p}{\partial y}. \quad (5.75)$$

Further if,

$$-\frac{1}{\rho}\frac{\partial p}{\partial x} = \frac{1}{2}\frac{\partial U_\infty^2}{\partial x}, \quad (5.76)$$

$$\frac{\partial}{\partial x}u + \frac{\partial}{\partial y}v = 0, \quad \frac{\partial u}{\partial t} + u\frac{\partial u}{\partial x} + v\frac{\partial u}{\partial y} = \frac{1}{2}\frac{\partial U_\infty^2}{\partial x} + v\frac{\partial^2 u}{\partial y^2}, \quad 0 = -\frac{\partial p}{\partial y}. \quad (5.77)$$

Introducing the stream function, $\psi = \psi(x,y)$ defines the velocity components

$$u = \frac{\partial \psi}{\partial y}, \quad v = -\frac{\partial \psi}{\partial x}. \tag{5.78}$$

In two dimensions, and if the pressure gradient term is neglected,

$$-(1/\rho)(\partial p/\partial x) = 0, \tag{5.79}$$

then starting with the boundary layer equations for continuity and momentum in two dimensions, in dimensional coordinates and dependent variables,

$$(\partial u/\partial x) + (\partial v/\partial y) = 0, \tag{5.80a}$$

$$\frac{\partial u}{\partial t} + u\frac{\partial u}{\partial x} + v\frac{\partial u}{\partial y} = v\frac{\partial^2 u}{\partial y^2}, \tag{5.80b}$$

the continuity equation is satisfied by the stream function. The momentum equation reduces in steady state to

$$\frac{\partial \psi}{\partial y}\frac{\partial^2 \psi}{\partial x \partial y} - \frac{\partial \psi}{\partial x}\frac{\partial^2 \psi}{\partial^2 y} = \frac{\partial^3 \psi}{\partial^3 y}. \tag{5.81}$$

The initial and boundary conditions are respectively given by

$$u(x, y > 0, t = 0) = U_\infty(x, 0), \quad v(x, y > 0, t = 0) = 0, \tag{5.82}$$

$$u(x, 0, t) = v(x, 0, t) = 0, \quad u(x, y \to \infty, t > 0) = U_\infty(x, t). \tag{5.83}$$

5.4.8 THE BLASIUS SOLUTION

Ludwig Prandtl's student H. Blasius obtained an exact solution for the boundary layer equations in two dimensions. First it is noted that the continuity equation is given by Equation (5.80a). The Blasius variable is introduced so the solution is self-similar for the flow and is,

$$\eta = y\sqrt{U/2vx}. \tag{5.84}$$

Then,

$$\partial \eta/\partial y = \sqrt{U/2vx}, \quad \partial \eta/\partial x = -(y/2x)\sqrt{U/2vx} = -\eta/2x. \tag{5.85}$$

Introducing the non-dimensional stream function,

$$F(\eta) = \psi(x,y)/\sqrt{2Uvx}. \tag{5.86}$$

Hence, it follows that,

$$\psi(x,y) = F(\eta)\sqrt{2Uvx}, \tag{5.87a}$$

$$u = \frac{\partial \psi}{\partial y} = \sqrt{2Uvx}\frac{\partial}{\partial \eta}F(\eta)\frac{\partial \eta}{\partial y} = \sqrt{2Uvx}\frac{\partial}{\partial \eta}F(\eta)\sqrt{\frac{U}{2vx}} = UF'(\eta). \tag{5.87b}$$

$$v = -\frac{\partial \psi}{\partial x} = -\sqrt{2Uvx}\frac{\partial}{\partial \eta}F(\eta)\frac{\partial \eta}{\partial x} - \frac{1}{2x}\sqrt{2Uvx}F(\eta)$$

$$= \frac{\sqrt{2Uvx}}{2x}\left(\eta\frac{\partial}{\partial \eta}F(\eta) - F(\eta)\right). \tag{5.87c}$$

Hence, since, $u = UF'(\eta)$, it follows that,

$$u/U = F'(\eta), \quad v = \sqrt{vU/2x}\left(\eta F'(\eta) - F(\eta)\right), \tag{5.88}$$

satisfy the continuity equation. Assuming $\partial p/\partial x = 0$, the steady state momentum equation is,

$$u\frac{\partial u}{\partial x} + v\frac{\partial u}{\partial y} = v\frac{\partial^2 u}{\partial y^2}, \tag{5.89a}$$

$$\frac{\partial u}{\partial x} = -\frac{1}{2x}U\eta F''(\eta), \quad \frac{\partial u}{\partial y} = UF''(\eta)\frac{\partial \eta}{\partial y} = UF''(\eta)\sqrt{\frac{U}{2vx}}, \tag{5.89b}$$

$$\frac{\partial^2 u}{\partial y^2} = UF'''(\eta)\left(\frac{\partial \eta}{\partial y}\right)^2 = \frac{U^2}{2vx}F'''(\eta), \tag{5.89c}$$

$$u\frac{\partial u}{\partial x} = -\frac{U^2}{2x}\eta F''(\eta)F'(\eta), \tag{5.89d}$$

$$v\frac{\partial u}{\partial y} = \frac{U^2}{2x}F''(\eta)\left(\eta F'(\eta) - F(\eta)\right), \quad v\frac{\partial^2 u}{\partial y^2} = \frac{U^2}{2x}F'''(\eta). \tag{5.89e}$$

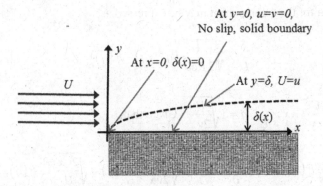

FIGURE 5.5 Illustration of boundary conditions.

Hence it follows that $F(\eta)$ satisfies the differential equation,

$$F'''(\eta) + F(\eta) F''(\eta) = 0, \qquad (5.90)$$

subject to the boundary conditions which are defined below. The physical boundary conditions are illustrated in Figure 5.5. One may write at $y=0$ and $u=v=0$ in terms of the transformed function and variables as: when $\eta = 0$, $F = F' = 0$. Similarly, as $y \to \infty$, $u = U$, which may be written as $\eta \to \infty$, $F' = 1$. Hence the Blasius equation is solved by numerical integration subject to the initial conditions, $\eta = 0$, $F = F' = 0$, $F'' = 0.4696$, where the last condition is found by interpolation to satisfy the boundary condition as $\eta \to \infty$ (White [3]).

Also, the following second derivatives are given by,

$$\frac{\partial^2 u}{\partial x^2} = \frac{U}{2vx} \left(\frac{y}{2x} \right)^2 U F''(\eta), \quad \frac{\partial^2 u}{\partial y^2} = \frac{U^2}{2vx} F'''(\eta) = -\frac{U^2}{2vx} F(\eta) F''(\eta). \quad (5.91)$$

$$\frac{\partial}{\partial y} v = \sqrt{\frac{vU}{2x}} \left(\sqrt{\frac{U}{2vx}} F'(\eta) - \sqrt{\frac{U}{2vx}} F'(\eta) + \sqrt{\frac{U}{2vx}} \eta F''(\eta) \right), \quad (5.92)$$

Thus,

$$\frac{\partial}{\partial y} v = \sqrt{\frac{\gamma U}{2x}} \left(\sqrt{\frac{U}{2vx}} \eta F''(\eta) \right) = \frac{U}{2x} \eta F''(\eta). \quad (5.93)$$

Hence, the equations for the velocity components are,

$$u/U = F'(\eta), \qquad (5.94a)$$

$$\frac{v}{U} = \sqrt{\frac{v}{2Ux}} \left(\eta F'(\eta) - F(\eta) \right)$$

$$\qquad (5.94b)$$

$$= \sqrt{\frac{1}{2 \operatorname{Re}_x}} \left(\eta F'(\eta) - F(\eta) \right), \quad \operatorname{Re}_x = Ux/v.$$

The equation for the v component may be expressed as,

$$\frac{v}{U} = \sqrt{\frac{1}{2\,\mathrm{Re}\,\bar{x}}}\left(\eta F'(\eta) - F(\eta)\right), \tag{5.95}$$

where $\mathrm{Re} = \dfrac{U\bar{c}}{v}$, $\bar{x} = \dfrac{x}{\bar{c}}$, $\eta = y\sqrt{\dfrac{U}{2vx}}$. Given that,

$$\frac{\partial}{\partial y}F(\eta) = F'(\eta)\frac{\partial \eta}{\partial y} = F'(\eta)\sqrt{\frac{U}{2vx}}, \quad F'(\eta) = \sqrt{\frac{2vx}{U}}\frac{\partial}{\partial y}F(\eta), \tag{5.96a}$$

$$F(\eta) = \sqrt{\frac{U}{2vx}}\int F'(\eta)\,dy = \sqrt{\frac{U}{2vx}}\int \frac{u}{U}\,dy, \tag{5.96b}$$

$$\frac{v}{U} = \sqrt{\frac{v}{2Ux}}\left(\eta F'(\eta) - F(\eta)\right) = \sqrt{\frac{1}{2\,\mathrm{Re}_x}}\left(\eta F'(\eta) - F(\eta)\right), \tag{5.96c}$$

reverting to the original variables,

$$\frac{v}{U} = \frac{1}{2x}\left(y\frac{u}{U} - \int \frac{u}{U}\,dy\right) = \frac{\delta}{2x}\left(\frac{y}{\delta}\frac{u}{U} - \int \frac{u}{U}d\frac{y}{\delta}\right). \tag{5.97}$$

If $\eta = y/\delta$, the equation for the v component may be expressed as,

$$\frac{v}{U} = \frac{\delta}{2x}\left(\eta F'(\eta) - F(\eta)\right). \tag{5.98}$$

(The change in the variable $\eta = y/\delta$, is made but it is not the Blasius variable).

The Blasius solution for a flat plate obtained numerically, is illustrated in Figure 5.6. Further details about the numerical computation may be found in White [3].

5.4.9 THE DISPLACEMENT, MOMENTUM AND ENERGY THICKNESSES

The displacement, momentum and energy thicknesses may now be defined:

i) *Displacement thickness*: The effect of the boundary layer on the external flow can be modeled by a quantity called the displacement thickness denoted by δ^*. In view of the no-slip condition, the streamwise velocity in the boundary layer is smaller than it would have been in an inviscid flow. As a consequence, the quantity of mass transported is also much smaller. The displacement thickness denotes the magnitude by which the wall has to be shifted out for an inviscid flow past the displaced wall to have the same

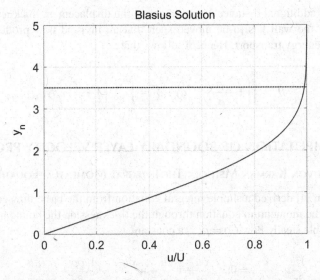

FIGURE 5.6 The Blasius solution for a flat plate; the dotted line is drawn horizontally where the horizontal velocity component $u(\delta) = 0.99U$ and $y_n = \eta = y\sqrt{U/2\nu x}$.

mass transport as the viscous flow along the original wall. Thus, assuming that the density is uniform and constant, the kinematic displacement thickness is defined as

$$\delta^* = \frac{1}{\rho_\infty U_\infty} \int_0^\delta \left(\rho_\infty U_\infty - \rho_\infty u \right) dy = \delta \int_0^1 \left(1 - \frac{u}{U_\infty} \right) d\eta. \tag{5.99}$$

ii) *Momentum thickness*: Similar to the displacement thickness which is based on the mass transported in the boundary layer, the viscous and inviscid flows based on the momentum transported in the boundary layer could also be compared. Thus, the momentum thickness is defined as the additional distance over and above the displacement thickness, over which the wall has to be moved such that an inviscid flow produces the same momentum transport. Thus, assuming that the density is uniform and constant, the kinematic momentum thickness is defined as

$$\theta = \frac{1}{\rho_\infty U_\infty^2} \int_0^\delta \left(\rho_\infty U_\infty^2 - \rho_\infty u^2 \right) dy - \delta^*$$

$$= \delta \int_0^1 \left(1 - \left(\frac{u}{U_\infty} \right)^2 \right) d\eta - \delta^* = \delta \int_0^1 \left(1 - \frac{u}{U_\infty} \right) \frac{u}{U_\infty} d\eta. \tag{5.100}$$

iii) *Energy Thickness*: Similar to the momentum thickness, assuming that the density is uniform and constant, the kinematic energy thickness is defined as the additional distance over and above the displacement thickness, over which the wall has to be moved such that an inviscid flow produces the same energy transport. Hence it follows that

$$\theta^* = \int_0^\delta \left(1 - \frac{\rho_\infty u^3}{\rho_\infty U_\infty^3}\right) dy - \delta^* = \int_0^\delta \frac{u}{U_\infty}\left(1 - \frac{u^2}{U_\infty^2}\right) dy. \qquad (5.101)$$

5.5   COMPUTATION OF BOUNDARY LAYER VELOCITY PROFILES

5.5.1   THE VON KARMAN METHOD: THE INTEGRAL MOMENTUM EQUATION

Von Karman [4] derived a simple integral equation from the basic flow equation by integrating the momentum equation through the thickness on the boundary layer. In incompressible steady flow ($\rho = \rho_0 =$ a constant),

$$\frac{\partial}{\partial x}u + \frac{\partial}{\partial y}v = 0, \quad u\frac{\partial u}{\partial x} + v\frac{\partial u}{\partial y} + w\frac{\partial u}{\partial z} = -\frac{1}{\rho}\frac{\partial p}{\partial x} + \frac{\mu}{\rho}\frac{\partial^2 u}{\partial y^2}. \qquad (5.102)$$

In two dimensions if,

$$-\frac{1}{\rho}\frac{\partial p}{\partial x} = \frac{1}{2}\frac{\partial U_\infty^2}{\partial x}, \qquad (5.103)$$

$$\frac{\partial}{\partial x}u + \frac{\partial}{\partial y}v = 0, \quad u\frac{\partial u}{\partial x} + v\frac{\partial u}{\partial y} = \frac{1}{2}\frac{\partial U_\infty^2}{\partial x} + v\frac{\partial^2 u}{\partial y^2}. \qquad (5.104)$$

The initial and boundary conditions are respectively given by,

$$u(x, y > 0, t = 0) = U_\infty(x, 0), \quad v(x, y > 0, t = 0) = 0, \qquad (5.105)$$

$$u(x, 0, t) = v(x, 0, t) = 0, \quad u(x, y \to \infty, t > 0) = U_\infty(x). \qquad (5.106)$$

If one assumes that $U_\infty(x,t)$ is steady and given by $U_\infty(x)$, it is possible to show that

$$u\frac{\partial u}{\partial x} + v\frac{\partial u}{\partial y} = \frac{1}{2}\frac{\partial U_\infty^2}{\partial x} + v\frac{\partial^2 u}{\partial y^2}, \qquad (5.107a)$$

$$\frac{\partial u^2}{\partial x} + \frac{\partial(vu)}{\partial y} = \frac{1}{2}\frac{\partial U_\infty^2}{\partial x} + v\frac{\partial^2 u}{\partial y^2}, \qquad (5.107b)$$

$$\frac{\partial(u^2 - U_\infty u)}{\partial x} + u\frac{\partial U_\infty}{\partial x} + U_\infty\frac{\partial u}{\partial x} + \frac{\partial(vu)}{\partial y} = \frac{1}{2}\frac{\partial U_\infty^2}{\partial x} + v\frac{\partial^2 u}{\partial y^2}. \qquad (5.107c)$$

Subtracting the expression

$$U_\infty \frac{\partial u}{\partial x} + U_\infty \frac{\partial v}{\partial y} = 0, \tag{5.108}$$

from the preceding two equations, Equations (5.107b) and (5.107c),

$$\frac{\partial(u^2 - U_\infty u)}{\partial x} + u \frac{\partial U_\infty}{\partial x} + \frac{\partial(v(u - U_\infty))}{\partial y} = \frac{1}{2} \frac{\partial U_\infty^2}{\partial x} + v \frac{\partial^2 u}{\partial y^2}, \tag{5.109a}$$

$$\frac{\partial(U_\infty u - u^2)}{\partial x} - u \frac{\partial U_\infty}{\partial x} + \frac{\partial(v(U_\infty - u))}{\partial y} = -\frac{1}{2} \frac{\partial U_\infty^2}{\partial x} - v \frac{\partial^2 u}{\partial y^2}. \tag{5.109b}$$

Let $\bar{u} = u/U_\infty$; $\bar{v} = v/U_\infty$;

$$\frac{\partial(U_\infty^2(\bar{u} - \bar{u}^2))}{\partial x} + \frac{1 - \bar{u}}{2} \frac{\partial U_\infty^2}{\partial x} + U_\infty^2 \frac{\partial(\bar{v}(1 - \bar{u}))}{\partial y} = -v U_\infty \frac{\partial^2 \bar{u}}{\partial y^2}. \tag{5.110}$$

Assuming that $u(x, y \to \infty, t > 0) = U_\infty(x)$, integrating through the boundary layer,

$$U_\infty^2 \frac{\partial \theta}{\partial x} + \frac{1}{2} (2\theta + \delta^*) \frac{\partial U_\infty^2}{\partial x} + U_\infty^2 \frac{\partial(\bar{v}(1 - \bar{u}))}{\partial y} = -v U_\infty \left. \frac{\partial \bar{u}}{\partial y} \right|_{\text{wall}}. \tag{5.111}$$

In the above equation, θ is the momentum thickness and δ^* is the displacement thickness which are respectively defined by,

$$\theta = \int_0^\delta \frac{u}{U_\infty} \left(1 - \frac{u}{U_\infty}\right) dy, \quad \delta^* = \int_0^\delta \left(1 - \frac{u}{U_\infty}\right) dy. \tag{5.112}$$

Since $v \approx 0$ in boundary layer, the wall shear stress is defined as

$$\tau_w = -\int_0^\delta \frac{\partial \tau_{xy}}{\partial y} dy = -\int_0^\delta \mu \frac{\partial^2 u}{\partial y^2} dy. \tag{5.113}$$

Neglecting the third term in the left-hand side, and identifying the right-hand side as proportional to the wall shear stress, the von Karman momentum integral equation is obtained as

$$\frac{d\theta}{dx} + (2 + H) \frac{\theta}{U_\infty} \frac{dU_\infty}{dx} = \frac{\tau_w}{\rho U_\infty^2}. \tag{5.114}$$

Further if $dU_\infty/dx = 0$,

$$\frac{d\theta}{dx} = \frac{\tau_w}{\rho U_\infty^2}. \tag{5.115}$$

The von Karman method requires assumed velocity profiles so one can evaluate θ the momentum thickness and δ^* the displacement thickness. Thus, typically a simple velocity profile may be assumed and obtain the displacement thickness, the momentum thickness and the shear stress. For example, if the velocity profile is assumed to be given by a second order polynomial,

$$u = Ay^2 + By + C, \tag{5.116}$$

where A, B and C are constants, to evaluate velocity profile constants A, B and C, boundary conditions must be applied. The following three boundary conditions may be used:

1) No-slip and no permeation adjacent to a wall boundary: $y=0$, $u=0$, $v=0$,
2) Boundary layer edge velocity is not affected by viscosity: $y=\delta$, $u=U$,
3) Boundary layer edge shear vanishes: $y=\delta$, $\partial u/\partial y = 0$.

The solution for the profile satisfying these minimal boundary conditions is

$$\frac{u}{U}(y) = 2\left(\frac{y}{\delta}\right) - \left(\frac{y}{\delta}\right)^2. \tag{5.117}$$

This represented a simple parabolic shape. The boundary layer velocity profile is only defined when $0 \le y \le \delta$. The use of this velocity profile may now be made to obtain the momentum thickness θ and τ_w

$$\theta = \int_0^\delta \frac{u}{U}\left(1 - \frac{u}{U}\right)dy = \delta \int_0^1 \frac{u}{U}\left(1 - \frac{u}{U}\right)d\left(\frac{y}{\delta}\right), \tag{5.118}$$

or,

$$\frac{\theta}{\delta} = \int_0^1 \left[2\frac{y}{\delta} - \left(\frac{y}{\delta}\right)^2\right]\left[1 - 2\frac{y}{\delta} + \left(\frac{y}{\delta}\right)^2\right]d\frac{y}{\delta}. \tag{5.119}$$

Note that defining a new variable $\eta = y/\delta$ makes the evaluation much easier.

$$\frac{\theta}{\delta} = \int_0^1 (2\eta - \eta^2)(1 - 2\eta + \eta^2)d\eta = \int_0^1 (2\eta - \eta^2 - 4\eta^2 + 2\eta^3 + 2\eta^3 - \eta^4)d\eta$$

$$\tag{5.120}$$

$$= \int_0^1 (2\eta - 5\eta^2 + 4\eta^3 - \eta^4)d\eta = 1 - \frac{5}{3} + 1 - \frac{1}{5} = \frac{15 - 25 + 15 - 3}{15} = \frac{2}{15}.$$

The displacement thickness is

$$\delta^* = \int_0^\delta \left(1 - \frac{u}{U}\right) dy = \delta \int_0^1 \left(1 - \frac{u}{U}\right) d\left(\frac{y}{\delta}\right) = \delta \int_0^1 \left(1 - 2\eta + \eta^2\right) d\eta$$

$$= \delta \left(\eta - \eta^2 + \frac{\eta^3}{3}\right)\Bigg|_0^1 = \frac{\delta}{3}.$$

(5.121)

Hence the shape factor, defined as the ratio of the displacement thickness to the momentum thickness is

$$H = \delta^*/\theta = 2.5.$$

(5.122)

Similarly, the wall shear stress is

$$\tau_w = \mu \frac{du}{dy}\bigg|_{y=0} = \frac{\mu U}{\delta} \times \frac{\partial(u/U)}{\partial(y/\delta)}\bigg|_{y=0} = \frac{\mu U}{\delta} \times \frac{\partial(u/U)}{\partial\eta}\bigg|_{\eta=0}$$

$$= \frac{\mu U}{\delta} \times \left[2 - 2\eta\right]\big|_{\eta=0} = \frac{2\mu U}{\delta},$$

(5.123)

for the assumed parabolic profile.

One could use the above results for θ and τ_w in the simplified momentum integral equation

$$\frac{d\theta}{dx} = \frac{\tau_w}{\rho U_\infty^2} = C_\tau.$$

(5.124)

The coefficient C_τ is an important parameter in the estimation of the drag.

Hence, since the ratio of θ/δ is a constant

$$\frac{\theta}{\delta} \frac{d\delta}{dx} = \frac{\tau_w}{\rho U^2}.$$

(5.125)

For a flat plate, it gives

$$\frac{2\mu U}{\delta} = \rho U^2 \times \left(\frac{2}{15}\right) \frac{d\delta}{dx}.$$

(5.126)

Rearranging the variables δ and x, and integrating gives,

$$\int_{\delta=0}^{\delta(x)} \delta \, d\delta = \int_{x=0}^{x} 15 \left(\frac{\mu}{\rho U}\right) dx.$$

(5.127)

Hence it follows that,

$$\frac{\delta^2}{2} = 15\left(\frac{\mu}{\rho U}\right) \Rightarrow \delta^2 = \frac{30\mu x}{\rho U}. \tag{5.128}$$

Expressing the δ^2 in a non-dimensional form, divide both sides by x^2,

$$\left(\frac{\delta}{x}\right)^2 = \frac{30\mu}{\rho U x} = \frac{30}{\mathrm{Re}_x}, \quad \mathrm{Re}_x = \frac{\rho U x}{\mu} = \frac{U x}{v}, \tag{5.129}$$

Where Re_x is the Reynolds number based upon the variable x. Thus,

$$\frac{\delta}{x} = \sqrt{\frac{30}{\mathrm{Re}_x}} = \frac{5.48}{\sqrt{\mathrm{Re}_x}}. \tag{5.130}$$

5.5.2 WALL SHEAR STRESS, MOMENTUM THICKNESS, DISPLACEMENT THICKNESS AND BOUNDARY LAYER THICKNESS FOR THE BLASIUS SOLUTION

Returning to the Blasius similarity solution for a two-dimensional boundary layer could compute several features defined earlier that characterize the solution. These are:

The displacement thickness δ^*:

$$\delta^* = \sqrt{\frac{2vx}{U}} \int_0^\infty \left(1 - F'(\eta)\right) d\eta = \sqrt{\frac{2vx}{U}} \left(\eta - F(\eta)\right)\Big|_{\eta\to\infty}. \tag{5.131}$$

The momentum thickness θ:

$$\theta = \sqrt{\frac{2vx}{U}} \int_0^\infty F'(\eta)\left(1 - F'(\eta)\right) d\eta$$

$$= \sqrt{\frac{2vx}{U}} \left(F(\eta)\Big|_{\eta=\infty} - F(\eta)F'(\eta)\Big|_{\eta=\infty} + \int_0^\infty F(\eta)F''(\eta)d\eta\right). \tag{5.132}$$

Taking into account the boundary conditions, the expression reduces to,

$$\theta = \sqrt{\frac{2vx}{U}} \int_0^\infty F(\eta)F''(\eta)d\eta. \tag{5.133}$$

Similarly, as the energy thickness δ^{**} is defined by Equation (5.101),

$$\delta^{**} = \sqrt{\frac{2\nu x}{U}} \int_0^\infty F'(\eta)\left(1 - F'(\eta)^2\right) d\eta$$

$$= 2\sqrt{\frac{2\nu x}{U} \int_0^\infty F(\eta)F'(\eta)F''(\eta)\,d\eta}. \qquad (5.134)$$

The skin-friction-related stress τ_w may be evaluated from the equations

$$\tau_w = \mu \frac{du}{dy}\bigg|_{y=0}, \quad \frac{\partial u}{\partial y} = UF''(\eta)\sqrt{\frac{U}{2\nu x}}. \qquad (5.135)$$

Hence,

$$\tau_w = \mu \frac{du}{dy}\bigg|_{y=0} = 0.4696 \mu U \sqrt{\frac{U}{2\nu x}} = 0.4696 \rho U^2 \sqrt{\frac{1}{2\,\mathrm{Re}_x}}. \qquad (5.136)$$

The dissipation integral D which is related to the rate at which energy is dissipated by the action of viscosity has been shown to be $\mu(du/dy)^2$ per unit time per unit volume, and is the integral of this energy dissipation rate across the boundary layer. It is

$$D = \int_0^\infty \mu \left(\frac{du}{dy}\right)^2 dy = \frac{U^3}{2\nu x} \int_0^\infty \left(F''(\eta)\right)^2 d\eta. \qquad (5.137)$$

Consequently D is the total dissipation in a cylinder of small cross-section with axis normal to the layer per unit time per unit area of the cross-section.

For the Blasius solution,

$$C_f = 2C_\tau = 0.4696 \frac{\sqrt{2}}{\sqrt{\mathrm{Re}_x}} = \frac{0.664}{\sqrt{\mathrm{Re}_x}}. \qquad (5.138)$$

Hence,

$$\frac{d\theta}{dx} = C_\tau, \quad \frac{\theta}{x} = 2C_\tau = \frac{0.664}{\sqrt{\mathrm{Re}_x}}, \quad \frac{\delta^*}{x} = \frac{1.721}{\sqrt{\mathrm{Re}_x}}, \quad H = 2.5919. \qquad (5.139)$$

But,

$$\frac{\theta}{\delta} \frac{d\delta}{dx} = C_\tau \quad \text{and} \quad 2\frac{d\delta}{\delta} = \frac{2C_\tau dx}{\theta} = \frac{dx}{x}. \qquad (5.140)$$

For $u(\delta)=0.99U$ and $F'(\eta)=0.99$, $\eta = \delta\sqrt{Re_x}\big/x\sqrt{2} = 3.47$. Hence,

$$\frac{\delta}{x} = \frac{4.91}{\sqrt{Re_x}} = 2.853\frac{\delta^*}{x}, \quad \frac{\delta^*}{\delta} = 0.35, \quad \frac{\theta}{\delta} = 0.1352, \quad \frac{\delta}{\theta} = 7.396. \quad (5.141)$$

Head's [5] shape factor is defined as $H_1 = (\delta/\theta)-(\delta^*/\theta)$ and is

$$H_1 = \frac{\delta}{\theta} - \frac{\delta^*}{\theta} = 7.396 - H = 4.8045 = 1.853 \times H, \quad (5.142)$$

$$\delta = \delta^* + H_1\theta = \delta^* + 1.853H\theta. \quad (5.143)$$

5.5.3 THE METHODS OF POHLHAUSEN AND HOLSTEIN AND BOHLEN

First it may be observed that the von Karman integral equation may be expressed as the von Karman momentum integral equation,

$$\frac{d\left(U_\infty^2\theta\right)}{dx} + \delta^* U_\infty \frac{dU_\infty}{dx} = \frac{\tau_w}{\rho}. \quad (5.144)$$

Pohlhausen's [6] method for solving the above Equation (5.144) was to assume that

$$\frac{u}{U_\infty} = \frac{y}{\delta}\left(a + \frac{y}{\delta}\left(b + \frac{y}{\delta}\left(c + d\frac{y}{\delta}\right)\right)\right). \quad (5.145)$$

The velocity u is subject to the boundary conditions which are at $y=0$, as $u=0$ and $vu_{yy} = -U_\infty\left(dU_\infty/dx\right)$. Similarly, as $y=\delta \to \infty$, $u=U_\infty$ and $u_y=u_{yy}=0$. Defining a parameter

$$\lambda = \frac{\delta^2 \times \rho}{\mu}\frac{\partial U}{\partial x} = \frac{\delta \times Re_\delta}{U}\frac{\partial U}{\partial x} \quad \text{with} \quad Re_\delta = \frac{\rho U\delta}{\mu}, \quad (5.146)$$

the velocity u is expressed as

$$\frac{u}{U_\infty} = F\left(\frac{y}{\delta}\right) + \lambda G\left(\frac{y}{\delta}\right), \quad (5.147)$$

with

$$F(\eta) = 2\eta - 2\eta^3 + \eta^4, \quad G(\eta) = \eta\left(1-\eta\right)^3\big/6 \quad \text{and} \quad \eta = y/\delta. \quad (5.148)$$

Thus, $u_y\big|_{y=0} = 0$ at $\lambda \geq 12$, and one can expect separation when $\lambda \leq -12$. Furthermore, when $u \geq U_\infty$, $\lambda \geq 12$. Hence the range of values for λ is restricted to $-12 \leq \lambda \leq 12$. Thus, the kinematic momentum thickness and the kinematic displacement thickness are respectively given by

$$\theta_k = \delta \int_0^\delta \frac{u}{U}\left(1-\frac{u}{U}\right) d\eta, \quad \delta_k^* = \delta \int_0^\delta \left(1-\frac{u}{U}\right) d\eta, \tag{5.149}$$

$$\delta_k^* \equiv \delta^* = \delta \int_0^1 \left(1-\frac{u}{U}\right) d\eta = \delta \int_0^1 \left(1-F(\eta)-\lambda G(\eta)\right) d\eta$$

$$= \delta\left(\frac{3}{10}-\frac{\lambda}{120}\right) = \frac{\delta}{120}(36-\lambda). \tag{5.150}$$

Similarly,

$$\theta_k \equiv \theta = \delta \int_0^1 \frac{u}{U}\left(1-\frac{u}{U}\right) d\eta = \delta\left(\frac{37}{315}-\frac{\lambda}{945}-\frac{\lambda^2}{9072}\right)$$

$$= \frac{\delta}{7530}\left(444-4\lambda-\frac{5\lambda^2}{12}\right). \tag{5.151}$$

The shear stress at the wall is given by,

$$\tau_w = \mu \frac{du}{dy}\bigg|_{y=0} = \frac{\mu U_\infty}{\delta}\left(2+\lambda/6\right). \tag{5.152}$$

The von Karman momentum integral equation reduces to

$$U_\infty \frac{dZ}{dx} = g(\lambda) + h(\lambda) U_\infty \frac{d^2 U_\infty}{dx^2} Z^2, \tag{5.153a}$$

with

$$Z = \delta^2/v = \lambda/(dU_\infty/dx), \tag{5.153b}$$

$$g(\lambda) = \frac{15120-2784\lambda+79\lambda^2+5\lambda^3/3}{(12-\lambda)(37+25\lambda/12)}, \tag{5.153c}$$

$$h(\lambda) = \frac{8+5\lambda/3}{(12-\lambda)(37+25\lambda/12)}. \tag{5.153d}$$

The equation for $Z(x)$ is solved numerically and one obtains

$$\frac{\delta}{x} = \sqrt{\frac{34.05}{Re_x}} = \frac{5.83}{\sqrt{Re_x}}, \quad \frac{\delta^*}{x} = \frac{1.75}{\sqrt{Re_x}}, \quad \frac{\theta}{x} = \frac{0.686}{\sqrt{Re_x}} \quad \text{and} \quad C_f = \frac{0.686}{\sqrt{Re_x}}. \quad (5.154)$$

Holstein and Bohlen [7] use a different dependent variable in place of $Z(x)$. Reconsider the von Karman momentum integral equation and express it as

$$\frac{U_\infty}{2v}\frac{d\theta^2}{dx} + (2+H)\frac{\theta^2}{v}\frac{dU_\infty}{dx} = \frac{\tau_w\theta}{\mu U_\infty} = \frac{\theta}{\delta}\left(2+\frac{\lambda}{6}\right). \quad (5.155)$$

If one lets $z = \theta^2/v = K/(dU_\infty/dx)$, where K was defined as a shape factor by Holstein and Bohlen. Hence,

$$\frac{\theta}{\delta} = \frac{37}{315} - \frac{\lambda}{945} - \frac{\lambda^2}{9072}, \quad K = \lambda\left(\frac{\theta}{\delta}\right)^2 = \lambda\left(\frac{37}{315} - \frac{\lambda}{945} - \frac{\lambda^2}{9072}\right)^2, \quad H(\lambda) = \frac{\delta^*}{\theta},$$

$$(5.156)$$

and

$$\frac{U_\infty}{2}\frac{dz}{dx} + (2+H(\lambda))K = \left(\frac{37}{315} - \frac{\lambda}{945} - \frac{\lambda^2}{9072}\right)\left(2+\frac{\lambda}{6}\right). \quad (5.157)$$

It follows that

$$U_\infty\frac{dz}{dx} = 2\left(\frac{37}{315} - \frac{\lambda}{945} - \frac{\lambda^2}{9072}\right)\left(2+\frac{\lambda}{6}\right) - 2(2+H(\lambda))K. \quad (5.158)$$

The above equation may be solved numerically and the reader is referred to Schlichting and Gersten [8] for further details. For a flat plate, one obtains

$$\theta/x = 0.686/\sqrt{Re_x}. \quad (5.159)$$

Alternately it was found by Walz [9] that the right-hand side may be approximated by a straight line in K as

$$U_\infty\frac{dz}{dx} = 2\left(\frac{37}{315} - \frac{\lambda}{945} - \frac{\lambda^2}{9072}\right)\left(2+\frac{\lambda}{6}\right) - 2(2+H(\lambda))K \approx a - bK, \quad (5.160)$$

where the constants, $a=0.47$, and $b=6$. Hence, it can be shown that

$$\theta^2 = \frac{0.47v}{U_\infty^6}\int_0^x U_\infty^5(x)\,dx. \quad (5.161)$$

5.5.4 REFINED VELOCITY PROFILES WITHIN THE BOUNDARY LAYER

Improved accuracy is obtained by using higher order polynomials in place of Pohlhausen's polynomial for the velocity u. Additional higher order boundary conditions are imposed at $y=0$, and at $y=\delta \to \infty$, based on physically valid assumptions. To obtain the velocity distribution $u(\eta)$ within the boundary layer, a general solution for the velocity profile is defined with two arbitrary constants. The method of defining the velocity profile is an extension of Pohlhausen's [6] method by Torda [10] and by Jain [11]. Jain [11] defines the velocity profile by a sixth-order polynomial. In general, one can define a complete set of polynomials and express the velocity as a linear combination of the first three polynomials in this set. Thus, the velocity profile in defined in terms of two arbitrary constants which are determined by satisfying Equations (5.99) and (5.100) for the momentum and displacement thicknesses obtained by numerically evaluating the equations. Hessenberger [12] uses a similar approach in principle and derives a seventh-order polynomial for the velocity profile, although in his case the polynomial is not a linear combination of polynomial functions. There were several other polynomial and exponential models of the flow velocity which were critically reviewed by Libby, Morduchow and Bloom [13]. The process employed in this section, which is similar, in principle, to the method adopted by Weighardt [14], is briefly described.

Consider the boundary conditions with no blowing, suction or pressure gradient, at

$$\eta = 0, \ \frac{u}{U} = 0, \ \frac{\partial^2}{\partial \eta^2} \frac{u}{U} = 0 \text{ and } \eta = 1, \ \frac{u}{U} = 1, \ \frac{\partial}{\partial \eta} \frac{u}{U} = 0, \ \frac{\partial^2}{\partial \eta^2} \frac{u}{U} = 0. \quad (5.162)$$

The velocity profile shape function is now a quartic in $\eta \equiv y/\delta$,

$$u = U_\infty F_0(\eta), \quad F_0(\eta) = \eta \left(2 - 2\eta^2 + \eta^3\right). \quad (5.163)$$

One may define two auxiliary functions which satisfy the boundary conditions,

i) For the boundary conditions at $\eta=0$, $\frac{u}{U} = 0$, $\frac{\partial^2}{\partial \eta^2} \frac{u}{U} = 1$ and at $\eta=1$,
$\frac{u}{U} = 0$, $\frac{\partial}{\partial \eta} \frac{u}{U} = 0$, $\frac{\partial^2}{\partial \eta^2} \frac{u}{U} = 0$,

$$G_1(\eta) = \frac{\eta}{6}(1-\eta)^3. \quad (5.164)$$

ii) For the boundary conditions at $\eta=0$, $\frac{u}{U} = 0$, $\frac{\partial}{\partial \eta} \frac{u}{U} = 0$, $\frac{\partial^2}{\partial \eta^2} \frac{u}{U} = 1$ and at
$\eta=1$, $\frac{u}{U} = 0$, $\frac{\partial}{\partial \eta} \frac{u}{U} = 0$, $\frac{\partial^2}{\partial \eta^2} \frac{u}{U} = 0$,

$$G_2(\eta) = \frac{\eta^2}{2}(1-\eta)^3, \quad (5.165)$$

Thus,

$$G_2'(\eta) = \eta(1-\eta)^3 - \frac{3}{2}\eta^2(1-\eta)^2 = \frac{(2-5\eta)}{2}\eta(1-\eta)^2, \qquad (5.166a)$$

$$G_2''(\eta) = (1-\eta)(1-\eta)^2 - (2-5\eta)\eta(1-\eta). \qquad (5.166b)$$

At the boundaries,

$$\eta = 0, \quad G_2(\eta) = 0, \quad G_2'(\eta) = 0, \quad G_2''(\eta) = 1; \qquad (5.167a)$$

$$\eta = 1, \quad G_2(\eta) = 0, \quad G_2'(\eta) = 0, \quad G_2''(\eta) = 0. \qquad (5.167b)$$

In general, an expression for the velocity profile may be expressed as,

$$\frac{u}{U} = F_0(\eta) + \lambda G_1(\eta) + g G_2(\eta), \qquad (5.168)$$

where,

$$F_0(\eta) = \eta(2 - 2\eta^2 + \eta^3), \quad G_1(\eta) = \eta(1-\eta)^3/6, \quad G_2(\eta) = \eta^2(1-\eta)^3/2. \qquad (5.169)$$

It follows that the following integral expressions are obtained,

$$\int \frac{u}{U} d\eta = \int F_0(\eta) d\eta + \lambda \int G_1(\eta) d\eta + g \int G_2(\eta) d\eta, \qquad (5.170a)$$

$$\int F_0(\eta) d\eta = \frac{\eta^2}{2}\left(2 - \eta^2 + \frac{2}{5}\eta^3\right), \qquad (5.170b)$$

$$\int G_1(\eta) d\eta = \frac{\eta^2}{6}\left(\frac{1}{2} - \eta + \frac{3}{4}\eta^2 - \frac{1}{5}\eta^3\right), \qquad (5.170c)$$

$$\int G_2(\eta) d\eta = \frac{\eta^3}{2}\left(\frac{1}{3} - \frac{3}{4}\eta + \frac{3}{5}\eta^2 - \frac{1}{6}\eta^3\right). \qquad (5.170d)$$

One may evaluate the normal velocity component corresponding to u. Hence, it is observed that:

$$\eta \frac{u}{U} = \eta F_0(\eta) + \lambda \eta G_1(\eta) + g \eta G_2(\eta), \qquad (5.171)$$

with

$$\eta F_0(\eta) = \frac{\eta^2}{2}\left(4 - 4\eta^2 + 2\eta^3\right), \quad \eta G_1(\eta) = \frac{\eta^2}{6}\left(1 - 3\eta + 3\eta^2 - \eta^4\right),$$

$$\eta G_2(\eta) = \frac{\eta^3}{2}\left(1 - 3\eta + 3\eta^2 - \eta^3\right). \tag{5.172}$$

Hence, it follows that

$$\frac{v}{U} = \frac{\delta}{2x}\left(\bar{F}_0(\eta) + \lambda\bar{G}_1(\eta) + g\bar{G}_2(\eta)\right), \tag{5.173}$$

with

$$\bar{F}_0(\eta) = \eta F_0(\eta) - \int F_0(\eta)\,d\eta = \frac{\eta^2}{2}\left(2 - 3\eta^2 + \frac{8}{5}\eta^3\right), \tag{5.174a}$$

$$\bar{G}_1(\eta) = \eta G_1(\eta) - \int G_1(\eta)\,d\eta = \frac{\eta^2}{6}\left(\frac{1}{2} - 2\eta + \frac{9}{4}\eta^2 - \frac{4}{5}\eta^3\right), \tag{5.174b}$$

$$\bar{G}_2(\eta) = \eta G_2(\eta) - \int G_2(\eta)\,d\eta = \frac{\eta^3}{2}\left(\frac{2}{3} - \frac{9}{4}\eta + \frac{12}{5}\eta^2 - \frac{5}{6}\eta^3\right). \tag{5.174c}$$

In the case of blowing and suction, following Jain [11], the streamwise velocity distribution in the boundary layer may be expressed as

$$u(\eta) = u_e\tilde{u}_6(\eta). \tag{5.175}$$

In Equation (5.175), $\tilde{u}_6(\eta)$ is a sixth-order polynomial in $\eta = y/\delta$, with $U = u_e$, $u = \tilde{u}_6(\eta)$ and the normal velocity given by $v = v(\eta)$ satisfies the boundary conditions:

$$\eta = 0, \quad \frac{u}{U} = 0, \quad \frac{v}{U} = v_0, \quad \text{Re}_\delta\, v_0 \frac{\partial}{\partial\eta}\frac{u}{U} = \frac{\delta \times \text{Re}_\delta}{U}\frac{\partial U}{\partial x} + \frac{\partial^2}{\partial\eta^2}\frac{u}{U},$$

$$\text{Re}_\delta\, v_0 \frac{\partial^2}{\partial\eta^2}\frac{u}{U} = \frac{\partial^3}{\partial\eta^3}\frac{u}{U}. \tag{5.176}$$

Let a suction parameter M be defined as

$$M = \text{Re}_\delta\, v_0, \tag{5.177}$$

and a pressure gradient parameter N be defined as

$$N = \frac{\delta \times \text{Re}_\delta}{U}\frac{\partial U}{\partial x} \quad \text{with} \quad \text{Re}_\delta = \frac{\rho U \delta}{\mu}. \tag{5.178}$$

Then, at $\eta=0$ with $\dfrac{u}{U}=0$, $\dfrac{v}{U}=v_0$,

$$M\frac{\partial}{\partial\eta}\frac{u}{U}=N+\frac{\partial^2}{\partial\eta^2}\frac{u}{U}, \quad M\frac{\partial^2}{\partial\eta^2}\frac{u}{U}=\frac{\partial^3}{\partial\eta^3}\frac{u}{U}, \tag{5.179}$$

At $\eta=1$ with $\dfrac{u}{U}=1$, $\dfrac{\partial}{\partial\eta}\dfrac{u}{U}=0$ and $\dfrac{\partial^2}{\partial\eta^2}\dfrac{u}{U}=0$,

$$\frac{\partial^3}{\partial\eta^3}\frac{u}{U}=0. \tag{5.180}$$

The polynomial, with $n=6$,

$$u=\tilde{u}_n(\eta)=\sum_{k=1}^{n}a_k\eta^k, \tag{5.181}$$

takes the form

$$\tilde{u}_6(\eta)=\frac{\left(120+N(12+M)\right)}{\left(60+12M+M^2\right)}\eta+\frac{\left(60M-30N\right)}{\left(60+12M+M^2\right)}\eta^2$$

$$+\frac{\left(20M-10N\right)M}{\left(60+12M+M^2\right)}\eta^3+\frac{\left(20N(3+M)-15\left(20+12M+3M^2\right)\right)}{\left(60+12M+M^2\right)}\eta^4$$

$$+\frac{\left(12\left(30+16M+3M^2\right)-15N(4+M)\right)}{\left(60+12M+M^2\right)}\eta^5$$

$$\tag{5.182}$$

$$+\frac{\left(2N(9+2M)-10\left(12+6M+M^2\right)\right)}{\left(60+12M+M^2\right)}\eta^6.$$

Similarly, one defines a seventh-order velocity profile as

$$\tilde{u}_7(\eta)=\frac{(45+4M)N+420}{4M^2+45M+216}\eta$$

$$+\frac{6\left(35M-18N\right)}{4M^2+45M+216}\eta^2+\frac{2M\left(35M-18N\right)}{4M^2+45M+216}\eta^3$$

$$+\frac{4N(16M+5))-35\left(4M^2+15M+24\right)}{4M^2+45M+216}\eta^4 \tag{5.183}$$

$$+\frac{84\left(M^2+5M+9\right)-9N(4M+15)}{4M^2+45M+216}\eta^5$$

$$-\frac{10\left(M^2+6M+12\right)-2N(2M+9)}{4M^2+45M+216}\eta^7.$$

A general solution to the velocity distribution may be expressed as

$$u(\eta) = u_e \tilde{u}_6(\eta) + c_1 \left(\tilde{u}_7(\eta) - \tilde{u}_6(\eta) \right). \tag{5.184}$$

When the eighth-order polynomial

$$\hat{u}_8(\eta) = \eta^4 (1-\eta)^4, \tag{5.185}$$

multiplied by an arbitrary constant, is added to $u(\eta)$, all of the boundary conditions continue to be satisfied. Hence, a general solution to the velocity distribution is

$$u(\eta) = u_e \tilde{u}_6(\eta) + c_1 \left(\tilde{u}_7(\eta) - \tilde{u}_6(\eta) \right) + c_2 \hat{u}_8(\eta), \tag{5.186}$$

the constants c_i may be chosen by satisfying Equations (5.99) and (5.100) for the momentum and displacement thicknesses obtained by numerically evaluating equations.

A higher order family of velocity profiles may be defined by imposing yet another boundary condition at the boundary layer edge in addition to Equations (5.179) and (5.180). Thus, at $\eta = 1$,

$$\frac{\partial^4}{\partial \eta^4} \frac{u}{U} = 0. \tag{5.187}$$

To maintain symmetry in the application of boundary conditions at the two ends, a compatible fourth-order boundary condition is applied at the wall, introducing another parameter F, and this is

$$\eta = 0, \quad \frac{Re_\delta^3 \delta}{4} C_f \frac{\partial}{\partial x} C_f = \frac{\partial^4}{\partial \eta^4} \frac{u}{U} = F. \tag{5.188}$$

Introducing the boundary condition given by Equation (5.188) requires an additional parameter, and is therefore avoided.

In general, the non-zero polynomial coefficients corresponding to Equations (5.184) and (5.185), with the highest order of the polynomial given by $n = 6, 7, \ldots$, are defined recursively. Defining the constants,

$$A = 10n/(n-3), \quad B = -3(n-2)/(n-3), \quad C = -6(n-1)/(n-3), \tag{5.189}$$

the non-zero coefficients a_k, $k = 1, \ldots, n$ are recursively defined as

$$a_1 = (6A + NM - 3NB)/D_n; \quad Ma_2 = 3a_3; \quad a_2 = (Ma_1 - N)/2;$$

$$a_3 = A + a_2 B + a_1 C = \left(10n - 3a_2(n-2) - 6a_1(n-1) \right)/(n-3);$$

$$a_4 = \left(-15n - 3a_2(n-2) - 8a_1(n-1) \right)/(n-4);$$

$$a_5 = \left(6n - a_2(n-2) - 3a_1(n-1)\right)/(n-5);$$

$$a_n = \left(6a_2 + 24a_1 - 60\right)/\left\{(n-5)(n-4)(n-3)\right\}. \tag{5.190}$$

When the additional boundary condition at the edge of the boundary layer, $\eta = 1$,

$$\frac{\partial^4}{\partial \eta^4} \frac{u}{U} = 0, \tag{5.191}$$

is also imposed, with the highest order of the polynomial given by $n = 7, 8, \ldots,$ the non-zero coefficients a_k, $k = 1, \ldots, n$ are recursively and implicitly defined as,

$$a_n = 360/\left\{(n-6)(n-5)(n-4)(n-3)\right\}$$
$$- 24\left(a_2 + 5a_1\right)/\left\{(n-6)(n-5)(n-4)(n-3)\right\},$$

$$a_6 = \left(-10n + a_2(n-2) + 4a_1(n-1)\right)/(n-6),$$

$$a_5 = 6 - 3a_1 - a_2 - 3a_6 - (n-4)(n-3)a_7/2,$$

$$a_4 = -3 + 2a_1 + a_2 - 2a_5 - 3a_6 - (n-3)a_7,$$

$$a_3 = 1 - a_1 - a_2 - a_4 - a_5 - a_6 - a_7,$$

$$Ma_2 = 3a_3, \quad Ma_1 = N + 2a_2. \tag{5.192}$$

However, the family of velocity profiles defined by Equations (5.184)–(5.186) were deemed to be adequate after comparing with the well solutions provided by Weighardt [14], Timman [15] and Cooke [16], and more recently by Mughal [17] and Bram [18]. While White [3] mentions that Pohlhausen's [6] fourth degree velocity profile is an inaccurate solution and should be discarded, Libby, Morduchow and Bloom [13] had concluded that the sixth-degree profile gives the most accurate velocity distribution throughout the boundary layer region. Although correlation-based methods may be superior, they do not provide a velocity profile. For this reason, the basic velocity profile could be assumed to be given by Equation (5.184), while Equation (5.186) could be used to model changes to the basic shape. The higher order velocity profiles are used to check for convergence, accuracy and consistency. The application of this methodology to the stability and transition assessment of boundary layers around morphing natural laminar flow airfoils is considered and reported by Vepa [19].

5.5.5 LAMINAR BOUNDARY LAYERS: INTEGRAL METHODS USING TWO EQUATIONS

When one lets the edge velocity be denoted as $u_e = U$ and $\partial u_e / \partial \xi = \delta\, \partial U / \partial x$, the integral momentum equation without blowing/suction could be written as

$$\frac{d\theta}{d\xi} + \left(2 + H - M_e^2\right)\frac{\theta}{u_e}\frac{du_e}{d\xi} = \frac{\tau_w}{\rho U_\infty^2} = \frac{C_f}{2}, \qquad (5.193)$$

where the shape parameter is $H = \delta^*/\theta$. The displacement thickness is defined to be

$$\delta^* = \int_0^\delta \left(1 - \frac{\rho u}{\rho_e u_e}\right) dy = \delta \int_0^1 \left(1 - \frac{\rho u}{\rho_e u_e}\right) d\eta. \qquad (5.194)$$

The momentum thickness is

$$\theta = \int_0^\delta \frac{\rho u}{\rho_e u_e}\left(1 - \frac{u}{u_e}\right) dy = \delta \int_0^1 \frac{\rho u}{\rho_e u_e}\left(1 - \frac{u}{u_e}\right) d\eta. \qquad (5.195)$$

The wall shear stress is

$$\tau_w = \int_0^\delta \frac{\partial \tau}{\partial \eta}\, d\eta = \frac{\rho u_e^2}{2} C_f$$

$$\equiv \mu \left.\frac{\partial u}{\partial y}\right|_{\text{wall}} \equiv \frac{u_e \mu}{\delta}\left.\frac{\partial}{\partial \eta}\left(\frac{u}{u_e}\right)\right|_{\text{wall}} = \frac{\rho u_e^2}{\mathrm{Re}_\delta}\left.\frac{\partial}{\partial \eta}\left(\frac{u}{u_e}\right)\right|_{\text{wall}}, \qquad (5.196)$$

In the preceding equation, the Reynolds number based on a length scale of δ, and the coefficient of friction which is related to the shear stress coefficient are respectively given by

$$\mathrm{Re}_\delta = \rho u_e \delta / \mu, \quad C_f = \tau_w / \left(\rho u_e^2 / 2\right) = 2 C_\tau. \qquad (5.197)$$

The other parameters, the kinematic shape parameter, the kinematic momentum thickness, the kinematic displacement thickness and Whitfield's [20] closure equation are respectively given by

$$H_k = \int_0^\delta \left(1 - \frac{u}{u_e}\right) dy \Big/ \int_0^\delta \frac{u}{u_e}\left(1 - \frac{u}{u_e}\right) dy, \qquad (5.198a)$$

$$\theta_k = \int_0^\delta \frac{u}{u_e}\left(1 - \frac{u}{u_e}\right) dy = \delta \int_0^1 \frac{u}{u_e}\left(1 - \frac{u}{u_e}\right) d\eta, \qquad (5.198b)$$

$$\delta_k^* = \int_0^\delta \left(1 - \frac{u}{u_e}\right) dy = \delta \int_0^1 \left(1 - \frac{u}{u_e}\right) d\eta, \tag{5.198c}$$

$$H_k = \left(H - 0.290 M_e^2\right) \big/ \left(1 + 0.113 M_e^2\right). \tag{5.198d}$$

The energy thickness θ^*, is defined by Equation (5.101) and obtained from

$$\frac{d\theta^*}{d\xi} + \left(\frac{\delta^{**}}{\theta^*} + 3 - M_e^2\right) \frac{\theta^*}{u_e} \frac{du_e}{d\xi} = 2C_D. \tag{5.199}$$

In the preceding equation, the density thickness δ^{**}, is

$$\delta^{**} = \int_0^\delta \frac{u}{u_e}\left(1 - \frac{\rho}{\rho_e}\right) dy = \int_0^\delta \left(1 - \frac{\rho u}{\rho_e u_e}\right) - \left(1 - \frac{u}{u_e}\right) dy$$

$$\tag{5.200}$$

$$= \delta^* - \int_0^\delta \left(1 - \frac{u}{u_e}\right) dy.$$

So it follows that the density thickness to the kinematic momentum thickness is given by $\delta^{**} = \delta^* - H_k \theta_k$.

It is useful to define a drag coefficient C_D by the equation

$$C_D(x) = \frac{\mu}{\rho_e u_e^3} \int_0^\delta \left(\frac{du}{d\eta}\right)^2 dy = \frac{2}{\rho_e u_e^2} \int_0^\delta \frac{\tau}{2u_e} \frac{du}{d\eta} dy. \tag{5.201}$$

It is related to the friction coefficient by

$$C_{Dc} = C_D\big|_{x=c} = \int_0^{x/c} C_f d\frac{x}{c} \bigg|_{x=c}, \tag{5.202}$$

Let $H^* = \theta^*/\theta$ be the kinetic energy shape parameter and $H^{**} = \delta^{**}/\theta$ the density shape parameter. H^{**} is given in terms of H_k by Whitfield [20] as

$$H^{**} = M_e^2 \left(0.251 + 0.064 \big/ \left(H_k - 0.8\right)\right). \tag{5.203}$$

The kinetic energy shape parameter equation may be shown, as done by Drela and Giles [21], to be directly related to drag coefficient C_D and the friction coefficient C_f and satisfies

$$\theta \frac{dH^*}{d\xi} + \left(2H^{**} + H^*\left(1 - H\right)\right) \frac{\theta}{u_e} \frac{du_e}{d\xi} = 2C_D - H^* \frac{C_f}{2}. \tag{5.204}$$

5.5.6 Effect of Suction, Blowing or Porosity

One method of controlling the energy thickness is by introducing blowing or suction. The integral momentum equation with blowing, suction or porosity is given by

$$\frac{d\theta}{d\xi} + \left(2 + H - M_e^2\right)\frac{\theta}{u_e}\frac{du_e}{d\xi} = \frac{C_f}{2} + m_s,$$ (5.205)

where m_s is a suction parameter defined by

$$m_s = \int_0^\delta \frac{\rho_s u_s}{\rho_e u_e}\, d\eta.$$ (5.206)

The energy thickness equation with blowing, suction or porosity is

$$\frac{d\theta^*}{d\xi} + \left(\frac{\delta^{**}}{\theta^*} + 3 - M_e^2\right)\frac{\theta^*}{u_e}\frac{du_e}{d\xi} = 2C_D + H^* m_s\left(1 - H_k\theta_k/\delta\right).$$ (5.207)

The kinetic energy shape parameter equation may be shown to be

$$\theta\frac{dH^*}{d\xi} + \left(2H^{**} + H^*\left(1 - H\right)\right)\frac{\theta}{u_e}\frac{du_e}{d\xi}$$

$$= \left(2C_D + H^*\left(m_s\left(1 - H_k\theta_k/\delta\right)\right)\right) - H^*\left(\frac{C_f}{2} + m_s\right).$$ (5.208)

Rearranging the right-hand side of Equation (5.208),

$$\theta\frac{dH^*}{d\xi} + \left(2H^{**} + H^*\left(1 - H\right)\right)\frac{\theta}{u_e}\frac{du_e}{d\xi} = 2C_D - H^*\left(\frac{C_f}{2} + H_k\theta_k m_s/\delta\right).$$ (5.209)

An alternate equation for Equation (5.209) is

$$\theta\frac{dH^*}{d\xi} + \left(2H^{**} + H^*\left(1 - H\right)\right)\frac{\theta}{u_e}\frac{du_e}{d\xi} = 2C_D - H^*\left(\frac{C_f}{2} + \frac{\delta_k^*}{\delta}m_s\right).$$ (5.210)

If the ratio of the density and the kinetic energy shape parameters is defined as

$$\bar{H}^{**} = H^{**}/H^*.$$ (5.211)

Then it follows that

$$\theta\frac{dH^*}{d\xi} + H^*\left(2\bar{H}^{**} + \left(1 - H\right)\right)\frac{\theta}{u_e}\frac{du_e}{d\xi} = 2C_D - H^*\left(\frac{C_f}{2} + \frac{\delta_k^*}{\delta}m_s\right).$$ (5.212)

For a flat plate it is assumed that $\left(\delta_k^*/\delta\right) \approx 1/3$. Ferreira [22] uses a slightly different approximation for this term. However, the numerical values obtained by him are similar, using the approximation

$$\theta \frac{dH^*}{d\xi} + H^*\left(2\bar{H}^{**} + \left(1-H\right)\right)\frac{\theta}{u_e}\frac{du_e}{d\xi} = 2C_D - H^*\left(\frac{C_f}{2} + \frac{1}{3}m_s\right). \qquad (5.213)$$

• In both equations, the integral momentum equation and the integral equation for the kinetic energy shape parameter, the skin-friction coefficient is modified independently. Considering the integral momentum equation, the skin-friction coefficient is modified to account for the effects of transpiration by using the relation by Kays and Crawford [23]. This is expressed implicitly as

$$\text{Re}_\theta \frac{C_{fpm}}{2} = \text{Re}_\theta \frac{C_f}{2}\left|\frac{\ln\left(1+B_{fm}\right)}{B_{fm}}\right|^{1.25}\left(1+B_{fm}\right)^{0.25},$$

$$B_{fm} = 2\,\text{Re}_\theta \frac{m_s}{2}\bigg/ \text{Re}_\theta \frac{C_{fpm}}{2}. \qquad (5.214)$$

The skin-friction coefficient is determined by the Newton Iteration. Considering the integral equation for the kinetic energy shape parameter, the skin-friction coefficient is again modified to account for the effects of transpiration by using the relation by Kays and Crawford [23] as

$$\text{Re}_\theta \frac{C_{fpe}}{2} = \text{Re}_\theta \frac{C_f}{2}\left|\frac{\ln\left(1+B_{fe}\right)}{B_{fe}}\right|^{1.25}\left(1+B_{fe}\right)^{0.25},$$

$$(5.215)$$

$$B_{fe} = 2\,\text{Re}_\theta \frac{m_s}{6}\bigg/ \text{Re}_\theta \frac{C_{fpe}}{2}.$$

Thus, the momentum and energy integral equations are expressed as

$$\frac{d\theta}{d\xi} + \left(2 + H - M_e^2\right)\frac{\theta}{u_e}\frac{du_e}{d\xi} = \frac{C_{fpm}}{2},$$

$$\theta \frac{dH^*}{d\xi} + H^*\left(2\bar{H}^{**} + \left(1-H\right)\right)\frac{\theta}{u_e}\frac{du_e}{d\xi} = 2C_D - H^*\frac{C_{fpe}}{2}. \qquad (5.216)$$

These equations are re-considered in Chapter 9.

5.5.7　REDUCTION OF THE EQUATIONS

Introducing the variable $\omega = \rho\theta^2/\mu$ and $\text{Re}_\theta = \rho u_e\theta/\mu = u_e\sqrt{\rho\omega/\mu}$, from the momentum integral equation, one has

$$\frac{u_e}{2}\frac{d\omega}{d\xi} = \frac{\rho u_e \theta}{\mu}\frac{d\theta}{d\xi} = \mathrm{Re}_\theta \frac{C_{fpm}}{2} - \left(2 + H - M_e^2\right)\frac{\rho \theta^2}{\mu}\frac{du_e}{d\xi}. \tag{5.217}$$

Hence,

$$\frac{u_e}{2}\frac{d\omega}{d\xi} + \left(2 + H - M_e^2\right)\frac{du_e}{d\xi}\omega = \mathrm{Re}_\theta \frac{C_{fpm}}{2}. \tag{5.218}$$

From the energy integral equation,

$$\frac{\theta}{H^*}\frac{dH^*}{d\xi} + \left(2\bar{H}^{**} + (1-H)\right)\frac{\theta}{u_e}\frac{du_e}{d\xi} = \frac{2C_D}{H^*} - \frac{C_{fpe}}{2}. \tag{5.219}$$

Thus Equation (5.219) may be expressed as,

$$\mathrm{Re}_\theta \theta \frac{d\log\left(H^*\right)}{dH}\frac{dH}{d\xi} + \left(2\bar{H}^{**} + (1-H)\right)\mathrm{Re}_\theta \frac{\theta}{u_e}\frac{du_e}{d\xi}$$

$$= \mathrm{Re}_\theta \frac{2C_D}{H^*} - \mathrm{Re}_\theta \frac{C_{fpe}}{2}, \tag{5.220}$$

which may also be expressed as

$$u_e \frac{d\log\left(H^*\right)}{dH}\omega \frac{dH}{d\xi} + \frac{du_e}{d\xi}\left(2\bar{H}^{**} + (1-H)\right)\omega = \mathrm{Re}_\theta \frac{2C_D}{H^*} - \mathrm{Re}_\theta \frac{C_{fpe}}{2}. \tag{5.221}$$

To further reduce these equations, some of the unknowns must be expressed in terms of two primary variables. These equations are known as "closure equations" and are obtained empirically and verified by experiment. Thus, the kinetic energy shape parameter $H^* = \theta^*/\theta$ may be expressed in terms $H = \delta^*/\theta$ by the relations

$$H^* = H^*(H) = 1.515 + 0.076(4-H)^2/H \quad \text{for} \quad H < 4$$

$$H^* = H^*(H) = 1.515 + 0.040(4-H)^2/H \quad \text{for} \quad H \geq 4. \tag{5.222}$$

The skin-friction and drag coefficients are then respectively expressed as

$$\mathrm{Re}_\theta \frac{C_f}{2} = f_1(H) = -0.067 + 0.01977(7.4-H)^2/(H-1) \quad \text{for} \quad H < 7.4,$$

$$= -0.067 + 0.022\left\{1 - 1.4/(H-6)\right\}^2 \quad \text{for} \quad H \geq 7.4,$$

$$\text{Re}_\theta \frac{2C_D}{H^*} = f_2(H) = 0.207 + 0.00205(4-H)^{5.5} \quad \text{for} \ \ H < 4,$$

$$\text{(5.223)}$$

$$= 0.207 - 0.003(4-H)^2 \Big/ \Big(1 + 0.02(H-4)^2\Big) \quad \text{for} \ \ H \geq 4.$$

The laminar closure equations reach a singularity at the point where $H_k = H$ reaches a value equal to 4, which is where the $H^*(H)$ reaches a minimum. This is referred to as the "Goldstein singularity" at a boundary layer separation point. The vanishing derivative of $H^*(H)$ causes a singularity in the equation for energy thickness, which can only be avoided if u_e adjusts to cause the rest of the equation to tend to zero as well. Therefore, any boundary layer method with a prescribed u_e that reaches separation will fail at this point. However, several methods have been proposed to get around this problem.

The boundary conditions are at a stagnation point, at $\xi = 0$, $d\omega/d\xi = 0$ and $dH/d\xi = 0$. With these conditions $\omega(0) = f_1(H_0) \Big/ \Big((2H_0 + H_0^2)(du_e/d\xi)\Big)$ and $H(0) \equiv H_0 \approx 2.24$ may be used to start the integration. After completing the integration, it is possible to estimate the displacement and momentum thickness, $\delta^*(\xi)$ and $\theta(\xi)$, along the streamwise direction above the boundary wall, respectively.

5.5.8   Special Cases

There are a number of special cases one could consider, including the case of constant momentum thickness θ, along with constant displacement thickness δ^* with uniform suction, the case of constant shape factor $H = \delta^*/\theta$, the case of prescribed $dH/d\xi$ and the case of prescribed $d\omega/d\xi$. The equations presented earlier could be specialized for each of these four special cases. These special cases are re-considered in Chapter 9.

5.5.9   Thwaites Correlation Technique

Thwaites [24] found by empirical correlation that most laminar boundary layers satisfy a simple differential equation. To obtain the differential equation, consider the momentum integral equation

$$\frac{C_f}{2} = \frac{\tau_w}{\rho U_\infty^2} = \frac{d\theta}{dx} + (2+H)\frac{\theta}{U_\infty}\frac{dU_\infty}{dx}. \tag{5.224}$$

Multiply both sides by $\theta U_\infty/\nu$ to obtain

$$\frac{U_\infty}{\nu}\frac{d\theta^2}{dx} = 2\left\{\theta\frac{U_\infty}{\nu}\frac{C_f}{2} - (2+H)\frac{\theta^2}{\nu}\frac{dU_\infty}{dx}\right\} \equiv 2\{L(\lambda) - (2+H)\lambda\}, \tag{5.225}$$

where

$$\lambda = \frac{\theta^2}{\nu}\frac{dU_\infty}{dx} \quad \text{and} \quad L(\lambda) = \theta\frac{U_\infty}{\nu}\frac{C_f}{2}. \tag{5.226}$$

Thwaites found a good fit for the above differential equation and expressed it as

$$\frac{U_\infty}{v}\frac{d}{dx}\left(\theta^2\right) = A - B\frac{\theta^2}{v}\frac{dU_\infty}{dx}, \quad A = 0.45 \text{ and } B = 6. \tag{5.227}$$

Thus

$$d\left(U_\infty^B\theta^2\right)/dx = vAU_\infty^{B-1}. \tag{5.228}$$

The solution is given by

$$\theta^2 = \frac{Av}{U_\infty^B}\int_{x=0}^{x} U_\infty^{B-1}dx + \left[\theta^2(x=0)\frac{U_\infty^B(x=0)}{U_\infty^B(x)}\right] = \frac{Av}{U_\infty^B}\int_{x=0}^{x} U_\infty^{B-1}dx. \tag{5.229}$$

Thwaites' integral correlation method [24] is used to calculate the laminar boundary layer parameters starting from the sharp-leading edge of the plate, up to transition onset specified by the point and determined by a transition occurrence criterion. Thwaites' method [24] enables calculation of the integral parameters θ, H and C_f of the laminar boundary layer in both the zero-pressure gradient and in the presence of a pressure gradient in an un-separated boundary layer, according to the following solution:

$$\theta^2 = \frac{0.45v}{U_\infty^6}\int_{0}^{x} U_\infty^5(x)dx. \tag{5.230}$$

After θ is found, the following closure equation are used to compute the shape factor $H(\lambda)$.

$$H(\lambda) = \begin{cases} 2.61 - 3.75\lambda + 5.24\lambda^2, & 0 \le \lambda \le 0.1 \\ 2.472 + 0.0147/(\lambda + 0.107), & -0.1 \le \lambda \le 0, \end{cases} \tag{5.231}$$

where

$$\lambda = \frac{\theta^2}{v}\frac{dU_\infty}{dx}. \tag{5.232}$$

The skin-friction coefficient and displacement thickness can respectively be calculated from the assumed one-parameter correlations [3]

$$\tau_w = \frac{\mu U_\infty}{\theta}S(\lambda), \quad \delta^* = \theta H(\lambda). \tag{5.233}$$

Thwaites suggested a correlation for $S(\lambda)$ and $H(\lambda)$, where $S(\lambda)$ is best represented by

$$S(\lambda) \approx (\lambda + 0.09)^{0.62}. \tag{5.234}$$

Then the friction coefficient can be calculated from the relation

$$C_f = \frac{2\tau_w}{\rho U_\infty^2} = \frac{2\mu}{\rho\theta U_\infty} S(\lambda). \tag{5.235}$$

5.6 TRANSITION AND SEPARATION

The general process by which transition of a laminar boundary layer to a turbulent one begins with a stable laminar boundary layer and with all linear instability
modes damped. As a critical Reynolds number is reached, modes of waves in the
boundary layer associated with particular frequencies become unstable and experience unbounded growth. The unstable waves grow in amplitude to a point when the
non-linear forces and moments become significantly large, and turbulent spots begin
to appear before the boundary layer ultimately transitions to turbulence. Empirical
criteria have been developed, based on linear wave amplification rates, to quantify
when the non-linear forces and moments become significant in the process of initiat-
ing transition, which will be discussed in Chapter 9.

Although the above description provides an apparently straightforward process
for the onset of transition, the transition to turbulence is affected by many external
factors, such as disturbances and noise in the freestream. When the disturbances
in the freestream are large, the entire linear amplification regime is practically and
rapidly bypassed and the transition to turbulence occurs almost instantaneously.
Freestream noise has an even more significant effect, and consequently, the transition
process can occur at Reynolds numbers an order of magnitude less than the values
observed in low-noise wind tunnels. Hence, the transition of an attached boundary
layer from a laminar to a turbulent state is broadly classified as being either a natural
type or a bypass type. Natural transition is the primary type of transition observed
in flows where the freestream turbulence level is relatively low, while the bypass
type of transition is present in flows with higher levels of freestream turbulence.
The former is therefore generally important in aircraft wings in steady flight while
the latter is observed in gas turbine engines. Key to the evolution of natural transition is receptivity, which is the process by which freestream velocity fluctuations
induce fluctuations within the boundary layer. Although the two types of transition are considered independent of each other, in natural transition, one observes
the development of the two-dimensional Tollmien–Schlichting waves as the first
stage of the transition process [8], which is followed by the development of three-
dimensional structures similar in nature to the streaky K-structures (or Klebanoff
modes) observed in bypass transition. Baines, Majumdar and Mitsudera [25] have
discussed the mechanics of the Tollmien–Schlichting wave, while Kachanov [26]
has given a succinct description of the physical mechanisms of laminar-boundary
layer transition. Tollmien–Schlichting waves, which propagate in the direction of
the flow, result from the instability of the laminar boundary layer and consequently
grow exponentially with the natural frequency for the boundary layer when the
damping is inadequate to attenuate them (i.e. when a critical Reynolds number is
exceeded). On the other hand, bypass fluctuations within the boundary layer seem

to correspond to the forced response due to the freestream fluctuations and are characterized both by an unstable transient response at the natural frequency as well
as a narrow band turbulent response near the resonant frequency of the boundary
layer, driven by the freestream turbulence. For this reason, the turbulence appears
almost instantaneously and is characterized by a superposition or convolution of
several frequencies. Characteristics of bypass transition are spanwise periodic,
three-dimensional flow patterns, and associated with these so-called Klebanoff patterns are the formation of longitudinal vortices. Furthermore, the forced response
to freestream turbulence is significantly affected by the non-linearities of the flow.
Thus the response features may include the secondary characteristics of a non-linear
resonance associated with a softening spring (resulting in a broad band amplification
at lower frequencies and selective amplification of certain structures at higher frequencies or C-structures), subharmonics associated with subharmonic instabilities
(H-structures) and in extreme cases may include tertiary features associated with
the response of chaotic dynamic systems, such as intermittency and period doubling.

An adverse pressure gradient occurs when the static pressure increases in the
direction of the flow i.e. the pressure gradient is positive. Boundary layer instability and transition are known to be related and are both strongly influenced by a
streamwise pressure gradient. When the static pressure increases in the downstream
direction, the margin of stability decreases and hence the possibility of transition to
turbulence becomes higher. Yet the reduction in the margin of stability or a slight
increase in instability of the boundary layer does not imply that turbulence. The
influence of an increase in the static pressure in the downstream direction on boundary layer separation, however, can be significant. Boundary layer separation is the
detachment of a boundary layer from the surface into a wake spreading over a larger
volume. Boundary layer separation occurs when the portion of the boundary layer
closest to the wall, or leading-edge, reverses in flow direction. Boundary layer separation generally occurs due to a sustained adverse pressure gradient, or a sharp or
sudden change in geometry, such as a sharp corner. For sharp edge airfoils, the onset
of separation is at the sharp-leading edge, since separation will start at sharp corners
due to the adverse pressure gradient. Sharp edge airfoils can experience the onset
of separation at an angle of attack as low as eight degrees. Separation of the boundary layer can occur prior to transition to turbulence or can occur after transition. A
separated shear layer which initially develops nominally under conditions of a zero-
pressure gradient entrains fluid both from the reversed flow region as well as from
the outer inviscid flow, external to the shear layer. Moreover, the velocity of the fluid
along the dividing stream surface increases and is followed by the reattachment of
the shear layer onto a downstream wall. Consequently, there is recompression associated with the confluence of top and bottom shear layers. The reattachment pressure
rises due to the recompression and the region over which it occurs may be expected
to depend primarily on the shear layer characteristics and the boundary conditions
at the boundary of reattachment. The mass entrained by the shear layers from the
reversed flow forms a recirculating bubble. The formation of the separation bubble
due to the reattachment process is explained by Hak [27]. The pressure field in a
separated flow is significantly influenced by the shear layer reattachment process and
therefore it is a key element in the dynamics of separated flows. The reattachment

process is quite uncertain and there are a number of reattachment criteria or models, which are not uniquely defined.

When the flow separates from a wall, the boundary layer theory no longer holds since a shear layer is formed. Vortices will appear and they will be shed from the separation points located at the leading and trailing edges in an alternate way. These vortices are energized by the interaction with each other so the ones that are shed from the leading edge are at a disadvantage since these leading-edge vortices are too weak to establish a formation, so they generally do not appear so until they reach the wake. Thus, the vortices are mainly observed in the wake. Thus, separation involves the mechanisms of the production, shedding, capture and enhancement of vortices at post-stall angles of attack and can be viewed as the resonance of a vortex layer following the increase in receptivity to disturbances and instability of the boundary layer. When flow separates, the result is the continuous shedding of a free shear layer. This shear layer, which is also a vortex layer in itself, is unstable to small disturbance perturbations (Kelvin-Helmholtz instabilities), and the instabilities will cause an interaction and coherent reinforcing of the vortices by a process akin to positive feedback. Thus, the resonance in the vortex layer continues to sustain the separation of the flow from the wall.

The point of separation may be defined as the point between forward and reverse flow in the layer very close to the wall i.e. at the point of separation

$$\left(du/dy\right)_{y=0} = 0. \tag{5.236}$$

Consequently, the shear stress at the wall, $\tau_w = 0$. The adverse pressure continues to exist downstream of the separation point and the flow is reversed, resulting in a back flow. From the dimensional form of the momentum equation at the wall, where $u = v = 0$, the following is obtained:

$$\left(d^2u/dy^2\right)_{y=0} = \left(dp/dx\right)/\mu. \tag{5.237}$$

Consider the situation with a favorable pressure gradient where $dp/dx < 0$, when, $\left(d^2u/dy^2\right)_{y=0} < 0$. As one traverses further into the freestream, the velocity u approaches $U_\infty = 0$ asymptotically, and du/dy decreases at a continuously at smaller rate in y-direction. This means that the rate of change of du/dy i.e. d^2u/dy^2 remains less than zero and is always negative as the edge of the boundary layer is approached. On the other hand, consider the case of adverse pressure gradient, $dp/dx > 0$ when $\left(d^2u/dy^2\right)_{y=0} > 0$. However, within the boundary layer $\left(d^2u/dy^2\right)_{y=0} < 0$. Thus, it is observed that for an adverse pressure gradient, there must exist a point for which $d^2u/dy^2 = 0$. This point is a "point of inflection" of the velocity profile in the boundary layer. Consequently, if there is a point of separation, there must exist a point of inflection in the velocity profile and vice versa.

5.6.1 WALZ–THWAITES' CRITERION FOR TRANSITION/SEPARATION

Since in Thwaites' method [24], it essentially assumed a shape for the profile, one could predict when the flow in the boundary layer reverses. This happens when:

$$\lambda = \left(\theta^2/v\right)\left(dU_\infty/dx\right) = -0.09. \tag{5.238}$$

The exact value is not very important, since λ changes quickly near the area of separation. This problem was also considered by Walz [9].

Another criterion that does not require numerical integration of the boundary layer equations is one due to Stratford [28]. This criterion asserts that laminar separation occurs when:

$$C_p'\left(x'\,dC_p'/dx\right)^2 = 0.0104. \tag{5.239}$$

C_p' and x' are the canonical pressure coefficient and effective boundary layer length. Stratford's laminar separation criterion appropriately reflects the destructive effect of the adverse gradient's severity and length.

Because the laminar boundary layer is prone to transition in an adverse gradient, it is difficult to predict whether the flow will transition or separate first. Sometimes the flow separates, transitions and then reattaches in what is called a laminar separation bubble. The length of the bubble is a function of the pressure gradient and Reynolds number, growing longer as the Reynolds number is reduced. In any case, laminar separation is to be avoided in airfoil design. This is done in several ways including forcing transition with surface roughness elements (grit) or building in a special transition region in the pressure distribution.

Once the flow is turbulent, an entirely different set of separation criteria must be applied. From the Thwaites expression for θ,

$$\theta^2 = \frac{3}{40}\frac{vL}{U_\infty}\left[\left(1-\frac{x}{L}\right)^{-6} - 1\right], \tag{5.240}$$

and the criterion for transition,

$$\lambda = \frac{\theta^2}{v}\frac{dU_\infty}{dx} = -0.09, \tag{5.241}$$

given the maximum pressure gradient $N=N_{max}$, where N is defined as,

$$N = \frac{\delta \times \mathrm{Re}_\delta}{U_\infty}\frac{\partial U_\infty}{\partial x}, \quad \mathrm{Re}_\delta = \frac{\rho U_\infty \delta}{\mu}, \tag{5.242}$$

it follows that,

$$\frac{\theta^2}{v} = \frac{3}{40}\frac{L}{U_\infty}\left[\left(1-\frac{x}{L}\right)^{-6} - 1\right], \tag{5.243}$$

and that,

$$\frac{\theta^2}{v}\frac{dU_\infty}{dx} = \frac{3L}{40}\frac{1}{U_\infty}\frac{dU_\infty}{dx}\left[\left(1-\frac{x}{L}\right)^{-6} - 1\right] = -0.09. \tag{5.244}$$

Rearranging the above equation,

$$\left(1-\frac{x}{L}\right)^{-6}=1-\frac{40}{3}\times\frac{0.09\delta\,\text{Re}_{\delta}}{NL}. \tag{5.245}$$

Thus, the first instance of transition is at $x=x_{tr}$ where x_{tr} is given by,

$$\frac{x_{tr}}{L}=1-\left(1-1.2\frac{\delta}{L}\frac{\text{Re}_{\delta}}{N_{\max}}\right)^{-1/6}. \tag{5.246}$$

5.6.2 THE TRANSITIONAL BOUNDARY LAYER

Numerous other correlations have been proposed in the literature for determining the location of the onset of transition. Among those correlations, the one proposed by Abu-Ghannam and Shaw [29] determines the starting of transition as a function of the freestream turbulence level Tu, and the local pressure gradient λ_{θ}. The empirical relation is based on extensive experimental data obtained in the highly turbulent flows likely to be encountered in turbomachinery, and could be adopted for use in flat plates and airfoils. The Reynolds number based on the momentum thickness at the start of the transition is given as

$$\text{Re}_{\theta S}=163+\exp\left\{f\left(\lambda_{\theta}\right)\left(1-\left(Tu/6.91\right)\right)\right\}, \tag{5.247}$$

where

$$f\left(\lambda\right)=\begin{cases}6.91+12.75\lambda+63.64\lambda^{2}, & \lambda<0\\6.91+2.48\lambda-12.27\lambda^{2}, & \lambda>0.\end{cases} \tag{5.248}$$

For case of flat plate with zero-pressure gradient the parameter λ_{θ} goes to zero. To calculate the development of the boundary layer during transition in this method, it is required to determine first, the ending point of transition X_{E} and the properties of the boundary layer at this point. The position of the end of transition X_{E} is specified by

$$\text{Re}_{XE}=\text{Re}_{XS}+\text{Re}_{L}, \tag{5.249}$$

where,

$$\text{Re}_{L}=16.8\left(\text{Re}_{XS}\right)^{0.8}, \tag{5.250}$$

and the momentum thickness Reynolds number at the end of transition $\text{Re}_{\theta E}$ is determined from

$$\text{Re}_{\theta E}=540+183.5\left(10^{-5}\,\text{Re}_{L}-1.5\right)\left(1-1.5\lambda_{\theta}\right). \tag{5.251}$$

The integral parameters at the end of transition (H_E, C_{fE}) are related to momentum thickness Reynolds number $(\mathrm{Re}_{\theta E})$ through the Ludwieg and Tillmann [30] correlation,

$$C_f = 0.246\left(10^{-0.678H}\right)\mathrm{Re}_{\theta}^{-0.268}. \tag{5.252}$$

Furthermore, the following empirical relationship between H and Re_{θ} proposed by Goksel [31] is considered

$$\log_e H/(H-1) = 0.112\left(\log_e \mathrm{Re}_{\theta}\right) + 0.375. \tag{5.253}$$

Specifying the momentum thickness Reynolds number at the end of transition, $\mathrm{Re}_{\theta E}$, Equations (5.253) and (5.252) can be solved for the shape factor, H_E, and friction coefficient, C_{fE}, at the end of transition respectively. Since $\mathrm{Re}_{\theta S}$, H_S and C_{fs} were known from Thwaites' prediction, the following equations are used to calculate the development of the different integral parameters within the transition region

$$\theta_{tr} = \eta^{1.35}, \quad H_{tr} = \sin\left(\pi\eta/2\right), \quad C_{ftr} = 1 - \exp\left(-5.645\eta^2\right), \tag{5.254}$$

where η is a non-dimensional length parameter during transition defined as

$$\eta = \frac{\mathrm{Re}_X - \mathrm{Re}_{XS}}{\mathrm{Re}_{XE} - \mathrm{Re}_{XS}}. \tag{5.255}$$

The momentum thickness ratio during transition region θ_{tr} is defined as

$$\theta_{tr} = \frac{\theta - \theta_S}{\theta_E - \theta_S}, \tag{5.256}$$

H_{tr} the shape factor ratio during transition is defined as

$$H_{tr} = \frac{H_S - H}{H_S - H_E}, \tag{5.257}$$

and C_{ftr}, friction coefficient during transition defined as

$$C_{ftr} = \frac{C_f - C_{fs}}{C_{fE} - C_{fs}} \tag{5.258}$$

5.7   TURBULENT BOUNDARY LAYERS

It is known that turbulent flow occurs if the flow velocity is large enough (or, the viscosity is small enough) to create a Reynolds number greater than the critical Reynolds number over an object. For spheres or circular cylinders, this critical Reynolds number is between 2 to 4×10^5. For flat plate flows this is around 500,000. There are implications of turbulent flows in the reduction of drag. What characterizes such flow is a flatter, fuller velocity profile. It is important to recognize that turbulent flows have two components: i) a mean, $\bar{u}$, and ii) a random one, u'. Thus, one could express a variable u as the sum of the two, that is, as $u = \bar{u} + u'$. Similarly, $v = \bar{v} + v'$ and $w = \bar{w} + w'$. The random u' cannot be determined without statistical means. Therefore, for turbulent fluid flows, with a *time averaged* mean flow $\bar{u}$ is usually considered first. When reference is made to a turbulent velocity profile it is the mean flow velocity profile $\bar{u}$ that is being considered. To avoid any confusion with this notation (the quantity $\bar{v}$ is also sometimes the *area averaged* velocity, and not only the time averaged velocity), turbulent flow velocities are written without the overbars.

In the case of turbulent flows, the mean flow must be properly represented. Turbulent flows in boundary layers over flat plates may be represented by the power law velocity profile:

$$\frac{u}{U}(y) = \left(\frac{y}{\delta}\right)^{1/n} = \eta^{1/n}, \quad \eta = \frac{y}{\delta}. \tag{5.259}$$

This profile covers a fairly broad range of turbulent Reynolds numbers for $6 < n < 10$. The most popular one is $n = 7$. For these profiles,

$$\theta = \theta_k = \int_0^\delta \frac{u}{u_e}\left(1 - \frac{u}{u_e}\right)dy = \delta\int_0^1 \frac{u}{u_e}\left(1 - \frac{u}{u_e}\right)d\eta. \tag{5.260}$$

The kinematic momentum and displacement thicknesses are, respectively,

$$\theta = \delta\int_0^1 \frac{u}{u_e}\left(1 - \frac{u}{u_e}\right)d\eta = \delta\int_0^1 \eta^{\frac{1}{n}}\left(1 - \eta^{\frac{1}{n}}\right)d\eta$$

$$= \delta\left(\frac{n}{n+1} - \frac{n}{n+2}\right) = \frac{\delta n}{(n+1)(n+2)} \tag{5.261}$$

$$\delta^* = \delta_k^* = \int_0^\delta \left(1 - \frac{u}{u_e}\right)dy = \delta\int_0^1 \left(1 - \frac{u}{u_e}\right)d\eta$$

$$= \delta\left(1 - \frac{1}{1+1/n}\right) = \frac{\delta}{(n+1)}. \tag{5.262}$$

The shape factor is given by

$$H = \delta^*/\theta = (n+2)/n = 1 + (2/n). \tag{5.263}$$

As $n = 1, 2, \ldots, 10$,

$$H = 3, \ 2, \ 1.6666, \ 1.5, \ 1.4, \ 1.3333, \ 9/7, \ 1.25, \ 11/9, \ 1.2. \tag{5.264}$$

5.7.1 PREDICTING THE TURBULENT BOUNDARY LAYER

In the turbulent region of the boundary layer, several integral methods have been proposed to predict the turbulent flow parameters. The first one was given by Head [5] is generally most popular. More recently, the method proposed by Drela [21] has gained popularity, as it can be used for both laminar and turbulent boundary layers and in unsteady flow. Using a method to predict the boundary layer in all three regimes, it is possible to validate the results for all the three types of boundary layers (laminar, transitional and turbulent) along with a transition prediction scheme (such as the Walz–Thwaites scheme (Section 5.6.1) or the Abu-Ghannam and Shaw [29] scheme). In all the turbulent boundary layer prediction methods, the boundary layer parameters are required at the end of the laminar region as a starting point. Since the boundary layer at the end of transition is fully turbulent, this point could be considered as a starting point for the prediction of the turbulent boundary layer.

5.7.2 THE ENTRAINMENT EQUATION DUE TO HEAD

In Head's method [5], Head suggested a new shape parameter H_1, given by the following relation

$$H_1 = \left(\delta - \delta^*\right)\big/\theta. \tag{5.265}$$

Head's concept of entrainment is illustrated in Figure 5.7. In the relation for H_1, $\delta = \delta(x)$ is the thickness of the boundary layer and the entrained mass transport into the boundary layer is given by the equation,

$$Q(x) = \int_0^{\delta(x)} u\,dy. \tag{5.266}$$

This quantity does not have to be constant, but usually increases significantly. An increase of $Q(x)$ has to be compensated by flow entering the boundary layer from outside. This flow can be expressed as

$$E = dQ(x)\big/dx, \tag{5.267}$$

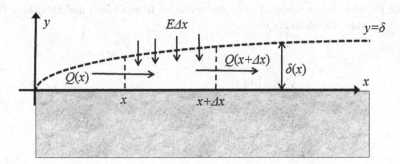

FIGURE 5.7 Concept of entrainment.

where E is known as the velocity of entrainment. Comparing the definition of $Q(x)$ with the definition of the displacement thickness δ^*,

$$U_\infty \delta(x) = Q(x) + U_\infty \delta^*(x). \tag{5.268}$$

Hence,

$$Q(x) = U_\infty \left(\delta(x) - \delta^*(x) \right) = U_\infty H_1 \theta. \tag{5.269}$$

Consequently,

$$E = dQ(x)/dx = d\left(U_\infty H_1 \theta \right)/dx. \tag{5.270}$$

Based on extensive experimental evidence, Head proposed the evolution of H_1 along the boundary layer to follow the entrainment equation

$$\frac{1}{U_\infty} \frac{d}{dx} \left(U_\infty \theta H_1 \right) = F(H_1) = 0.0306 \left(H_1 - 3 \right)^{-0.6169}. \tag{5.271}$$

In unsteady flow it is modified as

$$\frac{1}{U_\infty} \frac{d}{dx} \left(U_\infty \theta H_1 \right) = F(H_1) = 0.0306 \left(H_1 - 3 \right)^{-0.6169} + \frac{1}{U_\infty} \frac{d\delta}{dt}. \tag{5.272}$$

Equation (5.272) and the von Karman momentum integral equation, Equation (5.114), are solved by marching from transition location to trailing edge. The empirical closure relations suggested by Head is

$$H_1(H) = \begin{cases} 3.3 + 0.8234 \left(H - 1.1 \right)^{-1.287}, & H \leq 1.6, \\ 3.3 + 1.5501 \left(H - 0.6778 \right)^{-3.064}, & H > 1.6. \end{cases} \tag{5.273}$$

For closure, the friction coefficient C_f is approximated using the correlation from White [3] and H_1 is calculated using the relations from Cebeci and Bradshaw [32] which are valid for turbulent flow.

$$C_f = \frac{0.3 \exp(-1.33H)}{\left(\log_{10} \mathrm{Re}_\theta \right)^{1.74+0.31H}},$$

$$\tag{5.274}$$

$$H_1 = \begin{cases} 0.8234 \left(H - 1.1 \right)^{-1.287} + 3.3, & H < 1.6, \\ 0.5501 \left(H - 0.6778 \right)^{-3.064} + 3.3, & H \geq 1.6. \end{cases}$$

Using the closure relation from Cebeci and Bradshaw [32], and applying the product rule the entrainment equation is written as

$$\frac{dH}{dx} = \frac{1}{\gamma\theta}\left\{0.0306\left(H_1-3\right)^{-0.6169} - H_1\frac{d\theta}{dx} - \frac{\theta H_1}{u_e}\frac{du_e}{dx}\right\},$$

$$\gamma = \begin{cases} -1.0597\left(H-1.1\right)^{-2.287} + 3.3, & H < 1.6, \\ -1.6855\left(H-0.6778\right)^{-4.064} + 3.3, & H \geq 1.6. \end{cases} \tag{5.275}$$

Alternately, in Head's method the friction coefficient may be calculated by the Ludwig-Tillman correlation, given by Equation (5.252).

5.7.3 DRELA'S METHOD FOR A TURBULENT BOUNDARY LAYER

Drela [21] derived a different lag equation using the shear stress coefficient C_τ,

$$C_\tau = \frac{1}{u_e^2}\left(-\overline{u'v'}\right)_{max}, \tag{5.276}$$

where $\left(-\overline{u'v'}\right)_{max}$ is the maximum Reynolds stress. Drela defined the normal velocity gradient as

$$\frac{\partial u}{\partial \eta} = \frac{1}{L}\sqrt{\left(-\overline{u'v'}\right)_{max}}, \tag{5.277}$$

where L is the dissipation or mixing length. Substituting $\partial u/\partial \eta$ and C_τ in the Reynolds stress transport equation, neglecting normal convection and keeping only the terms with C_τ, Drela obtained the entrainment equation

$$\frac{\delta}{C_\tau}\frac{\partial C_\tau}{\partial x} = K_c\left(\sqrt{C_{\tau,eq}} - \sqrt{C_\tau}\right), \tag{5.278}$$

which is closed using equation set below. The original value for K_c is 5.6 but Drela [21] reported better results using $K_c=4.2$. C_τ can be initialized with a formulation used by Nishida [33] as discussed by Bram [18] in Section 3.3.2 of the reference.

While the initial estimate of C_τ in turbulent flow could be obtained using a formula due to Nishida [33], an alternate method is to estimate it from the laminar flow prior to transition. The calculated C_τ is used to estimate C_D. Two sets of closure equations are given in Bram [18], one for use with C_τ and the other without.

Drela's assumption that the higher order terms of the lag entrainment equation can be neglected was supported by Bhanderi and Babinksy [34]. The derivation of Drela's altered lag entrainment equation assumed the steady Reynolds stress transport equation. In the unsteady case, the lag entrainment equation can be written as in Hall [35], as discussed by Cebeci and Bradshaw [32].

$$\frac{\delta}{u_e^3 C_\tau U_{s,\max}} \frac{\partial\left(u_e^2 C_\tau\right)}{\partial t} + \frac{\delta}{u_e^2 C_\tau} \frac{\partial\left(u_e^2 C_\tau\right)}{\partial x} = K_c\left(\sqrt{C_{\tau,eq}} - \sqrt{C_\tau}\right). \tag{5.279}$$

In Equation (5.279) $U_{s,\max}$ is the equilibrium slip velocity at the point of maximum shear stress which is equivalent to the local slip velocity as given by Drela. The most general form of the unsteady entrainment equation, due to Hall [35], is

$$\frac{\delta}{u_e C_\tau U_{s,\max}} \frac{\partial C_\tau}{\partial t} + \frac{\delta}{C_\tau} \frac{\partial C_\tau}{\partial x} + \frac{2\delta}{u_e^2 U_{s,\max}} \frac{\partial u_e}{\partial t} + \frac{2\delta}{u_e} \frac{\partial u_e}{\partial x} - \left(\frac{2\delta}{u_e} \frac{\partial u_e}{\partial x}\right)_{EQ}$$

$$= K_c\left(\sqrt{C_{\tau,eq}} - \sqrt{C_\tau}\right), \tag{5.280}$$

where the additional term represents a measure of the equilibrium diffusion that is not ignored. Further details about Drela's method may be found in Drela's original paper [21] and in Bram [18].

5.8   STRATEGY FOR AIRCRAFT DRAG REDUCTION

In the preceding sections, first it was observed that there are two distinct types of drag that act to retard the aircraft when it is in forward flight: profile drag, consisting of form drag, due to pressure forces, and skin-friction drag, due to the shear stress or a boundary layer forms due to skin friction, and induced drag, due to vortex dissipation of the energy in the flow. A key feature of the flow over an airfoil, is the boundary layer, consisting of a laminar part towards the front of the airfoil and a turbulent boundary layer towards the aft of the airfoil. Moreover, the inviscid outer flow is displaced. Flow about the airfoil is characterized by an adverse pressure gradient and subsequent separation. The structure of the boundary layer is characterized by a laminar flow region where the boundary layer is defined by the Blasius theory, $\delta/x = 5/\sqrt{\mathrm{Re}_x}$ and turbulent flow when $\mathrm{Re}_{x,\mathrm{transition}} > 500000$ where $u(y)/U_\infty = (y/\delta)^{1/7}$, $\delta/x = 0.382/\mathrm{Re}_x^{1/5}$. Across the boundary layer there is a velocity gradient, $\partial u/\partial y$ which determines the shear stress. The velocity gradient at wall is significantly greater. It contributes to the drag; it must be reduced to reduce drag. The shear stress at the wall is proportional to the rate of change of momentum thickness with distance. It must be observed that the boundary layer envelope is not a streamline! The shear force, drag and skin-friction coefficients may be estimated from the velocity profiles in the boundary layer. Thus, there is a need to control the velocity profile. To reduce the skin-friction drag, for airfoils, laminar boundary layers are preferable. The smooth airfoil profile is altered by introducing surface roughness in the transition and in the turbulent boundary layer regions. This can prolong transition and results in reduced total drag. Various other mechanisms are available to change profiles and to inhibit transition and separation. Thus, one could adopt a particular strategy for drag reduction.

Clearly skin-friction drag is reduced by laminar flow due to a lower shear stress at the wall. Moreover, there is an increase in pressure drag when boundary layer separation occurs. Thus, pressure drag is reduced by turbulent flow by delaying boundary layer separation. There is an increase in skin-friction drag due to higher shear stresses at the wall. Naturally a trade-off is preferable, between increasing the skin friction drag and reducing the form drag or pressure drag. General laminar boundary layers are desirable for airfoils. So, one could spatially prolong the laminar region and reduce skin friction while also avoiding separation and prolonging transition when it happens.

CHAPTER SUMMARY

In this chapter, the basic equations of boundary layer theory were derived and approaches to solving the integral boundary layer equations, which are derived starting from the full Navier–Stokes equations, were also discussed. A brief description of the basic problems in boundary layer theory, in evaluating laminar boundary layer, transition and separation, turbulent boundary layers and the closure models used to obtain solutions for the quantities such as skin-friction and drag coefficients were also discussed. Further details may be found in the references provided.

REFERENCES

1. R. Vepa, *Flight Dynamics Simulation and Control of Aircraft: Rigid and Flexible*, CRC Press, Boca Raton, FL, August 18, 2014, p 660.
2. L. Prandtl, *Collected Works*, Vols. 1–3, Springer, Berlin, 1961.
3. F. M. White, *Viscous Fluid Flow*, McGraw Hill International Edition, New York, 2006.
4. T. von Kármán, Über laminare und turbulente Reibung, *Z. Angew. Math. Mech.*, 1(4):233–252, 1921 (English translation in NACA TM 1092).
5. M. R. Head. *Entrainment in the Turbulent Boundary Layer*, Ministry of Aviation, Aeronautical Research Council, London, A. R. C. R. & M. 3152, 1958.
6. K. Pohlhausen, Zur naherungsweisen Integration del' Differentialgleichung del' laminaren Grenzschicht. Zeitscher, *Z. Angew. Math. Mech.*, 1:115–120, 257–261, 1921.
7. H. Holstein, T. Bohlen, Ein einfaches Verfahren zur Berechnung laminarer Reibungsschichten, die dem Näherungsansatz von K. Pohlhausen genügen, *Lilienthal-Bericht*, S 10:5–16, 1940.
8. H. Schlichting, K. Gersten. *Boundary-Layer Theory*, Springer, Berlin Heidelberg, November 1999; 8th Revised and Enlarged Edition, 2008.
9. A. Walz, Ein Neuer Ansatz für das Geschwindig Keitsprofil der laminaren Reibungsschicht, *Lilienthal Bericht*, 141, 1941, S 8/21.
10. T. P. Torda, Boundary layer control by continuous surface suction or injection, *J. Math. Phys.*, 31(1–4):206, 1952.
11. A. C. Jain, On boundary layer control by continuous suction, *Proc. Natl. Acad. Sci. India A*, 26A(3):298–304, 1960.
12. K. Hessenberger, Programmierung eines Integralverfahrens zur Berechnung dreidimensionaler, laminarer, kompressibler Grenzschichten. Diplomarbeit, Institut für Aerodynamik und Gasdynamik, Universität Stuttgart, Dezember 1993.
13. P. A. Libby, M. Morduchow, M. Bloom, Critical study of integral methods in compressible laminar boundary layers, *Nat. Advisory Comm. Aeronaut.* (NACA) TN 2655, 1952.

14. K. Wieghardt. *On an Energy Equation for the Calculation of Laminar Boundary Layers*, B.I.G.S. 65, Joint Intelligence Objectives Agency, July 1946, Washington D.C.
15. R. Timman, A. One, *Parameter Method for the Calculation of Laminar Boundary Layer*, Re. F. 35, Nationaal Luchtvaartlaboratorium, pp F29–F46, 1949.
16. J. C. Cooke, Approximate calculation of three-dimensional laminar boundary layers, Technical Report 3201, Ministry of Aviation, Aeronautical Research Council, London, 1961.
17. B. H. Mughal. Integral methods for three-dimensional boundary layers, PhD Thesis, Massachusetts Institute of Technology, February 1998.
18. Van E. Bram, Comparison and Application of Unsteady Integral Boundary Layer Methods using various numerical schemes, Technical Report, Faculty of Aerospace Engineering, Energieonderzoek Centrum, Delft University of Technology, The Netherlands, November 2009.
19. R. Vepa, Stability and transition assessment of boundary layers around morphing natural laminar flow aerofoils, Paper presented at Advanced Aero Concepts, Design and Operations, Applied Aerodynamics Conference 2014, Bristol, 22–24 July 2014.
20. D. L. Whitfield, Analytical description of the complete turbulent boundary layer velocity profile, American Institute of Aeronautics and Astronautics Papers 78-1158, 11th Fluid and Plasma Dynamics Conference, July 12, 1978.
21. M. Drela, M. B. Giles, Viscous-inviscid analysis of transonic and low Reynolds number airfoils, *AIAA J.*, 25(10):1347–1355, 1987.
22. C. Ferreira, Implementation of boundary layer suction in XFOIL and application of suction powered by solar cells at high performance sailplanes, Master's Thesis, Delft University of Technology, 2002.
23. W. M. Kays, M. E. Crawford. *Convective Heat and Mass Transfer*, 3rd Edition, McGraw Hill, New York, 1993.
24. B. Thwaites, Approximate calculation of the laminar boundary layer, *Aeronaut. Q.*, 1(3):245–280, 1949.
25. P. Baines, S. Majumdar, H. Mitsudera, The mechanics of the Tollmien-Schlichting wave, *J. Fluid Mech.*, 312:107–124, 1996.
26. Y. S. Kachanov, Physical mechanisms of laminar-boundary-layer transition, *Annu. Rev. Fluid Mech.*, 26(1):411–482, 1994.
27. M. Gad-el Hak, *Flow Control-Passive, Active, and Reactive Flow Management*, 1st Edition, Cambridge University Press, Cambridge, UK, pp 150–203, 2000.
28. B. S. Stratford, *Flow in the Laminar Boundary Layer Near Separation*, Ministry of Aviation, Aeronautical Research Council, London, A. R. C. R & M No. 3002, November 1954.
29. B. J. Abu-Ghannam, R. Shaw, Natural transition of boundary layer: The effects of turbulence, pressure gradient, and flow history, *J. Mech. Eng. Sci.*, 22:213–228, 1980.
30. H. Ludwieg, W. Tillman. *Investigation of the Wall Shearing Stress in Turbulent Boundary Layers*, NACA T.M. 1284, 1950, Washington D.C.
31. O. T. Goksel, Some effects of spherical roughness upon the incompressible flow of a boundary layer with zero pressure gradient, PhD Thesis, Liverpool University, 1968.
32. T. Cebeci, P. Bradshaw, Momentum transfer in boundary layers, *Series in Thermal and Fluid Engineering*, Vol. 1, 1st Edition, Hemisphere Publishing, London, 1977.
33. B. A. Nishida. Fully simultaneous coupling of the full potential equation and the integral boundary layer equations in three dimensions, PhD Thesis, Massachusetts Institute of Technology, February 1996.
34. H. Bhanderi, H. Babinsky, *Improving the Lag Entrainment Method in the Case of Transonic Shock Wave / Boundary Layer Interaction*, AIAA 2004-2147, AIAA, 2004, Portland, OR.
35. M. G. Hall, A numerical method for calculating unsteady two-dimensional laminar boundary layers, *Ing. Arch.*, 38(2):97, 1969.

6. Electric Aircraft Propeller Design

6.1 INTRODUCTION

Propellers are basic elements essential for the propulsion of electric aircraft. They are very similar to helicopter rotors and could be modeled as thin rotating wings that produce a force in the same direction as the rotational axis. Unlike a wing, the propeller is continuous in a non-uniform flow field, which generally requires it to have a pre-twist and a non-uniform planform. Several theories have been developed for predicting the thrust generated by a propeller based on considerations of momentum balance, blade element-wise generation of lift, drag and pitching moment, hybrid blade element and momentum theories, lifting line approximations to the vortex distribution models and lifting surface approximations of steady and unsteady vortex and doublet distributions. Simplifying procedures have been developed to model and optimize the performance of the propellers, so one could estimate the chord and blade twist distribution as well as optimize the cross-sectional shape of the propeller across its span. The contours of a typical propeller cross-section have the shape of an airfoil, and usually the airfoil shape is not uniform across the span of the blade. As propellers for electric aircraft are generally driven by electric motors, hybrid blade element and momentum theories have emerged as the preferred methods for estimating the performance of such propellers. Moreover, as most propellers have two or more blades, the interaction of the pressure distribution across each blade with those on other blades needs to be modeled, at least empirically.

At any given point along a blade, the cross-section has all the characteristics of a typical airfoil section. These include the mean camber line, chord line and thickness distribution across the chord as well as leading and trailing edges. The chord line is a straight line connecting the leading edge to the trailing edge. The distance between the bottom and top sides, or the pressure and suction sides, of the airfoil along the chord line is the thickness distribution as shown in Figure 6.1. The line which is equally distant from the bottom and the top sides of the airfoil is the mean camber line and the camber distribution is the distance between the chord line and mean camber line. At the inner end, the blades connect to a hub which is usually directly attached to the shaft of an electric motor.

Three angles characterize the blade section location relative to the plane of rotation as shown in Figure 6.2. The first is the blade angle β, the angle that the blade section chord line makes with the plane of rotation. The blade section rotation velocity vector also lies in this plane of rotation. Given the vehicle's forward velocity, the net wind velocity is the vector sum of the blade's rotation velocity in the plane of rotation and the forward velocity vector which could include not only the wind velocity far upstream of the rotor, but also the induced wind velocity just ahead of the propeller

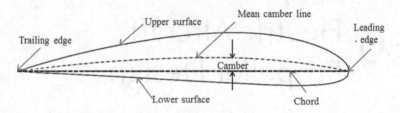

FIGURE 6.1 Geometry of a typical airfoil section.

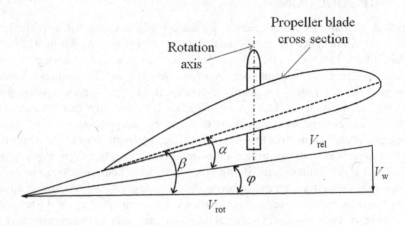

FIGURE 6.2 Definition of blade section characteristic angles.

disc. The angle that the total wind velocity vector makes to the plane of rotation is the wind angle φ, and it defines the direction of the wind vector relative to the plane of rotation. The angle which the blade chord makes with the relative wind velocity direction, which is the difference between the blade and wind angles, is the blade section angle of attack $\alpha = \beta - \varphi$. Another key parameter is the advance ratio J, which is the ratio between the distance the propeller moves forward through one rotation and the blade diameter. Given thrust force generated and the power required to drive the propeller, the non-dimensional thrust coefficient C_T, the non-dimensional power coefficient C_Q and the non-dimensional propeller efficiency η, are the key performance parameters characterizing a particular propeller. Basic propeller design procedures involve the geometric parametric design of the propeller given the number of blades, the flight speed, the propeller diameter, the desired distribution of airfoil lift and drag, the thrust force, the shaft power and the air density.

6.2 AEROFOIL SECTIONS: LIFT AND DRAG

Airfoils are generally designed to operate under specific flow conditions. Flow fields in which airfoils are typically designed to operate are characterized by a primary non-dimensional parameter called the Reynolds number. It is the ratio of inertia forces to viscous forces in a flow and is given by $R_e = \rho V d / \mu$, where ρ is the local flow density, V is the local flow velocity, d is a characteristic distance which for airfoils is assumed to be the semi-chord and μ is the coefficient of viscosity. Most propeller blades operate in a flow field characterized by $10000 \leq R_e \leq 100000$. Although these Reynolds numbers cannot be considered to be high, the coefficient of viscosity may be considered to be small and the flow field may be modeled as inviscid. The key features of the flow around the airfoil section are characterized by $C_\ell = C_\ell(\alpha)$ versus the angle of attack α characteristic, which were discussed in Chapter 5. It may be recalled that given the lift force L and the drag force D, per unit span and the chord length c, the coefficients of lift and drag may be defined as

$$C_\ell = \frac{L}{\frac{1}{2}\rho V^2 c}, \tag{6.1}$$

$$C_d = \frac{D}{\frac{1}{2}\rho V^2 c}. \tag{6.2}$$

Over the useful range of the angle of attack, α, which usually means the that $\alpha < \alpha_{stall}$, where α_{stall} is the angle of attack at which there is loss of lift due to the flow field stalling, so the corresponding lift coefficient is maximum, the lift coefficient c_l behaves linearly with α and could be expressed as,

$$C_\ell(\alpha) = a_0(\alpha - \alpha_0), \quad \alpha_0 < \alpha < \alpha_{stall}, \quad C_\ell(\alpha_{stall}) = C_{\ell,max}. \tag{6.3}$$

The angle α_0 is known as the zero-lift angle of attack and defines the lower limit of the range of α values over which the Equation (6.3) is valid. The stall-region of the $C_\ell = C_\ell(\alpha)$ characteristic, illustrated in Figure 6.3, occurs immediately after the maximum coefficient of lift is achieved. This region shows a dramatic decrease in lift and a corresponding increase in drag that is a result of separation of the flow when the flow is no longer attached over portions of the suction side of the airfoil.

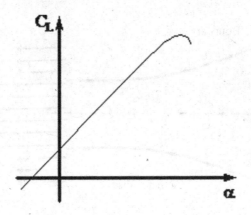

FIGURE 6.3   Lift coefficient characteristic.

6.3 MOMENTUM THEORY

Momentum theory is the simplest of all the propeller theories and is based on considerations of momentum and kinetic energy at certain key cross-sections along the flow. The theory was first introduced by Rankine [1] and Froude [2]. It is assumed that the propeller creates a uniform dynamic pressure differential between the front and the rear of the propeller's plane of rotation. Ignoring the effects of both compressibility and viscosity, the dynamic pressure and velocity before and after the disk are assumed to be respectively discontinuous and continuous as the flow passes through the propeller plane. The dynamic pressure far upstream and far downstream are both assumed to be equal.

As far as the velocity is concerned, the velocity field is assumed to smoothly, monotonically and asymptotically increase as it passes through the propeller plane and asymptotically reaches a steady value far downstream of the propeller. Furthermore, the static pressure is assumed to be uniform along the flow. The flow through the propeller disc region is illustrated in Figure 6.4. The mass flow at the propeller's plane of rotation may be expressed in terms of the local density of the

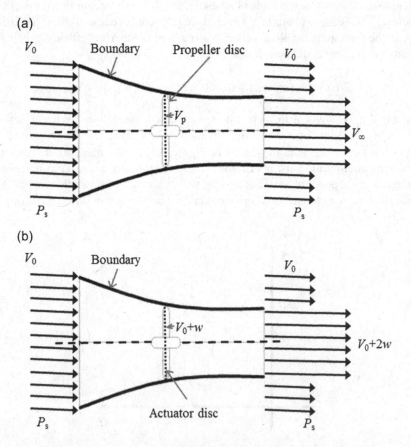

FIGURE 6.4   Flow through the propeller disc region and the actuator disc model.

flow ρ, the area of cross-section of the propeller disc and the velocity of the flow across the propeller's plane of rotation V_p. Thus, $\dot{m} = \rho A V_p$. Given the velocities of the flow far upstream and far downstream to be equal to V_0 and V_∞ respectively, the total change is momentum across the propeller's plane of rotation, which is also the net thrust acting on the propeller disc, is given by

$$\dot{m}\left(V_\infty - V_0\right) = F_T = \rho A V_p\left(V_\infty - V_0\right), \qquad (6.4)$$

The change in dynamic pressure across the propeller's plane of rotation, is given by energy considerations which is equivalent to applying the Bernoulli equation far upstream and far downstream of the propeller. Thus,

$$\Delta P = \left(P_s + \frac{1}{2}\rho V_\infty^2\right) - \left(P_s + \frac{1}{2}\rho V_0^2\right) = \frac{1}{2}\rho\left(V_\infty^2 - V_0^2\right), \qquad (6.5)$$

Thus, the thrust may also be expressed as

$$F_T = A\Delta P = \frac{1}{2}A\rho\left(V_\infty^2 - V_0^2\right). \qquad (6.6)$$

Comparing the Equations (6.4) and (6.6),

$$F_T = \rho A V_p\left(V_\infty - V_0\right) = \frac{1}{2}A\rho\left(V_\infty^2 - V_0^2\right). \qquad (6.7)$$

Hence, it follows that the velocity of the flow across the propeller's plane of rotation V_p is given by

$$V_p = \frac{1}{2}\left(V_\infty + V_0\right) = V_0 + \frac{1}{2}\left(V_\infty - V_0\right). \qquad (6.8)$$

It can be seen that half of the total increment in the velocity is added to the flow as it passes by the propeller disc.

6.4 ACTUATOR DISK

It is clear from the above analysis that as far as the generation of thrust is concerned, the propeller was modeled as a simple disc that provides a step increment in the pressure to generate a thrust as the flow passes by the disc. The momentum balance approach uses a simple analytical approach based on the principle of linear momentum conservation. It treats the propeller as a thin "actuator disk" which causes a step increase in pressure and an induced velocity across stations (1 and 2 in Figure 6.5). The principles of conservation of mass and energy are satisfied along the entire flow. If we allow the induced velocity at the propeller disc to be

$$w = \frac{1}{2}\left(V_\infty - V_0\right), \qquad (6.9)$$

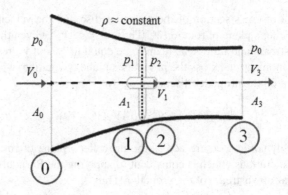

FIGURE 6.5 A stream tube enclosing the actuator disc.

it follows that,

$$V_p = V_0 + w, \quad V_\infty = V_0 + 2w. \tag{6.10}$$

With $A = A_1$, the thrust is given by,

$$F_T = 2\rho A(V_0 + w)w. \tag{6.11}$$

The ideal power required is given by

$$P = \frac{1}{2}\dot{m}V_3^2 - \frac{1}{2}\dot{m}V_0^2 = \frac{1}{2}\rho A(V_0 + w)\left[(V_0 + 2w)^2 - V_0^2\right] = 2\rho Aw(V_0 + w)^2. \tag{6.12}$$

The above expression for the power reduces to $P = F_T(V_0 + w)$ in terms of thrust F_T. Since power from an electric motor is constant at all altitudes, the thrust drops as the airplane picks up speed. Furthermore, if one wishes to find w as a function of F_T, from the earlier Equation (6.12),

$$(2\rho A)w^2 + (2\rho AV_0)w - F_T = 0. \tag{6.13}$$

The propeller-induced velocity ratio is

$$\frac{w}{V_0} = \frac{1}{2}\left(\sqrt{1 + \frac{2F_T}{\rho A_1 V_0^2}} - 1\right) \equiv \frac{1}{2}\left(\sqrt{1 + C_T} - 1\right). \tag{6.14}$$

With $V_0 = 0$, the static power is given by

$$P_o = P_{\text{ind},o} = F_o w_o = \frac{F_o^{3/2}}{\sqrt{2\rho A_1}}. \tag{6.15}$$

Ideal propeller propulsive efficiency is obtained as

$$\eta_{pr,i} = \frac{F_T V_0}{P} = \frac{F_T V_0}{F_T \left(V_0 + w\right)} = \frac{1}{1 + \dfrac{w}{V_0}} = \frac{2}{1 + \sqrt{1 + \dfrac{F_T}{qA}}}. \tag{6.16}$$

The ideal power consumed by the propeller is: $P = F_T(V_0 + w)$.

But using Equation (6.14), one obtains

$$P = \frac{F_T V_0}{2}\left(\sqrt{1 + \frac{2F_T}{\rho A V_0^2}} + 1\right). \tag{6.17}$$

From the propeller efficiency,

$$F_T V_0 = \eta_{pr,i} P = 2\eta_{pr,i}\rho A V_0^3 \frac{w}{V_0}\left(1 + \frac{w}{V_0}\right)^2 = 2\rho A V_0^3 \left(\frac{1 - \eta_{pr,i}}{\eta_{pr,i}^2}\right). \tag{6.18}$$

Hence, it follows that

$$\frac{P}{2\rho A V_0^3} = \left(\frac{1 - \eta_{pr,i}}{\eta_{pr,i}^3}\right). \tag{6.19}$$

Defining the advance ratio as

$$J = \frac{V_0}{nd_p} = \frac{V_0}{(\omega/(2\pi))(2R)} = \frac{\pi V_0}{\omega R}, \tag{6.20}$$

the efficiency and shaft horse power relation, in terms of $shp = SHP/\rho n^3 R^5$ is

$$\frac{1 - \eta_{pr,i}}{\eta_{pr,i}^3} = \frac{1}{16\pi}\frac{1}{J^3}\frac{P}{\rho n^3 R^5} = \frac{1}{16\pi}\frac{1}{J^3}\frac{SHP}{\rho n^3 R^5} = \frac{1}{16\pi}\frac{shp}{J^3} = \frac{1 - \eta_p}{\eta_p^3}. \tag{6.21}$$

One may use the relation Equation (6.21) to plot a set of propeller characteristics based on the actuator disc theory, as shown in Figures 6.6–6.8. An early version of the actuator disc theory was published by Hawthorne and Horlock [3].

6.5 BLADE ELEMENT THEORY

As with the actuator disc theory, in the blade element theory, the propeller blade is assumed to generate thrust force through the aerodynamic lift force component acting on the propeller section. The motor must deliver sufficient power to overcome the power absorbed by the aerodynamic drag force. Thus, it is essential that the section angle of attack must remain below the stall angle. If it does exceed the blade stall angle, it would result in a loss of lift and consequently, a reduction in the thrust force generated.

Consider a blade section at a radial position r, blade radius is R and hub radius is r_h. The blade is assumed to be rotating with a mechanical angular velocity, $\omega = \omega_m$.

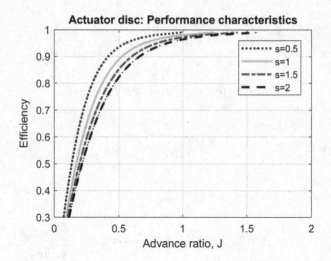

FIGURE 6.6 Propeller efficiency versus advance ratio.

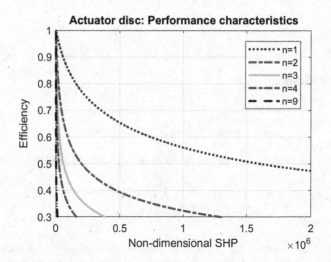

FIGURE 6.7 Propeller efficiency versus *shp*.

The blade section rotation velocity is then given by $\omega_m r$, while the inflow velocity far upstream of the rotor is assumed to be $V_0 = V_\infty$. Thus, the resultant velocity is:

$$V_R = \sqrt{\omega_m^2 r^2 + V_0^2}, \tag{6.22}$$

Furthermore, one defines propeller-induced inflow velocity or the induced downwash as w. Thus, there is an induced angle of attack which is given by:

$$\alpha_i = \sin^{-1}\left(w/V_R\right) \tag{6.23}$$

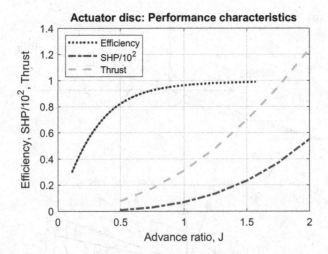

FIGURE 6.8 Typical propeller non-dimensional characteristics (efficiency, *shp* and thrust force $= \eta_p R \times \text{shp}/V_0$) versus advance ratio.

Before proceeding further, a simplifying assumption is made to avoid unnecessary and complex mathematics. The propellers are of a symmetric cross-section. Consequently, the chord line and the zero-lift-line would be coincident. Thus, the angle of attack as the angle made by the local wind vector relative to the zero-lift-line as shown in Figure 6.9(a). The incremental forces acting on a blade section are depicted in Figure 6.9(b). The overall resultant velocity including induced velocity is denoted as V_E and is given by

$$V_E = \sqrt{\left(\omega_m r - w\sin(\varphi + \alpha_i)\right)^2 + \left(w\cos(\varphi + \alpha_i) + V_0\right)^2} \qquad (6.24)$$

Thus, the increment of the blade section lift is given by

$$dL = \frac{1}{2}\rho V_E^2 c C_\ell dr. \qquad (6.25)$$

The corresponding increment of blade section drag is given by

$$dD = \frac{1}{2}\rho V_E^2 c C_d dr. \qquad (6.26)$$

The increment of thrust is obtained from

$$dF = dL\cos(\varphi + \alpha_i) - dD\sin(\varphi + \alpha_i). \qquad (6.27)$$

The increment in the torque is

$$dQ = rdF_Q = r\left[dL\sin(\varphi + \alpha_i) + dD\cos(\varphi + \alpha_i)\right]. \qquad (6.28)$$

(a)

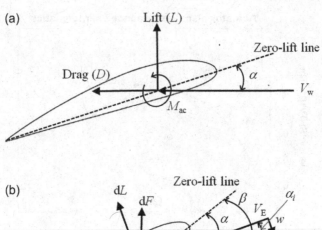

(b)

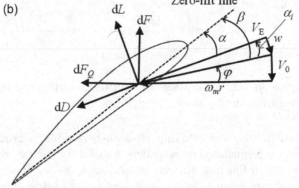

FIGURE 6.9   (a) Angle of attack definition. (b) Increment forces acting on a blade section.

The airfoil lift coefficient and lift curve slope are related by

$$C_\ell = \left(dC_\ell/d\alpha\right)\left(\beta - \alpha_i - \varphi\right) = a_o\left(\beta - \alpha_i - \varphi\right),\qquad (6.29)$$

where a_0 is the lift curve slope. The drag coefficients and the drag polar are defined by

$$C_{d0} = C_{d,\min},\quad C_d = C_{d0} + k\left(C_\ell - C_{\ell,\min}\right)^2.\qquad (6.30)$$

The increment in the thrust developed may be simplified and reduced to

$$dF \approx dL \cdot \cos\varphi = \left(B/2\right)\rho V_R^2 ca_o(\beta - \alpha_i - \varphi)dr \cdot \cos\varphi,\qquad (6.31)$$

where B is the number of blades in the propeller. In Equation (6.31), the induced angle of attack α_i is defined by Equation (6.23) and is a function of the induced downwash w, which is yet to be defined.

6.6   DYNAMICS AND MODELING OF THE INFLOW

There is a need to extend the momentum theory to so as to be able to estimate the induced downwash dynamically. The earliest dynamic extensions of momentum theory were due to Carpenter and Fridovitch [4], in a study on the thrust response during jump take-offs in an autogiro. They included an apparent mass term to the thrust equation that models the inertia of the air surrounding the rotor. To use this approach, it is first observed that with a sustained inflow, far upstream of the propeller, based on the actuator disc theory is given by Equation (6.11). Introducing the corrections by Carpenter and Fridovitch [4]

$$F_T = 2\rho A(V_0 + w)w + 0.637\rho A(4/3)R\dot{w} = 2\rho A\{(V_0 + w)w + 0.425R\dot{w}\}. \qquad (6.32)$$

Rather than use the time variable as the independent variable, it is useful to use a non-dimensional time variable based on the aerodynamic velocity which could be defined as

$$\tau = V_0 t / R. \qquad (6.33)$$

Thus, Equation (6.32) for the thrust is

$$F_T = 2\rho A\{V_0 w + w^2 + 0.425V_0 w'\}. \qquad (6.34a)$$

where w' is the derivative of w with respect to τ. It follows that the inflow dynamics satisfy the equation

$$0.425\frac{w'}{V_0} + \frac{w}{V_0} + \frac{w^2}{V_0^2} - \frac{F_T}{2\rho A V_0^2} = 0. \qquad (6.34b)$$

If one defines a non-dimensional induced inflow variable as $\lambda_i = w/V_0$, Equation (6.34) reduces to

$$\tau_i \lambda_i' + \lambda_i - \lambda_{st}(\lambda_i) + \lambda_i^2 = 0, \quad \lambda_{st} = F_T/(2\rho A V_0^2), \quad \tau_i = 0.425. \qquad (6.35)$$

To generate the inflow, in the blade element approach, Equation (6.35) must be integrated, where the thrust force F_T is found by integrating Equation (6.31). Thus, the thrust force F_T is a function of the induced angle of attack, $F_T = F_T(\alpha_i)$ with $\alpha_i = \sin^{-1}(\lambda_i V_0/V_R)$.

Observe that, with non-dimensional inflow,

$$V_R = V_E \cos\alpha_i, \quad V_0 = V_R \sin\varphi, \quad \alpha_i = \sin^{-1}(\lambda_i \sin\varphi). \qquad (6.36)$$

6.7   INTEGRATING THE THRUST AND TORQUE

From Equation (6.27), if one defines, $\tan\gamma = C_d/C_\ell$, it follows that

$$dF = \frac{1}{2}\rho V_E^2 cdr\left(C_\ell \cos(\varphi + \alpha_i) - C_d \sin(\varphi + \alpha_i)\right)$$

$$= \frac{1}{2}\rho V_E^2 cdr C_\ell \frac{\cos(\gamma + \varphi + \alpha_i)}{\cos\gamma}.$$

Thus, the differential thrust is given by

$$dF = \frac{1}{2}\rho V_0^2 cdrC_\ell \frac{\cos(\gamma+\varphi+\alpha_i)}{\cos^2\alpha_i \sin^2\varphi\cos\gamma}. \tag{6.37a}$$

Similarly, the differential torque is given by

$$dQ = \frac{1}{2}\rho V_0^2 crdrC_\ell \frac{\sin(\gamma+\varphi+\alpha_i)}{\cos^2\alpha_i \sin^2\varphi\cos\gamma}. \tag{6.38a}$$

Integrating Equations (6.37) and (6.38), and assuming the number of blades in the propeller to be B,

$$F_T = \frac{\rho V_0^2}{2}\int_0^R C_\ell \frac{\cos(\gamma+\varphi+\alpha_i)}{\cos^2\alpha_i \sin^2\varphi\cos\gamma}c(r)dr. \tag{6.37b}$$

$$Q = \frac{\rho V_0^2}{2}\int_0^R C_\ell \frac{\sin(\gamma+\varphi+\alpha_i)}{\cos^2\alpha_i \sin^2\varphi\cos\gamma}c(r)rdr. \tag{6.38b}$$

From Equation (6.16) the propulsive efficiency is,

$$\eta_{pr,i} = \frac{F_T V_0}{P} = \frac{F_T V_0}{F_T(V_0+w)} = \frac{1}{1+w/V_0} = \frac{1}{1+\lambda_i}. \tag{6.39}$$

6.8   BLADE ELEMENT MOMENTUM THEORY

The blade element momentum (BEM) theory, first proposed by Glauert [5], (see also Vepa [6]) combines the two methods of modeling the operation of a propeller. The first method of modeling uses the momentum balance on a rotating annular stream tube passing through a propeller disc. The second method of modeling, known as the blade element theory, uses the expressions for the airfoil lift and drag forces along various spanwise sections of the blade. The first and the second methods are used to obtain the induced flow velocities across the propeller, while the second is used to obtain the thrust and torque on the blades.

   To apply the blade element theory to a propeller blade, the blade is divided into a large number of sections or elements along its length. One makes two assumptions:

   i) There is no aerodynamic interference between different blade elements
   ii) The forces on the blade elements are solely determined by the two-dimen-
       sional section lift and drag forces.

Consider a blade divided into N elements along its length. Each of the blade elements will experience a slightly different flow as they have different rotational speeds, different chord lengths and different inflow angles. Blade element theory involves dividing the blade into a large number (usually about twenty or more) of elements and calculating the flow about each of the elements. The overall performance characteristics are determined by numerically integrating the forces and moments along the blade span.

Since the blade element momentum theory (BEMT) combines both the blade element and momentum balance theories, from the blade element theory the increment in thrust is:

$$dF \approx (B/4) 2\rho V_R^2 c a_o (\beta - \alpha_i - \varphi) dr \cdot \cos\varphi. \tag{6.40}$$

From the actuator disk theory,

$$dF \approx 2\rho \cdot dA \cdot (V_0 + w)w \approx 2\rho \cdot dA \cdot (V_\infty + w\cos\varphi)w\cos\varphi,$$

$$\approx 2\rho(2\pi r dr) \cdot (V_\infty + \alpha_i V_R \cos\varphi) \alpha_i V_R \cos\varphi. \tag{6.41}$$

Equating the two expressions for the incremental thrust given by Equations (6.40) and (6.41), one obtains

$$\alpha_i^2 + \left(\frac{V_\infty}{V_R \cos\varphi} + \frac{c a_o B}{8\pi r \cos\varphi} \right) \alpha_i - \frac{c a_o B}{8\pi r \cos\varphi} (\beta - \varphi) = 0. \tag{6.42}$$

In order to express Equation (6.42) in terms of meaningful non-dimensional parameters, several parameters characterizing a typical propeller are now defined. The overall propeller solidity is defined as

$$\sigma_{\text{ref}} = \frac{\text{blade area}}{\text{disk area}} = \frac{B \cdot c_{\text{ref}} R}{\pi R^2} = \frac{B c_{\text{ref}}}{\pi R}. \tag{6.43}$$

The local solidity may then be expressed as

$$\sigma = \frac{Bc}{\pi R} = x \frac{Bc}{\pi r}, \quad x = \frac{r}{R}. \tag{6.44}$$

The advance ratio, which was introduced earlier may also be expressed as

$$J = \frac{V_0}{n d_p} = \frac{V_0}{(\omega_m/(2\pi))(2R)} = \frac{\pi V_0}{\omega_m R}, \tag{6.45}$$

where the blade rotation speed in revs/s is related to the angular speed in rads/s by $n = \omega_m/2\pi$.

The inflow velocity ratio λ_f far away from the propeller disc is given by the ratio of the incoming wind velocity V_0 and the tip velocity of the blade $\omega_m R$, at the radius $r = R$. The non-dimensional inflow velocity ratio may be expressed as

$$\lambda_f = V_0/\omega_m R = J/\pi. \tag{6.46}$$

The inflow angle at any blade section is given by

$$\phi = \tan^{-1}\left(\frac{V_0}{\omega_m r}\right) = \tan^{-1}\left(\frac{\lambda_f}{x}\right). \tag{6.47}$$

Hence,

$$V_R \cos\phi = \omega_m r = x V_{\text{Tip}}, \quad V_{\text{Tip}} = \omega_m R. \tag{6.48}$$

In terms of the propeller's non-dimensional parameters, Equation (6.42) for the induced angle of attack is expressed as a quadratic given by:

$$\alpha_i^2 + \left(\frac{\lambda_f}{x} + \frac{\sigma a_o V_R}{8x^2 V_T}\right)\alpha_i - \frac{\sigma a_o V_R}{8x^2 V_T}\left(\beta - \varphi\right) = 0. \tag{6.49}$$

Solving the quadratic equation for the positive root,

$$\alpha_i = \frac{1}{2}\left\{\left[\left(\frac{\lambda_f}{x} + \frac{\sigma a_o V_R}{8x^2 V_T}\right)^2 + \frac{\sigma a_o V_R}{2x^2 V_T}\left(\beta - \varphi\right)\right]^{1/2} - \left(\frac{\lambda_f}{x} + \frac{\sigma a_o V_R}{8x^2 V_T}\right)\right\}. \tag{6.50}$$

Once the induced angle of attack α_i is determined, it is possible to estimate the induced inflow, propeller thrust and shaft power coefficients. Equation (6.50) may be cast in the same form as Equation (6.14). Figure 6.10 illustrates a typical distribution of the induced inflow along the blade span.

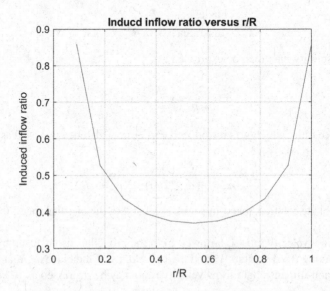

FIGURE 6.10 Typical distribution of the induced inflow along the blade span.

The propeller thrust and shaft power coefficients are respectively defined as:

$$C_T = \frac{F_T}{\rho n^2 d^4}, \quad C_{P_s} = \frac{P_S}{\rho n^3 d^5} = \frac{Q\omega}{\rho n^3 d^5} = \frac{2\pi Q}{\rho n^2 d^5} = 2\pi C_Q, \tag{6.51}$$

where Q is the propeller torque and $C_Q = Q/\rho n^2 d^5$.

The incremental thrust and power for all the blades in the propeller are respectively given by

$$dF = \frac{1}{2}\rho V_E^2 Bc[C_\ell \cos(\varphi + \alpha_i) - C_d \sin(\varphi + \alpha_i)]dr, \tag{6.52}$$

$$dP_S = \omega_m r \frac{1}{2}\rho V_E^2 Bc[C_\ell \sin(\varphi + \alpha_i) + C_d \cos(\varphi + \alpha_i)]dr, \tag{6.53}$$

where

$$V_E^2 \approx V_R^2 = V_0^2 + \omega_m^2 r^2 = \frac{\omega_m^2 r^2}{\pi^2}\left(J^2 + \pi^2 x^2\right). \tag{6.54}$$

The propeller thrust and shaft power coefficients are respectively given by

$$C_T = \frac{\pi^2}{4\rho\omega_m^2 R^4}F_T,$$

$$C_T = \frac{\pi^2}{4\rho\omega_m^2 R^4}\int dF = \frac{\pi}{8}\int_{x_h}^{1}\sigma\left(J^2 + \pi^2 x^2\right)\left[C_\ell \cos(\varphi + \alpha_i) - C_d \sin(\varphi + \alpha_i)\right]dx, \tag{6.55}$$

and

$$C_{P_s} = \frac{\pi^3}{4\rho\omega_m^3 R^5}P_S,$$

$$C_{P_s} = \frac{\pi^3}{4\rho\omega_m^3 R^5}\int dP_S$$

$$= \frac{\pi^2}{8}\int_{x_h}^{1}\sigma x\left(J^2 + \pi^2 x^2\right)\left[C_\ell \sin(\varphi + \alpha_i) + C_d \cos(\varphi + \alpha_i)\right]dx. \tag{6.56}$$

Equations (6.55) and (6.56) can form the basis for the propeller design procedure. Equations (6.55) and (6.56) for the thrust and shaft power coefficients, C_T and C_{P_s}, are integrated from the hub station ($x = x_h$) to the blade tip ($x = 1$) using a numerical approach. The integration is along the blade, which could be of varying chord and twist, assuming a closed form solution for linear or other twist variation is available. The various pertinent coefficients of lift, drag, inflow angle, induced angle of attack and the propeller parameters are estimated. The optimum (desired) twist distribution is chosen to keep the net angle of attack increments nearly uniform, and just below the stall angle to avoid stall. The shaft power coefficient C_{P_s} is as desired (usually a maximum). A typical optimum desired distribution, for a blade with a uniform chord distribution, is compared with the linear variation of the twist in Figure 6.11.

The power dissipated by the blade section drag is estimated by the following expression

$$P_{\text{drag}} \approx \frac{2\eta_{pr}}{3} \frac{V_T}{V_0} \left(\frac{C_d}{C_l} \right) \Bigg|_{\alpha=0} P_S.$$ (6.57)

If one defines useful thrust power over overall shaft power as

$$\eta_{sp} = \frac{FV_0}{P_S},$$ (6.58)

then it follows that

$$\eta_{sp} = \frac{C_T \rho n^2 d^4 V_0}{C_P \rho n^3 d^5} = \frac{C_T}{C_P} J.$$ (6.59)

In the preceding analysis, the induced velocity components in the radial direction were ignored. For larger propellers these important components must be accounted for. In a case when the induced velocity in the radial direction needs to be included,

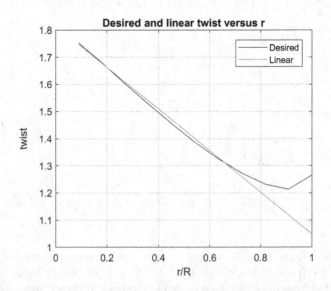

FIGURE 6.11 Optimum twist distribution compared with a linear twist variation.

the above analysis is modified. The inflow at the propeller disc is modified additionally by the axial velocity induction factor a and the angular velocity induction factor a' by,

$$\lambda_f + \lambda_0 = \frac{V_0(1+a)}{\omega_m R(1-a')}.$$

(6.60)

Thus,

$$\lambda_0 = \frac{V_0(1+a)}{\omega_m R(1-a')} - \frac{V_0}{\omega_m R} = \frac{V_0(a+a')}{\omega_m R(1-a')} = \lambda_f \frac{a+a'}{1-a'}.$$

(6.61)

The local flow velocity magnitude for the blade section is thus

$$V_1 = \sqrt{V_0^2(1+a)^2 + (\omega_m R)^2(1-a')^2} = \omega_m R\sqrt{\lambda_f^2(1+a)^2 + (1-a')^2}.$$

(6.62)

To obtain expressions for a and a', one obtains expressions for the differential thrust normal to the rotor disc dF_T and the differential torque dQ using blade element theory. Comparing the equations for dF_T and dQ obtained by using the momentum theory and by using the blade element theory, expressions for a and a' may be obtained.

The aerodynamic forces and moments are developed in terms of the non-dimensional velocity components at a point r along the radius of the blade, where the velocity normal to the disc is denoted as u_P and in-plane components in the tangential and radial directions are denoted as u_T and u_R respectively. These are then given by

$$u_T = r(1-a')/R \equiv \bar{r}(1-a'), \quad u_R = 0 \text{ and } u_P = \lambda_f(1+a),$$

(6.63)

where $\bar{r} = r/R$ is the non-dimensional position of a section along the blade. The sectional aerodynamic lift L, and drag forces D, are respectively defined in terms of the lift coefficient $C_\ell = C_L$ and the drag coefficient $C_d = C_D$ as

$$L = (1/2)\rho\omega_m^2 R^2(u_T^2 + u_P^2)cC_L, \quad D = (1/2)\rho\omega_m^2 R^2(u_T^2 + u_P^2)cC_D$$

(6.64)

where ρ is the local density of the flow and c the local chord at the section. The sectional angle of attack is $\alpha = \theta - \phi$ where θ is blade section pitch angle and ϕ is the local inflow angle, defined as

$$\varphi = \tan^{-1}(u_P/u_T) = \tan^{-1}(\lambda_f(1+a)/\bar{r}(1-a')).$$

(6.65)

The differential thrust normal to the rotor disc acting on a rotor blade section is

$$dF_T = dL\cos\varphi + dD\sin\varphi$$

$$= \frac{1}{2}\frac{\rho\omega_m^2 R^2}{\sin^2\varphi}\lambda_f^2(1+a)^2 c(C_L\cos\varphi + C_D\sin\varphi)dr.$$

(6.66)

The differential thrust acting tangential to the blade disc due to a section of the rotor blade is

$$dF_\theta = -dL\sin\varphi + dD\cos\varphi$$

$$= \frac{1}{2}\frac{\rho\omega_m^2 R^2}{\sin^2\varphi}\lambda_f^2(1+a)^2 c(-C_L\sin\varphi + C_D\cos\varphi)dr. \tag{6.67}$$

The torque acting on the disc is

$$dQ = rdF_\theta = \frac{1}{2}\frac{\rho\omega_m^2 R^2}{\sin^2\varphi}\lambda_f^2(1+a)^2 c(-C_L\sin\varphi + C_D\cos\varphi)rdr. \tag{6.68}$$

For N blades, introducing the local solidity ratio, $\sigma' = Nc/\pi r$

$$dF_T = \frac{1}{2}\pi\sigma'\frac{\rho\omega_m^2 R^2}{\sin^2\varphi}\lambda_f^2(1+a)^2(C_L\cos\varphi + C_D\sin\varphi)rdr, \tag{6.69}$$

$$dQ = \frac{1}{2}\pi\sigma'\frac{\rho\omega_m^2 R^2}{\sin^2\varphi}\lambda_f^2(1+a)^2(-C_L\sin\varphi + C_D\cos\varphi)r^2dr. \tag{6.70}$$

Based on the changes in the linear and angular momentum, the differential thrust and torque are given by

$$dF_T = (2\pi rdr)\rho V_{\text{disc}}(V_{ds} - V_0), \quad V_{\text{disc}} = (V_0 + V_{ds})/2, \tag{6.71}$$

and

$$dQ = (2\pi rdr)\rho V_{\text{disc}}(2a'\omega_m r \times r), \tag{6.72}$$

where V_{disc} is flow velocity at the propeller disc and V_{ds} is the flow velocity downstream. From the momentum balance analysis, V_{ds} is given by

$$V_{ds} = V_w(1+2a). \tag{6.73}$$

The differential thrust and torque are respectively expressed as [30]:

$$dF_T = 4\pi\rho Q_{\text{tip}}\omega_m^2 R^2\lambda_f^2(1+a)ardr, \quad dQ = 4\pi\rho Q_{\text{tip}}\omega_m^2 R^2\lambda_f(1+a)a'\bar{r}r^2dr \tag{6.74}$$

where Q_{tip} is Prandtl's tip flow correction factor defined as

$$Q_{\text{tip}} = (2/\pi)\cos^{-1}(f_{\text{loss}}), \quad f_{\text{loss}} = N(\bar{r}^{-1} - 1)/(2\sin\varphi). \tag{6.75}$$

Comparing the expressions obtained by applying the momentum and the blade element theories, the induction factors satisfy the relations

$$\frac{a}{(1+a)} = \frac{\sigma'}{8Q_{tip}\left(\sin^2\varphi\right)}\left(C_L\cos\varphi + C_D\sin\varphi\right), \tag{6.76}$$

$$\frac{a'}{(1+a)} = \frac{\sigma'}{8Q_{tip}\bar{r}\left(\sin^2\varphi\right)}\lambda_f\left(-C_L\sin\varphi + C_D\cos\varphi\right). \tag{6.77}$$

Solutions for a and a' respectively may be obtained as

$$a = \left(\left\{\frac{\sigma'}{8Q_{tip}\left(\sin^2\varphi\right)}\left(C_L\cos\varphi + C_D\sin\varphi\right)\right\}^{-1} - 1\right)^{-1}, \tag{6.78}$$

$$a' = \frac{\sigma'(1+a)}{8Q_{tip}\bar{r}\left(\sin^2\varphi\right)}\lambda_f\left(-C_L\sin\varphi + C_D\cos\varphi\right), \tag{6.79}$$

where a and a' are functions of $\bar{r}$ and not assumed to be constant. Hence, the factor $1+a$ may be expressed as

$$1+a = \left\{1 - \frac{\sigma' C_L}{8Q_{tip}\left(\sin\varphi\tan\varphi\right)}\left(1 + \frac{C_D\tan\varphi}{C_L}\right)\right\}^{-1}. \tag{6.80}$$

The thrust F_T and torque Q, respectively, are given by

$$F_T = \frac{1}{2}\pi\rho\omega_m^2 R^4 \int_{r_h}^1 \frac{\lambda_f^2\sigma'(1+a)^2 C_D}{\sin\varphi\tan\varphi}\left(\frac{C_L}{C_D} + \tan\varphi\right)\bar{r}d\bar{r}$$

$$\equiv \rho\frac{\omega_m^2}{4\pi^2}\left(2R\right)^4 C_F, \tag{6.81}$$

$$Q = \frac{1}{2}\pi\rho\omega_m^2 R^5 \int_{r_h}^1 \frac{\sigma'\lambda_f^2(1+a)^2 C_D}{\sin\varphi\tan\varphi}\left(1 - \frac{C_L}{C_D}\tan\varphi\right)\bar{r}^2 d\bar{r}$$

$$\equiv \rho\frac{\omega_m^2}{4\pi^2}\left(2R\right)^5 C_T. \tag{6.82}$$

The power consumed by an annular element is given by

$$dP = \omega_m dQ. \tag{6.83}$$

Hence the total power consumed by the propeller disc is

$$P_c = \frac{1}{2}\pi\rho\omega_m^3 R^5\lambda_f^2 \int_{r_h}^1 \frac{\sigma'(1+a)^2 C_D}{\sin\varphi\tan\varphi}\left(1 - \frac{C_L\tan\varphi}{C_D}\right)\bar{r}^2 d\bar{r}. \tag{6.84}$$

The power of the wind is given by

$$P_{\text{wind}} = \frac{1}{2}\pi\rho\omega_m^3 R^5 \lambda_f^3 = \rho\frac{\omega_m^3}{8\pi^3}(2R)^5 \frac{\pi^4}{8}\lambda_f^3. \tag{6.85}$$

Thus, relative to the wind, the coefficient of power consumed based on the BEM theory may be expressed as

$$C_{P_c} = \frac{P_c}{P_{\text{wind}}} = \frac{1}{\lambda_f}\int_{\bar{r}_h}^{1}\frac{\sigma'(1+a)^2 C_D}{\sin\varphi\tan\varphi}\left(1-\frac{C_L\tan\varphi}{C_D}\right)\bar{r}^2 d\bar{r}. \tag{6.86}$$

The power consumed by the propeller is:

$$P_c = \frac{1}{2}\pi\rho\omega_m^3 R^5 \lambda_f^2 \int_{\bar{r}_h}^{1}\frac{\sigma'(1+a)^2 C_D}{\sin\varphi\tan\varphi}\left(1-\frac{C_L\tan\varphi}{C_D}\right)\bar{r}^2 d\bar{r}$$

$$= \frac{1}{2}\pi\rho\omega_m^3 R^5 \lambda_f^3 C_{P_c}. \tag{6.87}$$

It reduces to

$$P_c = (1/2)\pi\rho\omega_m^3 R^5 \lambda_f^3 C_{P_c} = (1/2)\pi\rho V_0^3 R^2 C_{P_c}. \tag{6.88}$$

where C_{P_c} is expressed as

$$C_{P_c} = \frac{P_c}{P_{\text{wind}}} = \frac{1}{\lambda_f}\int_{\bar{r}_h}^{1}\frac{\sigma'(1+a)^2}{\sin\varphi}\left(\frac{C_D}{\tan\varphi}-C_L\right)\bar{r}^2 d\bar{r}, \tag{6.89}$$

with

$$\varphi = \tan^{-1}\left(\frac{u_P}{u_T}\right) = \tan^{-1}\left(\frac{\lambda_f}{\bar{r}}\frac{(1+a)}{(1-a')}\right), \quad \alpha = \theta - \varphi. \tag{6.90}$$

The propulsive efficiency with respect to the shaft power is given by

$$\eta_{sp} = V_0 F_T/\omega_m Q = J C_F/(2\pi C_Q) = J C_F/C_{P_s}, \tag{6.91}$$

where $C_F = 4\pi^2 F_T/\rho\omega_m^2 (2R)^4$, $C_Q = 4\pi^2 Q/\rho\omega_m^2 (2R)^5$. In the above expression, J, is the advance ratio which is a dimensionless term defining the forward speed of the aircraft V_0, relative to the angular speed of the propeller and is given by

$$J = 2\pi V_0/2R\omega_m. \tag{6.92}$$

The electrical energy conversion efficiency is given by

$$\eta_{\text{elec}} = \omega_m T / EI.$$ (6.93)

An alternate way of expressing a power coefficient is to express the power required to drive the propeller in terms of the torque and the torque coefficient. Hence, the power consumed by the propeller is expressed as

$$P_c = \left(1/2\right)\pi\rho\omega_m^3 R^5 \lambda_f^3 C_{P_c} = \left(1/4\pi^2\right)\rho\omega_m^3 \left(2R\right)^5 C_Q.$$ (6.94)

Hence, it follows that the coefficient of the shaft power required to drive the propeller is related to the coefficient of power consumed relative to the wind and the torque coefficients by

$$C_{P_s} = \left(1/8\right)\pi^4 \lambda_f^3 C_{P_c} = 2\pi C_Q.$$ (6.95)

The typical characteristics of a propeller blade obtained using the modifications to both the axial and radial inflow are shown in Figure 6.12. The corresponding lift and drag coefficients and their derivatives are shown in Figure 6.13. The optimum non-linear twist in this case is chosen, with the angle of attack just under the stall angle so that no section of the blade will stall at the design speed, the speed at which the propulsive efficiency is a maximum. These characteristics could also be estimated under stalled conditions over a part of the blade span, as well as for a variety of lift and drag models.

6.8.1 APPLICATION TO DUCTED PROPELLERS

One application of the actuator disc and blade element momentum theories are to a ducted propeller. Given the equation for the conservation of mass and the Navier–Stokes equation for steady incompressible laminar flow, over a thin cylindrical cell including momentum sources, which represent the magnitudes of momentum sources in the three coordinates at a point in the flow one may define an equivalent actuator disc to model a ducted propeller. To define an equivalent actuator disc, the momentum sources must be specified. It can be shown that the equivalent actuator disc model can successfully simulate the propeller slipstream as long as the flow velocity distribution function across the momentum source is representative of the actual flow field in the actuator disc plane.

Extending the non-linear actuator disk theory of Conway [7, 8], Bontempo and Manna [9] have obtained the exact solution of the flow around a ducted actuator disk. Based on the earlier model of Bontempo and Manna [9], Bontempo [10] and Bontempo et al. [11, 12] have presented an extension of the actuator disc theory of propellers to ducted rotors. The method simultaneously accounts for the proper shape of the slipstream, the rotation of the wake, a variable radial distribution of the load and ducts of general shape. Bontempo et al. [11] present a generalized semi-analytical actuator disk model as applied to the analysis of the flow around ducted

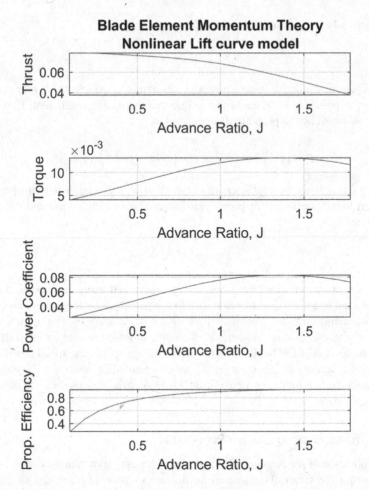

FIGURE 6.12 Typical characteristics of a propeller blade using the modifications to both the axial and radial inflow.

propellers in different operating conditions. They have shown that power and thrust coefficients for a ducted propeller have two contributions: one from the actuator disc representing the propeller, and the second from the duct. When properly optimized the two add up to maximize the performance of the ducted propeller, and the duct functions as an accelerating duct. Bontempo and Manna [13] have found that an optimized duct with augmentation of both the camber and thickness of the duct leads to an increase in the ideal propulsive efficiency of the ducted propeller.

The axisymmetric flow field around a ducted propeller could be analyzed by a simplified analytical model such as the blade element momentum approach, with appropriate modifications for the presence of the shroud, provided one is able to capture the critical aspects of shrouded systems like the interaction between the propeller and the duct, and the slipstream contraction and rotation. The blade element momentum theory could be modified and applied to a ducted propeller by assuming

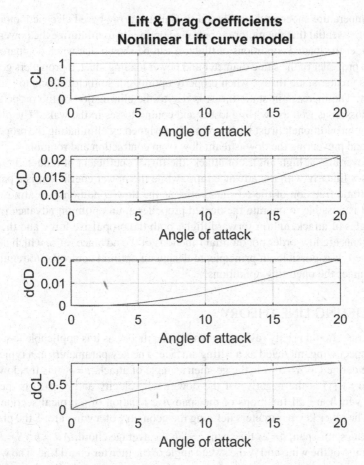

FIGURE 6.13 Models of lift and drag coefficient and their derivatives used in computing the characteristics shown in Figure 6.12.

that the axial velocity induction factor a is enhanced by the induced velocity due to the duct. Weir [14] combined an expression for the increment of velocity induced by the duct in forward flight, and the duct induced velocity due to the interaction of the duct and the propeller disc in forward motion with blade element theory to predict the performance of a ducted propeller. It is also possible to assume that both the axial velocity induction factor and the angular velocity induction factor a' are functions of the radial position where these functions are approximated based on experimentally obtained distributions of the axial and tangential velocity distributions, as has been done by Wang, Chen and Guo [15]. Thus, to design a ducted propeller, the duct is modeled by a distribution of ring vortices. The strengths of ring vortices modeling the duct geometry and the axial velocity components induced at propeller disc by these vortices are estimated. Then axial velocities induced by duct/hub are added to the induced inflow at the propeller disc.

Ducted propellers, or shrouded propellers can produce high static thrust and are suitable as propulsion systems for electric aircraft, as they could by driven by either high temperature superconductor-based or Halbach array-based electric motors. It is, however, essential that the motor is controlled optimally to minimize the power drawn by it under the loaded conditions of flight. It can be shown that when compared to an isolated propeller of the same diameter and power loading, ducted propellers generally produce greater static thrust, when properly optimized, so the duct acts to accelerate the flow. The increase in static thrust is due to the elimination of the propeller's tip vortex losses, as well as swirling losses and coning losses in the wake. The duct provides for an additional thrust when optimally designed by eliminating the propeller tip vortex and preventing the downstream flow from contraction and rotation.

In hover and at high angles of attack, the thrust coefficient of a ducted propeller decreases linearly with the advance ratio, while the power coefficient is a parabolic or quadratic function of the advance ratio, as the flow conditions are almost static. Thus, it is possible to operate the ducted propeller at an optimum advance ratio. At low angles of attack, and in forward flight, both the propulsive force and the power coefficients are low-order polynomials functions of the advance ratio, which could be linearized. It is possible, in principle, to design an optimal controller to regulate the motor under the operating conditions.

6.9 LIFTING LINE THEORY

Firstly, one should briefly review the lifting line theory as it is applicable to a lifting, large aspect wing, modeled as a lifting surface. The key parameters at a typical section are defined as $\alpha = \alpha(y)$ is the geometric angle of attack, $\varepsilon = \varepsilon(y)$ is the downwash angle, $w = w(y)$ is the negative of the downwash velocity and $c = c(y)$ is the chord length, which are all functions of the spanwise location of a typical section of the wing. The other key parameters defining the geometry of a wing are: S the planform area, s the semi-span, $b = 2s$ the span, $c = S/b$ the average chord, $AR = b^2/S = b/c$ the aspect ratio of the wing and Λ the sweep angle of the quarter-chord line. The wing lift and drag coefficient are respectively defined as $C_L = 2L/\rho V_\infty^2 S$ and $C_D = 2D/\rho V_\infty^2 S$. The key assumptions are that the roll-up of the wake could be ignored, that the wake remains in a plane parallel to the freestream and that it is modeled using a single vortex sheet starting at the quarter-chord line. The vortex distribution along the span of the wing has a strength, $\Gamma = \Gamma(y)$, which is also the circulation around the particular wing section. Thus, the strength of the vortex shed and the corresponding downwash at a location $y = y_1$ are respectively given by

$$d\Gamma_{\text{shed}} = -\left.\frac{d\Gamma}{dy}\right|_{y=y_1} dy_1. \tag{6.96}$$

According to the Biot–Savart law, the incremental velocity field $d\bar{u}$, induced by an infinitesimally small vortex line segment of unit strength, length dl, directed towards the unit vector $\bar{s}$ and at a position vector $\bar{r} = xi + yj + zk$ from it, is given by line integral

$$d\bar{u} = \frac{1}{4\pi} \int_{dl} \frac{\bar{s} \times \bar{r}}{|\bar{r}|^3} dl. \tag{6.97}$$

Thus, the components of the induced incremental velocity are given by

$$du_x = \frac{1}{4\pi} \int\limits_{dl} \frac{s_y z - s_z y}{|\vec{r}|^3} dl, \quad du_y$$

$$= \frac{1}{4\pi} \int\limits_{dl} \frac{s_z x - s_x z}{|\vec{r}|^3} dl, \quad du_z = \frac{1}{4\pi} \int\limits_{dl} \frac{s_x y - s_y x}{|\vec{r}|^3} dl. \tag{6.98}$$

Hence, it can be shown, from the Biot–Savart law, that the induced downwash velocity component due to an element of the spanwise directed vortex line segment is

$$-dw(y) = -\frac{1}{4\pi(y_1 - y)} \left. \frac{d\Gamma}{dy} \right|_{y=y_1} dy_1. \tag{6.99}$$

The induced downwash due to the entire vortex line along the span is given by

$$w(y) = \int\limits_{-s}^{s} \frac{1}{4\pi(y_1 - y)} \left. \frac{d\Gamma}{dy} \right|_{y=y_1} dy_1. \tag{6.100}$$

The integral is evaluated by the method adopted by Glauert [16] in thin airfoil theory and this requires a transformation of the variable of integration to an angle varying from 0 to π. Now considering a typical section of the wing, and assuming the flow over each section to be two-dimensional, the sectional lift coefficient and the circulation may be related downwash at the quarter-chord in accordance with thin airfoil theory by the equations

$$dC_L = \frac{2dL}{\rho V_\infty^2 c} = \frac{2\rho V_\infty \Gamma}{\rho V_\infty^2 c} = 2\pi(\alpha - \alpha_0 - \varepsilon), \quad \Gamma = \pi V_\infty c(\alpha - \alpha_0) + \pi w(\Gamma)c. \tag{6.101}$$

In deriving the above relations, the following definitions were adopted: the sectional lift and induced drag are $dL = \rho V_\infty \Gamma$, $dD_i = \rho V_\infty \Gamma \varepsilon = -\rho \Gamma w$. Thus, the total wing lift and induced drag are given by

$$L = \rho V_\infty \int\limits_{-s}^{s} \Gamma \, dy = \rho V_\infty \int\limits_{-s}^{s} \left(\pi V_\infty c(\alpha - \alpha_0) + \pi w(\Gamma)c \right) dy, \tag{6.102}$$

$$D_i = -\rho \int\limits_{-s}^{s} w \Gamma \, dy = -\rho \int\limits_{-s}^{s} w \left(\pi V_\infty c(\alpha - \alpha_0) + \pi w(\Gamma)c \right) dy. \tag{6.103}$$

Thus, the lift and induced drag coefficients are given by

$$C_L = \frac{2L}{\rho V_\infty^2 S} = \frac{2}{V_\infty S} \int\limits_{-s}^{s} \Gamma \, dy, \quad C_{D_i} = \frac{2D_i}{\rho V_\infty^2 S} = -\frac{2}{V_\infty^2 S} \int\limits_{-s}^{s} w \Gamma \, dy. \tag{6.104}$$

The final lifting line equation, which is an integral equation that must be solved is

$$\Gamma = \pi V_\infty c (\alpha - \alpha_0) + \pi w(\Gamma) c, \qquad (6.105)$$

with

$$w(y) = \int_{-s}^{s} \frac{1}{4\pi (y_1 - y)} \left. \frac{d\Gamma}{dy} \right|_{y=y_1} dy_1. \qquad (6.106)$$

To apply the above method to a propeller, observe the following:

i) It is no longer possible to assume that the wake is planar; for a propeller the wake is either assumed to be helical in shape and the induced velocity is obtained by applying the Biot–Savart law, or a suitable potential function is assumed in the wake region and the induced velocities are obtained from the potential function;

ii) The induced velocities and forces of relevance have two components; one in the axial direction, and other a tangential component at a blade section;

iii) The freestream velocity, relative to the blade section, is no longer uniform, as the propeller itself is in rotation, while the propeller disc is also moving forward.

Figure 6.14 illustrates the velocities and forces (per unit radius) at a typical blade section. The propeller angular velocity is assumed to be given by ω. The axial and tangential inflow velocities are given by V_a and V_t; the axial and tangential induced velocities are assumed to be u_a^* and u_t^*. Hence, the total resultant inflow velocity has the magnitude

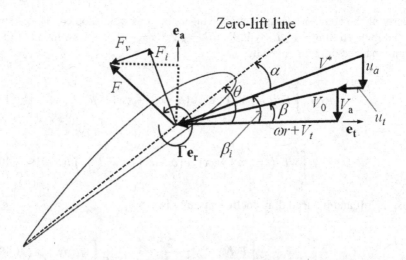

FIGURE 6.14 Propeller velocities and forces diagram at a radius r.

$$V^* = \left| \mathbf{V}^* \right| = \sqrt{\left(V_a + u_a^* \right)^2 + \left(\omega r + V_t + u_t^* \right)^2}. \tag{6.107}$$

The total resultant inflow velocity vector makes an angle β_i to the direction of the tangential unit vector $\mathbf{e}_t$ given by,

$$\sin \beta_i = \left| \mathbf{V}^* \right|^{-1} \left(V_a + u_a^* \right), \quad \cos \beta_i = \left| \mathbf{V}^* \right|^{-1} \left(\omega r + V_t + u_t^* \right). \tag{6.108}$$

Hence, the angle of attack is related to the blade pitch angle by $\alpha = \theta - \beta_i$. Furthermore, the blade section circulation is Γe_r and the corresponding lift force is $F_i = \rho V^* \Gamma$. The total sectional drag force is, $F_D = \frac{1}{2} \rho V^{*2} c C_D$. The coefficient of lift and drag are related to the circulation by

$$C_L = \frac{2F_i}{\rho V^{*2} c} = \frac{2\rho V^* \Gamma}{\rho V^{*2} c} = \frac{2\Gamma}{V^* c}, \tag{6.109}$$

$$C_D = \frac{2F_D}{\rho V^{*2} c} = C_{D0} + \frac{2\rho V^* u_t^* \Gamma}{\rho V^{*2} c} = C_{D0} + \frac{2 u_t^* \Gamma}{V^* c}. \tag{6.110}$$

The total thrust in the e_a and torque in the $-e_t$ directions, acting on a N bladed propeller are obtained by integrating the sectional thrust and torque from the hub radius r_h, to the tip radius R, and are respectively given by

$$F_T = N \int_{r_h}^{R} \left(F_i \cos \beta_i - F_D \sin \beta_i \right) dr, \quad Q = N \int_{r_h}^{R} \left(F_i \sin \beta_i + F_D \cos \beta_i \right) r dr. \tag{6.111}$$

Substituting for the lift and drag forces as well as for angle β_i,

$$F_T = N \rho \int_{r_h}^{R} \left(\Gamma \left(\omega r + V_t + u_t^* \right) - \frac{1}{2} c C_D V^* \left(V_a + u_a^* \right) \right) dr, \tag{6.112}$$

$$Q = N \rho \int_{r_h}^{R} \left(\Gamma \left(V_a + u_a^* \right) + \frac{1}{2} c C_D V^* \left(\omega r + V_t + u_t^* \right) \right) r dr. \tag{6.113}$$

The power required to drive the propeller is given by $P = Q\omega$. The useful power produced by the propeller is $P_r = F_T V_\infty$ where the propeller's forward speed at infinity (i.e. freestream speed) is V_∞. Thus, the shaft power efficiency is

$$\eta_{sp} = \frac{F_T V_\infty}{Q\omega}. \tag{6.114}$$

All that remains now is to relate the blade section circulation, Γe_r to the axial and tangential induced velocities, u_a^* and u_t^*. There are broadly two classes of method for doing this; in the first class, a wake model is assumed and the relations between Γe_r and, u_a^* and u_t^* are derived using the Biot–Savart law. The wake model or shape of the wake is chosen to satisfy certain requirements. In the second approach, the relations between Γe_r and, u_a^* and u_t^* are derived using a suitable assumed potential function. This approach was introduced by Goldstein [17], refined by Kawada [18] and discussed in some detail by Flood [19].

The earliest propeller theories were due to Betz [20] who considered the propeller problem for the case of uniform inflow ($V_a = V_\infty$) and no viscous forces. Betz used a vortex model of the rotating blades based on the lifting-line technique in which the vortex strength varies along the wingspan. Assuming that $\Gamma = \Gamma(r)$ is the optimum circulation distribution, the circulation is incremented at some arbitrary radius r, causing incremental changes in thrust and torque. Betz argued that for the circulation to be optimum, the change in the "shaft power efficiency" should be independent of the radius. Joukowsky, in a series of papers published in 1912–1918 [21, 22, 23, 24] defined the optimum rotor as one having constant circulation along the blade span, such that the vortex system for a multi-bladed rotor consists of multiple helical tip vortices, each with a predefined and equal strength and an axial hub vortices of equal and opposite strength. Lerbs [25] extended the work of Betz to the case of non-uniform axial inflow $V_a(r)$. Kerwin, Coney and Hsin [26] sought to produce a specified thrust with the requirement that the torque must be minimum. Kerwin, Coney and Hsin [26] solved the torque optimization problem by adopting the Lagrange multiplier approach for both single- and multi-bladed propellers. In the single-bladed case, they discretized the vortex distribution along the radius of the propeller blade to find a set of M vortex panel circulations that produce the least torque for a specified thrust, $F_T = F_{Ts}$. They form an auxiliary Hamiltonian function, $H = Q + \lambda_Q(F_T - F_{Ts})$, where λ_Q is a Lagrange multiplier, and they find the optimum $\Gamma = \Gamma(r)$ by setting the partial derivatives of

$$H = Q + \lambda_Q\left(F_T - F_{Ts}\right), \qquad (6.115)$$

with respect to Γ and λ_Q to zero.

$$\frac{\partial H}{\partial \Gamma(j)} = 0, \quad j = 1, 2, \ldots, M, \quad \frac{\partial H}{\partial \lambda_Q} = \left(F_T - F_{Ts}\right) = 0. \qquad (6.116)$$

The second condition leads to the constraint equation, while the first results in an additional system of M non-linear equations for the unknown circulation strengths at the collocation points on the vortex panels, which can be solved iteratively.

One is interested in approximating the velocities induced on the lifting line due to the influence of the helical vortex sheets in the wake of each blade. The computation of these induced velocities is governed by the Biot–Savart Law which relates the magnitude, direction and proximity of a single vortex filament in the wake of a section of the blade to the velocity at the lifting line, as the vortex sheet is made up of individual line vortices shed from each foil section, and the influence of each section's wake vortex affects the velocity induced at any point on the lifting line. As the vortex is shed from each of the propeller blade section's wakes, it is essential to compute the influence of each of these shed vortices on every other section. The aerodynamic influence coefficients may be found by the application of the Biot–Savart law. The axial and tangential velocities induced at radius r_m on a typical propeller blade lifting line by a set of N unit strength helical vortices shed at a radius r_j can be expressed as integrals using the law of Biot–Savart respectively as

$$\bar{u}_a\left(r_m,r_j\right)=\frac{1}{4\pi}\sum_{k=1}^{N}\int_0^\infty \frac{r_j\left(r_j-r_m\cos\left(\phi+\delta_k\right)\right)d\phi}{\left\{\left(r_j\phi\tan\beta_h\right)^2+r_j^2+r_m^2-2r_jr_m\cos\left(\phi+\delta_k\right)\right\}^{3/2}}, \quad (6.117)$$

$$\bar{u}_t\left(r_m,r_j\right)=\frac{1}{4\pi}\sum_{k=1}^{N}\int_0^\infty \frac{r_j\tan\beta_h\left\{\left(r_j-r_m\cos\left(\phi+\delta_k\right)\right)-\left(r_j\phi\sin\left(\phi+\delta_k\right)\right)\right\}d\phi}{\left\{\left(r_j\phi\tan\beta_h\right)^2+r_j^2+r_m^2-2r_jr_m\cos\left(\phi+\delta_k\right)\right\}^{3/2}}. \quad (6.118)$$

In the above expressions, β_h is the pitch angle of the helix at r_j. The variable of integration, ϕ, is the angular coordinate of a general point on the helix shed from a typical propeller blade. The corresponding angular coordinate of a point on the kth blade is found by adding the blade indexing angle, δ_k, given by

$$\delta_k=2\pi\frac{k-1}{N}, \quad k=1,\,2,...,\,N. \quad (6.119)$$

The total induced velocity components on the lifting line can now be obtained by integrating the contributions of the helical vortices over the radius

$$u_a^*\left(r_m\right)=-\int_{r_h}^{R}\bar{u}_a\left(r_m,r_j\right)\frac{\partial\Gamma\left(r_j\right)}{\partial r_j}dr_j,\quad u_t^*\left(r_m\right)=-\int_{r_h}^{R}\bar{u}_t\left(r_m,r_j\right)\frac{\partial\Gamma\left(r_j\right)}{\partial r_j}dr_j, \quad (6.120)$$

where the integrals are evaluated in the sense of Cauchy or the singular part of the integral is evaluated independently as discussed below. The aerodynamic influence functions u_a^* and u_t^* must generally be evaluated numerically. In the propeller lifting line methods, the axial and tangential induced velocities, u_a^* and u_t^*, are related to the discrete vortex strengths $\Gamma(j)$ by

$$u_a^*\left(m\right)=\sum_{j=1}^{M}\bar{u}_a^*\left(m,j\right)\Gamma\left(j\right),\quad u_t^*\left(m\right)=\sum_{j=1}^{M}\bar{u}_t^*\left(m,j\right)\Gamma\left(j\right),\quad m=1,\,2,...,M. \quad (6.121)$$

The aerodynamic influence coefficients are defined by $\bar{u}_a^*\left(m,j\right),\bar{u}_t^*\left(m,j\right)$. Hence it follows that

$$\frac{\partial u_a^*\left(m\right)}{\partial\Gamma\left(j\right)}=\bar{u}_a^*\left(m,j\right),\quad \frac{\partial u_t^*\left(m\right)}{\partial\Gamma\left(j\right)}=\bar{u}_t^*\left(m,j\right) \quad (6.122)$$

and

$$\frac{\partial V^*(m)}{\partial \Gamma(j)} = \sin \beta_i(m) \bar{u}_a^*(m,j) + \cos \beta_i(m) \bar{u}_t^*(m,j). \tag{6.123}$$

Using the expressions given by Kawada [18] for the assumed potential function for the induced velocity components, Wrench [27] derived a set of closed form approximations for the aerodynamic influence coefficients $\bar{u}_a^*(m,j)$, $\bar{u}_t^*(m,j)$ from expansions of modified Bessel functions of the first and second kind. Wrench's approximations and the velocity integrals are computed using the following equations and conditions [28], [29]. The velocity components are essentially obtained from the inflow velocity. The control point radius r_m is the radial location on the lifting line where the velocity is induced, and the radius r_j is the radial location of the particular vortex filament of concern. When $r_j < r_m$,

$$\bar{u}_a^*(m,j) = \frac{N}{4\pi r_m}(y - 2Nyy_0 F_1), \quad \bar{u}_t^*(m,j) = \frac{N^2}{2\pi r_m} y_0 F_1. \tag{6.124}$$

When $r_j < r_m$,

$$\bar{u}_a^*(m,j) = \frac{N^2}{2\pi r_m} yy_0 F_2, \quad \bar{u}_t^*(m,j) = \frac{N}{4\pi r_m} y(1 + 2Ny_0 F_2), \tag{6.125}$$

with

$$F_1 \approx \frac{-1}{2Ny_0}\left(\frac{1+y_0^2}{1+y^2}\right)^{\frac{1}{4}}\left\{\frac{U}{1-U} + \frac{1}{24N}\left(\frac{2+9y_0^2}{\left(\sqrt{1+y_0^2}\right)^3} - \frac{2-3y^2}{\left(\sqrt{1+y^2}\right)^3}\right)\ln\left(\frac{1}{1-U}\right)\right\},$$

$$F_2 \approx \frac{1}{2Ny_0}\left(\frac{1+y_0^2}{1+y^2}\right)^{\frac{1}{4}}\left\{\frac{1}{U-1} - \frac{1}{24N}\left(\frac{2+9y_0^2}{\left(\sqrt{1+y_0^2}\right)^3} - \frac{2-3y^2}{\left(\sqrt{1+y^2}\right)^3}\right)\ln\left(\frac{U}{U-1}\right)\right\},$$

$$U = \left\{\frac{y_0\left(\sqrt{1+y^2}-1\right)}{y\left(\sqrt{1+y_0^2}-1\right)}\exp\left(\sqrt{1+y^2}-\sqrt{1+y_0^2}\right)\right\}^N,$$

$$y = \frac{r_m}{r_j \tan \beta_i}, \quad y_0 = \frac{1}{\tan \beta_i}. \tag{6.126}$$

The computation of the induced velocities involves a singularity when r_m is equal to r_j. To account for this singularity, it is important to factor out the singular part, leaving a regular function that depends on the geometry of the wake. The residue of the singular integral is assumed to be a constant in the region very close to the singularity [28]. This is done with the induction factors proposed by Lerbs [25]. The singular integrals, representing the self-induced velocities due to the vortex filaments, are then evaluated by Glauert's method. Once the aerodynamic influence coefficients, $\bar{u}_a^*(m, j)$, $\bar{u}_t^*(m, j)$, are evaluated, all that remains is the solution of the non-linear system equations for $\Gamma(j)$ and λ_O. All of the propeller's performance characteristics may then be found.

An alternative, but equivalent, approach is to use the Biot–Savart law and write the axial and tangential induced velocities as

$$\frac{u_a^*(r_m)}{V_\infty} = \int_{r_h}^{R} \frac{i_a}{2(r_m - r)} \frac{dG}{dr} dr, \quad \frac{u_t^*(r_m)}{V_\infty} = \int_{r_h}^{R} \frac{i_t}{2(r_m - r)} \frac{dG}{dr} dr, \quad (6.127)$$

where $G(\Gamma) = \Gamma(r)/2\pi R V_\infty$.

In Equation (6.127), i_a and i_t are the induction factors proposed by Lerbs [25] for correcting the induced velocities for a helical wake and are given by

$$i_a(r_m, r_j) = -\frac{\bar{u}_a^*(m, j)}{1/(4\pi(r_m - r_j))}, \quad i_t(r_m, r_j) = \frac{\bar{u}_t^*(m, j)}{1/(4\pi(r_m - r_j))}, \quad (6.128)$$

where $\bar{u}_a^*(m, j)$, $\bar{u}_t^*(m, j)$ are given by Equations (6.124) and (6.125). As the radius of the vortex, r_j, approaches the radius of the control point, r_m, the velocity induced by the helical vortices will approach the value induced by a semi-infinite vortex oriented in a direction tangent to the helix at its starting point on the lifting line. Therefore, as $r_m \to r_j$ one finds that, $i_a(r_m, r_j) = \cos\beta_i$, $i_t(r_m, r_j) = \sin\beta_i$. The complete solutions for the non-dimensional circulation $G(\Gamma)$ as a function of the radial coordinate are given by Eastbridge [29].

The lifting line theory discussed in this section could be extended to model propeller blades as lifting surfaces. The lifting surface model which results in the vortex lattice method consists of layers of constant strength, quadrilateral shaped vortex line elements along the mean camber lines of the blades, or covering both surfaces of the blades when the thickness effect is not to be excluded. At the trailing edge, the helical horseshoe vortices extend from the last vortex ring, downstream, to infinity, to form a constant radius cylindrical wake. Since the adjacent sides of the quadrilateral vortex ring elements are coincident, the strength of each vortex element in the lattice is given by the difference in the strength of adjacent vortex ring elements and is determined from the application of the boundary conditions on blade surfaces. The strengths of the induced velocities are determined by the application of the Biot–Savart law. Typical applications of the vortex lattice method are to ducted propellers as well as to low aspect ratio propeller blades. Runyan [30], Tsakonas, Jacobs and Ali [31], Kerwin and Kinnas [32] and Kerwin [33] have discussed the application of lifting surface theories to propellers and ducted propellers.

6.10   BLADE CIRCULATION DISTRIBUTION: POTENTIAL FLOW-BASED SOLUTIONS

The flow field due to an advancing propeller in space with a fixed or inertial frame of reference and away from it and its wake, may be considered to be irrotational and isentropic. The viscous forces acting on a propeller blade may be ignored. Thus, the flow field may be represented by a velocity potential, Φ and the corresponding velocity field may be assumed to be $\vec{V}$. Thus, the governing equation for the velocity potential, in general compressible flow, may be shown to be

$$\nabla^2\Phi = \frac{1}{a^2}\left(\vec{V}\cdot\frac{D\vec{V}}{Dt}+\frac{1}{2}\frac{\partial}{\partial t}\left|\vec{V}\right|^2+\frac{\partial^2\Phi}{\partial t^2}\right). \tag{6.129}$$

The freestream velocity and the perturbation velocity vector are respectively assumed to be V_∞ and u, while a is the local speed of sound. When u is small relative to V_∞, it is also small relative to $\sqrt{\omega^2R^2+V_\infty^2}$, where ω is the propeller angular velocity and R is the propeller disc radius, Equation (6.129) could be linearized and written in terms of a perturbation velocity potential, φ. Assume further a cylindrical coordinate system, $r,\tilde{\theta},z$ where the propeller is advancing in the direction of z and t is the independent time variable. Given that the freestream Mach number is $M_\infty=V_\infty/a_\infty$, and that $a_\infty=a$ is a constant, the linearized equation governing the perturbation potential φ is given by

$$\nabla^2\varphi - M_\infty^2\frac{\partial^2\varphi}{\partial z^2}=\frac{1}{a^2}\left(2M_\infty a\frac{\partial^2\varphi}{\partial z\partial t}+\frac{\partial^2\varphi}{\partial t^2}\right). \tag{6.130}$$

The boundary conditions requiring the flow to be tangential to the surface of the body applies on the surface of the propeller blades, while there should be no pressure discontinuity across the wake. Moreover, the disturbance potential must vanish at infinity.

Transforming to a blade fixed rotating frame of reference, where $\theta = \tilde{\theta}+\omega_m t$, the linearized equation governing the perturbation potential φ is given by

$$\nabla^2\varphi - M_\infty^2\frac{\partial^2\varphi}{\partial z^2}-2\frac{M_\infty M_\theta}{r}\frac{\partial^2\varphi}{\partial z\partial\theta}-\frac{M_\theta^2}{r^2}\frac{\partial^2\varphi}{\partial\theta^2}$$
$$=\frac{1}{a^2}\left(2M_\infty a\frac{\partial^2\varphi}{\partial z\partial t}+2\frac{M_\theta a}{r}\frac{\partial^2\varphi}{\partial\theta\partial t}+\frac{\partial^2\varphi}{\partial t^2}\right), \tag{6.131}$$

where $M_\theta = \omega_m r/a$ is the radial flow Mach number. In terms of the downwash velocity in the slipstream, the boundary condition may be expressed as

$$w = \frac{\partial\varphi}{\partial z}-\frac{V_\infty}{\omega_m r}\frac{\partial\varphi}{r\partial\theta}. \tag{6.132}$$

Introducing the dimensionless variables, $\rho = \omega r/V_\infty$ and $\bar{z} = \omega z/V_\infty$, and considering the steady flow case, the linearized equation governing the perturbation potential φ is given by

$$\frac{\partial^2 \varphi}{\partial \rho^2} + \frac{1}{\rho}\frac{\partial \varphi}{\partial \rho} + \beta^2 \frac{\partial^2 \varphi}{\partial z^2} + \frac{1}{\rho^2}\left(1 - M_\infty^2 \rho^2\right)\frac{\partial^2 \varphi}{\partial \theta^2} - 2M_\infty^2 \frac{\partial^2 \varphi}{\partial z \partial \theta} = 0, \qquad (6.133)$$

where $\beta^2 = 1 - M_\infty^2$. Equation (6.133) must solved for the flow through the propeller disc, subject to the boundary conditions across the blades, the wake and at infinity. In what follows, the bars on $\bar{z}$ are dropped.

Following Runyan [30], one may introduce helical coordinates, defined by $\sigma = \theta + z$ and $\zeta = \theta - z$,

$$\frac{\partial^2 \varphi}{\partial \rho^2} + \frac{1}{\rho}\frac{\partial \varphi}{\partial \rho} + \left(1 + \frac{1}{\rho^2}\right)\frac{\partial^2 \varphi}{\partial \zeta^2} + \left(1 + \frac{1}{\rho^2} - 4M_\infty^2\right)\frac{\partial^2 \varphi}{\partial \sigma^2}$$

$$-2\left(1 + \frac{1}{\rho^2} - 2M_\infty^2\right)\frac{\partial^2 \varphi}{\partial \zeta \partial \sigma} = 0. \qquad (6.134)$$

The trailing vortices behind a propeller form a helical vortex sheet with a constant diameter. This is a massless surface of discontinuity which cannot be penetrated by the fluid particles.

Consider a two-bladed propeller and the geometric equations of the two rigid vortex sheets or helical surfaces of the wake are given by $\theta = z$ and $\theta - \pi = z$. Thus, $\theta - z = 0$, π and the helix axis coincides with the z-axis which is also the propeller axis. Moreover, when a helical surface moves along its axis with a velocity w and rotates with an angular velocity $\omega w/V_\infty$ about its axis, its displacement is entirely with its own surface. Thus $\zeta = \theta - z$ represents a helical surface that starts after an angle ζ after the helical surface defined by $0 = \theta - z$, where θ and ζ are angles defined in the same direction and the spiral lines along the helical surfaces represent lines of constant velocity potential. The velocity potential across the helical vortex sheet is constant along the helical vortex line of a given radius and is equal to the circulation at the corresponding point on the propeller where the vortex line is shed. Thus, the total velocity potential is only a function of ρ and $\zeta = \theta - z$. Due to the helical nature of the wake, Goldstein [17] assumed that the derivatives along the helix $\partial/\partial\sigma = 0$ and found a solution to the simplified potential equation obtained from Equation (6.134).

Wells [34] has expressed the solution to Equation (6.133) in the form

$$\varphi_c^\pm (\rho, \theta, z) = \int_0^\infty C_0(\gamma) J_0(\gamma \rho) e^{\pm(\gamma/\rho z)} dz$$

$$+ \sum_{n-1}^\infty e^{inN\theta} \int_0^\infty C_n(\gamma) \gamma J_{nN}(\gamma \rho) \exp\left\{f\left(\beta, n, N, M_\infty, \gamma\right) z/\beta^2\right\} d\gamma, \qquad (6.135)$$

where $f\left(\beta,n,N,M_{\infty},\gamma\right)=inNM_{\infty}^{2}\pm\sqrt{\beta^{2}\gamma^{2}-n^{2}N^{2}M_{\infty}^{2}}$ and $C_{n}(\gamma)$ are "constants" that are to be determined. Prior to stating the expressions defining $C_{n}(\gamma)$, it is essential to define a number of related functions. Given the spanwise circulation distribution, $\Gamma(\rho)$, and with $\rho_{0}=\rho\big|_{tip}$ let,

$$\Gamma_{n}'(\gamma)=\int_{0}^{\rho_{0}}\Gamma\left(s\right)sJ_{nN}\left(\gamma s\right)ds,\quad R_{n}'(\gamma)=\int_{0}^{\rho_{0}}R_{n}\left(s\right)sJ_{nN}\left(\gamma s\right)ds,\qquad(6.136)$$

$$R_{n}\left(\rho\right)=(-1)^{n}\frac{N}{n\pi}K_{nN}\left(nN\rho\right)\int_{0}^{\rho}I_{nN}\left(nN\xi\right)\xi\frac{d\Gamma}{d\xi}d\xi$$

$$+(-1)^{n}\frac{N}{n\pi}I_{nN}\left(nN\rho\right)\left(\int_{0}^{\rho_{0}}K_{nN}\left(nN\xi\right)\xi\frac{d\Gamma}{d\xi}d\xi-K_{nN}\left(nN\xi\right)\xi\frac{d\Gamma}{d\xi}\bigg|_{\xi=\rho_{0}}\right),$$

$$(6.137)$$

$$I_{1}=\int_{0}^{\infty}\gamma J_{nN}\left(\gamma s\right)J_{nN}\left(\gamma\rho\right)e^{-\gamma z}d\gamma=\frac{1}{2z}\exp\left(-\frac{\rho^{2}+s^{2}}{4z}\right)I_{nN}\left(\frac{\rho s}{2z}\right),\quad(6.138)$$

$$I_{2}=\int_{0}^{\infty}\gamma^{nN-1}J_{nN}\left(\gamma\rho\right)e^{-\gamma z}d\gamma$$

$$(6.139)$$

$$=(-1)^{nN-1}\rho^{-nN}\frac{d^{nN-1}}{dz^{nN-1}}\left(\frac{\left(\sqrt{z^{2}+\rho^{2}}-z\right)^{nN}}{\sqrt{z^{2}+\rho^{2}}}\right).$$

Then, using Equations (6.138) and (6.139), one obtains

$$C_{0}=\frac{N}{2\pi}\frac{\beta}{2\gamma}\Gamma_{0}'(\gamma),\quad C_{n}=A_{n}+B_{n},\qquad(6.140)$$

with,

$$A_{n}=\frac{N}{2\pi}\frac{2(-1)^{n}}{nN}\frac{\Gamma_{n}'(\gamma)}{2}\left(\text{sign}(z)i-\frac{nNM_{\infty}^{2}}{\sqrt{\gamma^{2}\beta^{2}-n^{2}N^{2}M_{\infty}^{2}}}\right),\qquad(6.141)$$

$$B_{n}=-\frac{R_{n}'(\gamma)}{2}\left(\text{sign}(z)i-\frac{nN}{\sqrt{\gamma^{2}\beta^{2}-n^{2}N^{2}M_{\infty}^{2}}}\right),\qquad(6.142)$$

where $\text{sign}(z)=-1$ for $z<0$ and $\text{sign}(z)=1$ for $z\geq0$.

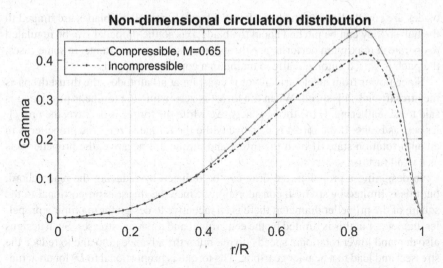

FIGURE 6.15 Typical circulation distribution computed using Equation (6.143).

Defining the non-dimensional variables, $\varphi^* = \omega\varphi/V_\infty^2$ and $\Gamma^* = \omega\Gamma(\rho)/V_\infty^2$, $\gamma_{\rho s} = nN/\rho s$, $\Gamma_n'^*(\gamma) = \omega\Gamma_n'(\gamma)/V_\infty^2$, $R_n'^*(\gamma) = \omega R_n'(\gamma)/V_\infty^2$ and simplifying the boundary conditions, Wells [34] was able to show that

$$\rho\frac{\partial\varphi^*}{\partial z} - \frac{1}{\rho}\frac{\partial\varphi^*}{\partial\theta} + \frac{2\Gamma^*}{\rho\bar{c}C_{L_\alpha}} = (1+\rho^2)\alpha_g, \quad z\to 0, \ \theta\to\frac{\pi}{N}, \tag{6.143}$$

where α_g is the geometric angle of attack, $\bar{c} = c/R$ and C_{L_α} is the section lift curve slope (Figure 6.15).

Equation (6.15) represents an integral equation for the unknown circulation distribution Γ^* which can be obtained by assuming a shape function of the form,

$$\Gamma(\rho) = \sum_{k=1}^{K} g_k(\rho_0 - \rho)\rho^{k+2}. \tag{6.144}$$

The non-dimensional circulation distribution function obtained on a typical uniform propeller blade is shown in Figure 6.13. Wells [34] has also obtained expressions for the axial and tangential perturbation velocities.

6.11　STANDARD PROPELLER FEATURES AND DESIGN CONSIDERATIONS

Propellers are either tractor or pusher propellers. A tractor propeller is generally placed in front of the prime mover while the aircraft structure is downstream of it, so the propeller pulls the aircraft. A pusher propeller is placed behind the prime mover while the aircraft structure is upstream of it, so the propeller pushes the aircraft. Propellers are classed as fixed or variable pitch propellers. A fixed pitch propeller's blades are rigidly fixed to the hub. A variable pitch propeller's blades are hinged to the hub so they can be pitched about the blade axis so the propeller can be regulated to operate at maximum performance throughout its operational range. In some cases, the pitch is varied automatically to maintain a constant tip speed.

Since power from an electric motor is constant at all altitudes, the thrust drops as the aircraft picks up speed. When the thrust is zero while the torque is positive, it is said to be feathering. If the thrust is negative while the torque is positive, the propeller acts as brake. If the thrust is negative while the torque is zero, the propeller is in an auto-rotation state. If both the thrust and torque are negative, the propeller acts as a wind turbine.

In designing a propeller, the drag can be reduced by reducing the blade chord, but this is limited by strength considerations. Since the thrust is proportional to the square of the propeller diameter, there is an incentive to design for maximum propeller diameter, but this is limited by the centrifugal and torsional stresses. Such designs also demand lower rotational speeds (to maintain the advance ratio and to reduce the stresses) and lead to a heavier gear-box. The torque is proportional to D^5 for an actuator disc, but this is clearly reduced in practice (along with propulsive efficiency), as the number of blades is reduced and not infinite.

To balance engine torque characteristics, two, three or four blades are normally chosen. The materials used are aluminum, hollow steel, wood or polymer composites, as fatigue and fracture strength is crucial. Propeller blades are known to fracture quite frequently. Centrifugal forces are much lower than for fans. Bird-strikes and other sources of damage must be considered at the design stage.

6.12   PROPELLERS FOR DISTRIBUTED PROPULSION

Electric motors have vastly different performance features and characteristics when compared with conventional fossil fuel-powered motors, including incredibly higher efficiencies, much smaller volumes, lower mass, fewer moving parts, independence of motor shaft power with air density and scalability over wide range power requirements. Electric motors are relatively easily controllable and thus facilitate the incorporation of distributed propulsion with a multiplicity of smaller motors driving highly efficient but smaller propellers to deliver large levels of thrust to the aircraft. This is the basis of a new and emerging strategy generally called distributed electric propulsion (DEP) so as to facilitate full propulsion control of an aircraft. DEP requires two types of propellers, low-speed, high-lift propellers used during low-speed maneuvering of the aircraft and high-speed, low-lift propellers used during cruising. The requirement is that they generate a higher axial induced velocity for the same level of thrust force generated. The high-lift propeller system is defined by selecting the number of propellers, the propeller disc diameter, the shape of the blade, including the twist distribution and chordwise blade section profile, and the number of blades. The blade section lift coefficient is an important parameter that should generally be higher. Patterson [35] and Patterson, Derlaga and Borer [36] have considered and proposed suitable methods for the design of such propeller systems, wherein they give due importance to propeller—wing interactions.

CHAPTER SUMMARY

Propellers are the most important means of propulsion of electric aircraft that are expected to be flying at low speeds. For this reason, the underlying theories of propeller design for the distributed propulsion of electric aircraft are discussed at some length in this chapter.

REFERENCES

1. W. J. M. Rankine, On the mechanical principles of the action of propellers, *Trans. Inst. Nav. Architects*, 6:13–39, 1865.
2. R. E. Froude, On the part played in propulsion by differences of fluid pressure, *Trans. Inst. Nav. Architects*, 30:390–405, 1889.
3. W. R. Hawthorne, J. H. Horlock, Actuator disc theory of the incompressible flow in axial compressors, *Proc. Inst. Mech. Eng. Lond.*, 176(30.1): 789–814, June 1, 1962.
4. P. J. Carpenter, B. Fridovich, *Effect of a Rapid Blade-Pitch Increase on the Thrust and Induced-Velocity Response of a Full-Scale Helicopter Rotor*, TN 3044, National Advisory Committee for Aeronautics, USA, 1953.
5. H. Glauert, Airplane propellers, pp 169–269. In Durand W. F. (Ed.). *Aerodynamic Theory*, Vol. 4(Division L), Dover, New York, 1935; (Dover Ed., 1963).
6. R. Vepa (Ed.), Wind power generation and control, Chapter 4. In *Dynamic Modelling Simulation and Control of Energy Generation*, Lecture Notes in Energy Series, No. 20, Springer Verlag, London, UK, 2013.
7. J. T. Conway, Analytical solutions for the actuator disk with variable radial distribution of load, *J. Fluid Mech.*, 297:327–355, 1995.
8. J. T. Conway, Exact actuator disk solutions for non-uniform heavy loading and slipstream contraction, *J. Fluid Mech.*, 365:235–267, 1998.
9. R. Bontempo, M. Manna, Solution of the flow over a nonuniform heavily loaded ducted actuator disk, *J. Fluid Mech.*, 728:163–195, 2013.
10. R. Bontempo, The nonlinear actuator disk as applied to open and ducted rotors, Ph.D. Thesis, University of Naples Federico II, Naples, Italy, 2014.
11. R. Bontempo, M. Cardone, M. Manna, G. Vorraro, Ducted propeller flow analysis by means of a generalized actuator disk model, *Energy Procedia*, 45C:1107–1115, 2014.
12. R. Bontempo, M. Cardone, M. Manna, G. Vorraro, A comparison of nonlinear actuator disk methods for the performance analysis of ducted propellers, Proceedings of the 11th European Conference on Turbomachinery Fluid Dynamics and Thermodynamics (ETC '15), Madrid, Spain, March 2015.
13. R. Bontempo, M. Manna, Effects of duct cross section camber and thickness on the performance of ducted propulsion systems for aeronautical applications, *Int. J. Aerosp. Eng*, 2016(8913901):1–9, 2016, (Hindawi), doi:10.1155/2016/8913901.
14. R. J. Weir, Ducted propeller design and analysis, Sandia Report, SAND87-2118 UC-32, Sandia National Laboratories, October 1987.
15. Z. Wang, L. Chen, S. Guo, Numerical analysis of aerodynamic characteristics for the design of a small ducted fan aircraft. *Proc IMechE Part G, J. Aerosp. Eng.*, 227(10):1556–1570, 2012.
16. H. Glauert. *The Elements of Aerofoil and Airscrew Theory*, Cambridge University Press, Cambridge, 1926.
17. S. Goldstein, On the vortex theory of screw propellers, *Proc. R. Soc. Lond. A*, 123:440–465, 1929.
18. S. Kawada, Induced velocity by helical vortices, *J. Aeronaut. Sci.* 3:86–87, 1936. Also: Report of the Aeronautical Research Institute, Tokyo, Imperial University, No. 172, 1939.

19. K. M. Flood, Propeller performance analysis using lifting line theory, Master of Science (in Mechanical Engineering) Thesis, Massachusetts Institute of Technology, June 2009.

20. A. Betz. Schraubenpropeller mit geringstem energieverlust, *K. Ges. Wiss. Gottingem Nachr. Math.-Phys.*, 193–217, 1919. Also: Dissertation, Gottingen Nachrichten, Gottingen.

21. N. E. Joukowsky, Vortex theory of screw propeller, I. *Trudy Otdeleniya Fizicheskikh Nauk Obshchestva Lubitelei Estestvoznaniya* 16(1):1–31, 1912 (in Russian). French translation in: *Theorie tourbillonnaire de l'helice propulsive* (Gauthier-Villars, Paris, 1929) 1–47.

22. N. E. Joukowsky, Vortex theory of screw propeller, II. *Trudy Otdeleniya Fizicheskikh Nauk Obshchestva Lubitelei Estestvoznaniya* 17(1):1–31, 1914 (in Russian). French translation in: *Theorie tourbillonnaire de l'helice propulsive* (Gauthier-Villars, Paris, 1929) 48–93.

23. N. E. Joukowsky, Vortex theory of screw propeller, III. *Trudy Otdeleniya Fizicheskikh Nauk Obshchestva Lubitelei Estestvoznaniya* 17(2):1–23, 1915 (in Russian). French translation in: *Theorie tourbillonnaire de l'helice propulsive* (Gauthier-Villars, Paris, 1929) 94–122.

24. N. E. Joukowsky, Vortex theory of screw propeller, IV. *Trudy Avia Raschetno-Ispytatelnogo Byuro*, 3:1–97, 1918 (in Russian). French translation in: *Theorie tourbillonnaire de l'helice propulsive* (Gauthier-Villars, Paris, 1929) 123–198.

25. H. Lerbs, Moderately loaded propeller with a finite number of blades and an arbitrary distribution of circulation, *Trans. Soc. Nav. Architects Mar. Eng. (SNAME)*, 60:73–123, 1952.

26. J. E. Kerwin, W. B. Coney, C. Y. Hsin, Optimum circulation distributions for single and multi-component propulsors, Twenty-First American Towing Tank Conference, pp 53–62, 1986.

27. J. W. Wrench, The calculation of propeller induction factors, Technical Report 1116, David Taylor Model Basin, 1957.

28. J. E. Kerwin, J. B. Hadler. *Principles of Naval Architecture: Propulsion*. Society of Naval Architects and Marine Engineers (SNAME), Jersey City, NJ, 2010.

29. J. R. Eastridge, Investigation and implementation of a lifting line theory to predict propeller performance, Senior Honors Theses, Paper 72, School of Naval Architecture and Marine Engineering, University of New Orleans, USA, 2016.

30. H. L. Runyan, Unsteady lifting surface theory applied to a propeller and helicopter rotor, Doctoral Thesis, Submitted in partial fulfilment of the requirements for the award of Doctor of Philosophy by Loughborough University, Loughborough, UK, 1973.

31. S. Tsakonas, W. R. Jacobs, M. R. Ali, Propeller-duct interaction due to loading and thickness effects, Proceedings of the Propellers/Shafting '78 Symposium, Society of Naval Architects and Marine Engineers (SNAME), 1978.

32. J. E. Kerwin, S. A. Kinnas, A surface panel method for the hydrodynamic analysis of ducted propellers, *Trans. Soc. Nav. Architects Mar. Eng.*, 95:93–122, 1987.

33. J. E. Kerwin. *Hydrofoils and Propellers*, MIT Course 13.04 Lecture Notes, Cambridge, MA, 2001.

34. V. L. Wells, Propellers in compressible flow, ICAS-84-5.6.1, Proceedings of the '14th Congress of the International Council of the Aeronautical Sciences, Toulouse, France, September 9–14, 1984.

35. M. D. Patterson. Conceptual design of high-lift propeller systems for small electric aircraft, PhD Thesis, Georgia Institute of Technology, 2016.

36. M. D. Patterson, J. M. Derlaga, K. Borer, *High-Lift Propeller System Configuration Selection for NASA's SCEPTOR Distributed Electric Propulsion Flight Demonstrator*, AIAA-2016-3922, AIAA Aviation, Washington, DC, June 2016.

7 High Temperature Superconducting Motors

7.1 HIGH TEMPERATURE SUPERCONDUCTORS (HTS)

Superconducting materials with characteristically low resistance to electron flow were first discovered by the Dutch physicist Heike Kammerlingh Onnes in 1911. High-temperature superconductors (HTS) were discovered in 1986 by two physicists, Johannes Georg Bednorz and Karl Müller, working at the International Business Machines (IBM) Labs. Normally the property of superconductivity is only found at very low temperatures around 0 K and conventional superconductors only work when cooled near absolute zero. However, Bednorz and Müller [1] detected superconductivity in a sample of LaBaCuO material at the much higher temperature of 36 K. Subsequently, high-temperature superconductivity was also identified by a group led by Chu [2], in structured materials containing planes of copper oxide, such as $YBa_2Cu_3O_7$. These materials had a transition temperature beginning at 93 K and had zero resistance at 80 K. The warmest of the superconducting materials discovered until recently, is hydrogen sulfide, which functions at 203 K [3]. Recently several researchers have announced the discovery of room temperature superconductivity. A room-temperature superconductor is a material that is capable of exhibiting superconductivity at room temperatures of about around 298 K. In one case the material is excited and stressed by a PZT actuator which renders it a superconductor at room temperatures. More recently a team of scientists at the Indian Institute of Science have reported having synthesized a room-temperature superconductor using particles of gold and silver [4]. Another group of physicists [5] have synthesized a room-temperature superconductor using a pressurized superhydride, LaH_{10}, a compound of lanthanum and hydrogen. Superconducting materials are known to have problems with flux creep, weak links and poor mechanical properties, which have also been identified by Goyal [6] and are currently being resolved.

Superconductivity was thought to have been completely explained in 1957 by the Bardeen, Cooper and Schrieffer (BCS) theory [7, 8] based on quantum mechanics. According to the BCS theory, when a single negatively charged electron slightly distorts the lattice of atoms in the superconductor, it attracts toward it a small excess of positive charge. This excess was considered to be responsible for attracting a second electron. This weak, indirect attraction was said to bind the two electrons together into what is known as a Cooper pair. The concept of a Cooper pair was put forward by Cooper [9] in 1956. Cooper pairs function as supercurrent carriers, thus facilitating superconductivity. However, the BCS theory also predicted a theoretical maximum to Tc of around 30–40 K, and it was shown that above this temperature, Cooper pairs were not sustainable. Although the BCS theory completely failed to predict the existence of HTS, it did provide a framework for characterizing HTS, and it can be credited with several other successes. A superconducting material is in the superconducting state when a point determined by the operating temperature, current density and magnetic field characteristics lies below a surface defined by the material's critical characteristics. Critical surface characteristics are defined in terms of three variables: temperature, current density and magnetic field intensity. Thus, superconductivity is generally observed below a critical temperature at a given critical (magnetic) field strength and a critical current density. A superconductor needs to be kept below a critical temperature T_c and the applied field must be less than the critical field H_c. When a magnet is placed above the superconductor, there is a force of repulsion which could overcome gravity, allowing the magnet to levitate above the superconductor. However, it is not a completely stable arrangement, giving the magnet the freedom to spin or rotate above the superconductor while it tries to orient its magnetic poles. If the magnetic field is removed or the temperature of the superconductor is raised above the critical value T_c, the surface currents and magnetization disappear, and the magnet will no longer levitate. Alternatively, if an applied magnetic field is stronger than the critical field, it can penetrate the superconductor, causing a quenching of the superconducting state. Even though the temperature may be below the critical temperature, it will no longer be superconducting. Materials exhibiting this behavior are known as type I superconductors, and the BCS theory applies to these materials.

7.1.1 THE MEISSNER STATE AND THE MEISSNER EFFECT

HTS materials are known as type II, which exhibit two critical magnetic field states before losing the property of superconductivity. For type II superconductors, there is an additional state that occurs between a Meissner state and the normal state. This state is known as a vortex state. There are now two critical magnetic fields and the transition from the vortex state to the Meissner state occurs at the lower critical field. Within the vortex state, a phenomenon called flux pinning occurs which allows small amounts of the magnetic field to pass via flux tubes or vortices. However, within the Meissner state all of the magnetic field is excluded. While the most well-known characteristic of superconductivity is zero resistance, the magnetic field inside the Meissner state of the superconductor, which is proportional to the sum of the magnetization and the applied field, adds up to zero. The exclusion of magnetic fields within the Meissner state of the superconductor is the defining characteristic of a superconductor and is due to the phenomenon known as the Meissner effect. Generally, superconductors are known not to permit magnetic fields to penetrate through them, a phenomenon known as the Meissner effect. It is not known yet if room temperature superconductors also share this property, so it is not clear if room temperature superconductors can be used to replace high temperature superconductors in all applications.

7.1.2 FEATURES OF SUPERCONDUCTING MATERIALS

The disappearance of the electrical resistance and the Meissner effect when the magnetic field inside the superconductor is shielded by a lossless current flowing in a thin layer on the surface of the superconducting material are key properties of superconductors. Such materials can be considered to be the perfectly diamagnetic, which occurs when an external magnetic field penetrates only a finite, outer region of the material and does not influence the remaining inner domain of the material. Furthermore, in a cooled superconductor after the magnetization and the applied field add up to zero, if the applied field is removed, the magnetization remains and the internal field is now acting in a direction opposite to the originally applied field. Thus, the flux is essentially "trapped" within the superconductor. These superconductors are also referred to as flux pinned superconductors and the trapped field could be as high as 15+ Tesla. They are also known as trapped field superconductors. The trapped field can be maintained, in principle, by a quantum process known as flux pumping, whereby small amounts of magnetic flux are pumped in by the repeated application of magnetic transients, which require relatively low power. Such magnetized superconducting materials have an advantage over field windings, since no electrical connections are necessary. Magnetic flux can be trapped within an HTS by cooling through the transition temperature in the presence of an applied external magnetic field. Currents continue to flow in the superconductor and re-create the field that previously existed, even after the removal of the external field. The material with its trapped magnetic field can be used like a permanent magnet. Another method of trapping a magnetic field in an HTS while the material is in the superconducting state, is to apply a large magnetic pulse. Long after the pulse has been applied, a field remains trapped in the superconductor. Although the concept has not yet been exploited in the design of a real motor, it is a very promising concept.

Yttrium barium copper oxide (YBCO) is a family of crystalline chemical compounds known for exhibiting high-temperature superconductivity. This family of superconductors includes the first material that was discovered to become superconducting above the boiling point of liquid nitrogen (77 K) at about 93 K. Typically, YBCO has a layered structure consisting of copper–oxygen planes with yttrium and barium atoms in the crystal structure as well. The resulting crystal structure is similar to a perovskite, with a unit cell consisting of stacked cubes of $BaCuO_3$ and $YCuO_3$.

7.2 HTS MOTORS

An electrical motor can be termed an HTS motor if its design contains elements that are HTS and these elements are working in the superconducting state at much higher temperatures than elements exhibiting the normal features of superconductivity. The motor usually contains an HTS element, such as an armature or a field winding or both. It is common practice to use a high current density and low-loss HTS wire for winding the armature or field coils or both. The HTS-field winding in the rotor of a synchronous motor creates a higher magnetic field in the air gap than iron-core copper windings. Thus, the use of HTS windings can reduce the size of the motor, the losses in the motor including the iron losses and the weight of the motor due to the useless material in its design. Aircraft propulsion systems can benefit from electric HTS machines in two distinct ways. In the first instance, power can be generated by a conventional gas turbine and converted into electrical energy in an HTS generator, which in turn is used to power a controlled HTS motor driving a propeller. Due to the onboard availability of electric power, an electric motor may be used to drive the compressor in a high bypass ratio gas "turbine" engine, thus making the turbine redundant. Thus, such a hybrid propulsion system uses both HTS generators and motors. This turbo-generator motor drive has several advantages over a direct drive gas turbine propulsion system. The turbine main shaft speed is not limited by the fan and therefore can be operated at higher efficiencies. The use of a turbo-generator allows for greater redundancy, which in turn improves the overall reliability of the propulsion system. Moreover, a turbo-generator motor drive facilitates greater control of thrust and consequently makes propulsion-based flight control feasible, particularly when the gas turbine can be distributed along the lifting surfaces.

On the other hand, an all-electric propulsion system would use only HTS motors driven by batteries or fuel cells without the need for an onboard generator. The YBCO family of compounds have been used as conductors which are manufactured as wire assemblies, and as tapes that are now available from several industrial suppliers and are very promising conductors for the design of HTS electric motors and their applications. Many YBCO compounds have the general formula $YBa_2Cu_3O_{7-x}$ (also known as Y123), although materials with other Y:Ba:Cu ratios also exist, such as $YBa_2Cu_4O_y$ (Y124) and $Y_2Ba_4Cu_7O_y$ (Y247). The bismuth strontium calcium copper oxide HTS superconductor family (BSCCO tapes) which include the following materials: Bi-2201 or $Bi_2Sr_2CuO_6$, Bi-2212 or $Bi_2Sr_2CaCu_2O_8$, Bi-2223 or $Bi_2Sr_2Ca_2Cu_3O_{10}$, have also been successfully used in the design and construction of HTS motors.

Kalsi [10] has described in some detail the different types of superconducting machines and has provided models that could be used for simulation. Masson et al. [11], Luongo et al. [12] and Armstrong et al. [13] have outlined the requirements of HTS motors for aircraft propulsion. A wide variety of HTS reluctance motors with diamagnetic rotors, HTS permanent magnets, wound field synchronous motors and HTS brushless direct current (DC) motors are being developed for aircraft and other applications [14–17]. Thus, several superconducting field synchronous machines with room-temperature or cryogenic high purity metal armatures are under development. In these applications, it is advantageous to reduce the number of moving parts in the HTS machines. For this reason, for most aircraft drive systems, rotary machines with HTS stator windings and permanent magnets or wound and trapped-field rotors are preferred. The stator is generally constructed by winding HTS tape, about 4–6 mm wide, flat on spools of the same width, over each of the poles, which are quite large. Such windings are referred to as pancake coils. Alternately, the windings are inserted into circular tracks on a disc, and these types of windings are referred to as race-track windings.

Most superconducting machines have employed liquid neon or cold gaseous helium for cooling HTS coils. Liquid nitrogen-cooled HTS motors have also been developed for ship propulsion by Okazaki, Sugimoto and Takeda [18]. The coolant is generally transferred from an external refrigerator to the rotor through a rotating coupling. The field winding consists of several HTS coils that are conduction-cooled through the support structure or exposed to the tubes carrying the coolant to enhance the cooling rate. The torque tube is mainly responsible for transferring torque from the "cold" (cryogenically-cooled) environment to the "warm" shaft end. The pole pairs and the support structure are enclosed in the vacuum-sealed cryostat to minimize heat input due to radiation and to provide an insulated operating environment for the HTS field coils. A refrigeration system or cylinder, which uses the cold circulating gas in a closed loop, continuously maintains the HTS field winding at the desired low cryogenic temperature.

One approach to keeping the field coils magnetized is to use flux-trapped superconductors to maintain the flux in the field coils. Such a superconductor is also known as a flux pump, a concept which has been discussed in the preceding section. A flux pump enables a DC current to be driven within a closed superconducting circuit without the need for a direct electrical connection to a power supply. This eliminates the thermal losses associated with metal leads conducting current. Such machines are modeled as high-field and low-resistance machines, but the torque constant is relatively high. They need minimal energy for super-cooling over an entire flight, and continuous magnetization is provided by periodic flux pumping.

7.2.1 HTS DC Motors

The first HTS motors began appearing just before the end of the last millennium and in the following years. In 1992, the Electric Power Research Institute in the US in collaboration with the Reliance Electric Company manufactured a small-scale prototype 25 W HTS DC motor [19]. Although it was initially believed that a synchronous motor with an HTS field winding in the rotor and the traditional copper armature winding in the stator was the best topology for commercial application, DC motors were also designed and found to be particularly suitable for propulsion applications in ships. Eckels and Snitchler [20] designed and tested a large HTS motor for ship propulsion. Okazaki, Sugimoto and Takeda [18] built a large HTS DC motor for ship propulsion. Siemens has developed Europe's first motor using windings formed from HTS. The motor has been operated continuously at 400 kW in both motor and generator modes [21, 22].

A major problem in the design of DC motor is the transfer of large DC currents across a rotating joint, in order to pass the current to the HTS rotor coils. Conventional commutation techniques such as slip rings and high-frequency brushless commutators all reduce performance by increasing the losses, and also increase cost and complexity. Arc erosion at high currents is a major issue with slip rings making contact with brushes, particularly at high speeds when the wear increases substantially. Inductive brushless-commutators use stabilized switched-mode power transistors built into the rotor. They can not only be unsuitable at high speed and high currents, but also reduce the reliability of the motor, due to the increase in the number of failure modes. HTS flux pumps therefore provide a suitable alternative. Another alternative is to use liquid metal current collectors. Sodium potassium (NaK) liquid metal current collectors could be used to pass the current from the stator of the machine and the rotating armature, in place of conventional solid brushes. Liquid metal current collectors can operate at higher current densities and at higher rotational speeds with better power conversion efficiencies and have a higher life expectancy.

7.2.2 HTS Synchronous and Induction Motors

In 1995, the Electric Power Research Institute in the US, in collaboration with the Reliance Electric Company, built two small-scale synchronous HTS AC motors generating about 5 Kw of power [23]. Since then a number of large synchronous motors have been built for various applications [24, 25]. A cross-sectional view of a typical HTS synchronous motor is shown in Figure 7.1. Nick, Grundmann and Frauenhofer [26] at Siemens have built and tested a 400 MW HTS motor. Hu et al. [27] have considered the 3D modeling of all-superconducting synchronous electric machines, where both the stator and rotor are designed using HTS coils, by the finite element method.

Most superconducting motors have been assembled with superconducting coils wound by using a superconducting wire. In place of the wound coil, several groups are developing large motors employing a large size magnet composed of bulk HTS material. Bause et al. [28] developed an analytical model for the electromagnetic field in the HTS equivalent of a radial-gap permanent magnet synchronous machine in which bulk HTS material is used to produce the magnetic field, instead of conventional permanent magnets in the rotor. Inácio et al. [29] have presented a lumped parameter equivalent circuit model of a bulk HTS hysteresis motor. Unlike a conventional synchronous motor, a hysteresis motor, which is a synchronous motor with

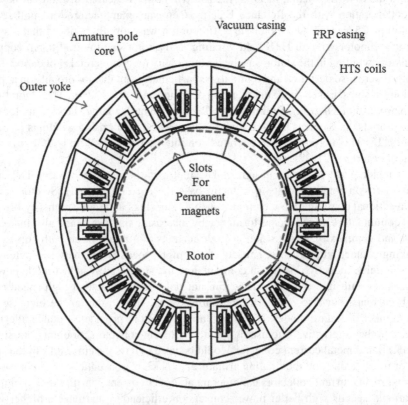

FIGURE 7.1 Cross-sectional view of a typical HTS permanent-magnet synchronous motor.

a uniform air gap, does not require DC excitation. It can be designed to operate both in a single and in a three-phase supply. The torque of such a motor is produced due to the hysteresis and eddy current induced in the rotor by the action of the rotating flux of the stator windings. Vepa [30] has discussed the modeling and dynamics of HTS motors, including synchronous motors, for aircraft propulsion applications. González-Parada [31] have discussed in detail the design and development of axial flux HTS induction motors. Their paper presents the design and optimization of a superconductor induction motor in axial flux configuration, using HTS BSCCO-2223 tapes, which are optimized to have a high critical temperature T_c. Xiao et al. [32] have discussed the modeling and analysis of a linear stator permanent-magnet machine which uses bulk HTS material.

The rotor of a synchronous HTS motor generally uses several HTS race-track or pancake coils which can generate relatively large fields in the stator-rotor air gap. These fields can only be generated by injecting currents of large magnitudes into the rotor coils. The rotor cryostat used to cool the windings to the required low temperatures must be penetrated by and/or encapsulate two or more metal leads which must carry the required current from an environment at an ambient temperature to the environment within the cryostat maintained at the cryogenic temperature. Cryocoolers generally use either liquid nitrogen or liquid helium to provide the required level of cooling. Thus, the current leads will generate a major thermal load due to both Ohmic losses and thermal conduction on the cryogenic cooling system, leading to much larger cryostats.

7.2.3 CRYOSTATS FOR HTS MOTORS

For cooling a typical HTS motor armature or field coil, typically liquid nitrogen coolant, supplied by a cryostat, is most commonly used. Although cryostats could be designed based on open or closed thermodynamic cycles, the emerging designs for aircraft propulsion seem to rely mainly on closed cycles involving either recuperative type cryostats based on the Brayton cycle or regenerative cryostats such as Gifford-McMahon designs, based on the Stirling cycle. The design of a typical sub-cooled liquid nitrogen circulation cooling system which operates at a temperature below 70 K is presented by Chen et al. [33]. A flow diagram of the sub-cooled liquid nitrogen cooling system designed by Chen et al. [33] is shown in Figure 7.2. The system consists of the HTS coils, a cryostat that houses the coils, a cryo-refrigerator, a heat exchanger in the sub-cooler bath, a cryogenic pump immersed in a liquid nitrogen reservoir which pumps the liquid nitrogen to the sub-cooled heat exchanger, a helium gas bottle for pressurizing the liquid nitrogen in the reservoir, a high-strength liquid nitrogen storage Dewar for storing and supplying the liquid nitrogen as and when required and several flow control valves for regulating the flow of liquid nitrogen and the helium gas. Figure 7.2 shows 1) the liquid nitrogen Dewar, 2) the cryostat, 3) the cryo-refrigerator, 4) the liquid nitrogen pump, 5) the helium gas bottle, 6) the cryogenic liquid nitrogen reservoir, 7) the heat exchanger and 8) the liquid nitrogen transfer lines.

Chen et al. [33] have shown that the major power loss is in cooling the coils due to the heat generated by AC current in the coils, while a further 20–25% of power is lost

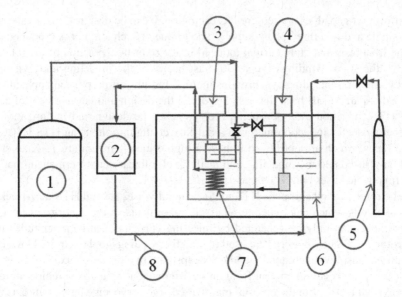

FIGURE 7.2   Flow diagram of the sub-cooled liquid nitrogen cooling system.

due to the presence of the current leads, conduction, radiation, flow control valves and ports and due to the flow in the transfer lines.

7.2.4   CONTROL OF 3-PHASE HTS PMSM

The space-vector pulse width modulation (SV PWM) technique is the preferred pulse width modulation technique of a three-phase voltage-source inverter (VSI) for the control of HTS PMSM (permanent-magnet synchronous motor) motors. The PMSM control system usually employs the field-oriented control method for decoupling the torque and the flux to ensure a fast response. The objective of the SV PWM technique is to approximate the reference output voltage instantaneously by a combination of the switching states corresponding to the basic space vectors which define the relative magnitudes of the outputs in the three phases. A typical design may be implemented by using either insulated gate bipolar and metal oxide semiconductor field effect transistors or low-cost microcontroller or custom digital signal processing hardware. The structure of a typical three-phase VSI is shown in Figure 7.3. As shown in the figure, there are three output voltages of the inverter and six power transistors that shape the outputs, controlled by six corresponding inputs. However, when one of the upper transistors is switched on, the corresponding lower transistor is switched off. Thus, the on-off states of the upper transistors alone are sufficient to define the output voltage.

The output voltages are applied to the star-connected motor windings, and the continuous inverter input voltage is a DC voltage. Since the corresponding lower transistor is switched off when an upper transistor is switched on, there are only eight different combinations of the three transistor switching states, which could be represented as (000), (100), (110), (010), (011), (001), (101) and (111). The first and last

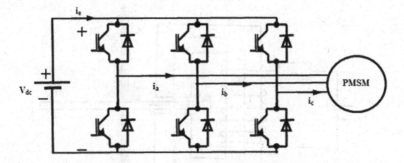

FIGURE 7.3 Three-phase VSI inverter circuit used for controlling motors.

states do not cause a current to flow to the motor, and hence, the line-to-line voltages are zero. The other six states can produce voltages to be applied to the motor. The transformation relating these switching states to the output voltages can be defined as a matrix transformation. The eight transformed vectors arising from the switching vectors are referred to as the basic space vectors, representing the output voltages. In the SV PWM, the switching times could be directly related to the magnitudes of the required output voltages and, assuming that the switches operate almost instantaneously, the outputs to the three phases are continuously generated. Usually the outputs are also filtered to remove the harmonics.

7.3 HOMOPOLAR MOTORS

Probably the most highly researched HTS machine, in the early years of the development of HTS motors, is the homopolar motor. The DC homopolar motor converts electrical energy into mechanical rotational energy by generating a Lorentz force, tangential to a coil rotating about a static permanent magnet. In a homopolar motor, like any DC electric motor with two magnetic poles, the conductors always cut unidirectional lines of magnetic flux. But unlike a conventional DC motor, this is done by rotating the conductor around a fixed axis so that the conductor is at right angles to a static magnetic field generated by the non-rotating poles. Figure 7.4 illustrates the principle of operation of Faraday's disc, invented by Michael Faraday, working as a motor. The disc, in a disc type homopolar machine, is made of aluminum or copper and is placed in the air gap of the electromagnet. It illustrates the principle of the disc type of a homopolar motor.

DC current is supplied to the disc via the brushes. The interaction of the disc current and the magnetic field between the two poles of the electromagnet generates a Lorentz force tangential to the disc, and the resulting electromagnetic torque drives the disc. One of the major issues is the high current that must be supplied to the disc in order to generate a torque. On the other hand, the advantage is the absence of a commutator although the machine does use brushes to make contact with the disc.

The structure of a modern drum type homopolar machine is shown in Figure 7.5. The stator consists of a ferromagnetic cylinder with two coils embedded within it.

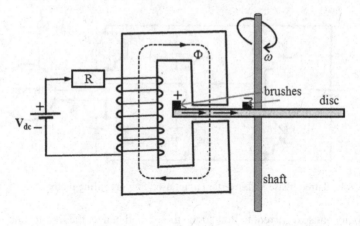

FIGURE 7.4   Principle of the operation of a homopolar motor.

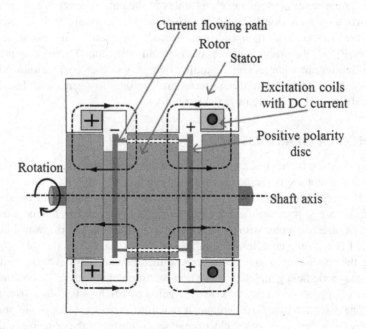

FIGURE 7.5   Structure of a modern homopolar motor.

The rotor is in the form of a ferromagnetic core on a shaft and has also a copper cage attached to it. The cage has bars placed axially within the core which are short-circuited by two rings at both ends. The two stator coils are excited by the DC current flowing in opposite directions with respect to one another. The DC current generates the magnetic flux in the stator coils that passes to the rotor through the air gap in a radial direction. The rotor rings are supplied through the brushes that are distributed uniformly around each ring. The currents that flow through the bars embedded in the rotor core are almost at right angles to the magnetic flux, giving rise to the tangential Lorentz force and the resulting torque that drives the rotor.

7.3.1 SUPERCONDUCTING HOMOPOLAR MOTORS

In 1995, United States Naval Surface Warfare Center [34] developed an HTS homopolar DC motor with a field winding construct with BSCCO HTS wire. The motor produced 125 kW of output power with its HTS field winding operating in liquid helium at a temperature of 4.2 K, and produced 91 kW of output power with the field winding cooled by liquid neon to a temperature of 28 K. In 2013, the first fully HTS homopolar motor, with both the stator and rotor consisting of HTS coils was built and tested by Lee et al. [35]. It appears that pancake coils were used for the stator winding and racetrack coils for armature winding. The HTS coils were cooled in liquid nitrogen. In a disk type machine, the disc itself is made from HTS materials such as BSCCO-2223.

7.4 DESIGN OF HTS MOTORS FOR AIRCRAFT PROPULSION

Several research and development groups worldwide are involved in the design of HTS motors for aircraft propulsion. HTS motors i.e. motors with armature and field windings made of HTS wires or tapes that are encased within a vacuum-sealed cryostat, so the temperature of the stator and rotor are in the superconductivity domain, may be used for aircraft propulsion. The cryostat is essentially a refrigeration system or cylinder which uses the cold circulating gas in a closed loop, and continuously maintains the HTS field winding at the desired low cryogenic temperature. Cryostats required for aerospace propulsion have been shown to require relatively low power as they need to mainly compensate for the power losses in the HTS coils. The design of an integrated, low power-consuming cryostat is the key to the use of HTS motors for aircraft applications. Pienkos et al. [36] describe the design of a small HTS motor, including the required cryostat that is integrated within the hub of the aircraft's propeller (see also Masson et al. [11] and Luongo et al. [12]). Vratny et al. [37] presented the design of an optimal HTS motor and the associated controller architecture with regard to efficiency and mass for all electric aircraft propulsion. They have estimated that the specific power requirements for the HTS motor could vary between 10 to 40 kW/kg with very high propulsive efficiencies, while the power required for the cryostat would be less than 0.33 kW/kg. The study clearly points to the large requirements of stored electric power in the form of batteries. Several research groups are also investigating the feasibility of different superconducting technologies not only for the storage of electric power, but also for electromagnetic launch to assist civil aircraft to take off. This would reduce the total in-flight power requirements which in turn would reduce the required size of the batteries for electric power storage. Haran et al. [38] have presented a technology roadmap for the development of superconducting machines for aircraft propulsion and other applications.

CHAPTER SUMMARY

In this chapter, the principles governing the design of HTS electric motors are briefly reviewed. The advantages of HTS motors that would become key enablers for the development of propulsive systems for electric aircraft which are not currently practical with conventional technology are considered. It is clear that all-electric aircraft flight would be feasible in the not too distant future.

REFERENCES

1. J. G. Bednorz, K. A. Müller, Possible high T_c superconductivity in the Ba-La-Cu-O system, *Z. Phys. B*, 64:189–193, 1986.
2. M. K. Wu, J. R. Ashburn, C. J. Torng, P. H. Hor, R. L. Meng, L. Gao, Z. J. Huang, Y. Q. Wang, C. W. Chu, Superconductivity at 93K in a new mixed-phase Y-Ba-Cu-O compound system at ambient pressure, *Phys. Rev. Lett.*, 58(9):908–910, 1987.
3. A. P. Drozdov, M. I. Eremets, I. A. Troyan, V. Ksenofontov, S. I. Shylin, Conventional superconductivity at 203 kelvin at high pressures in the sulfur hydride system, *Nature*, 525:73–76, 2015.
4. D. K. Thapa, S. Islam, S. K. Saha, P. S. Mahapatra, B. Bhattacharyya, T. P. Sai, R. Mahadevu, S. Patil, A. Ghosh, A. Pandey, *Coexistence of Diamagnetism and Vanishingly Small Electrical Resistance at Ambient Temperature and Pressure in Nanostructures*, eprint arXiv:1807.08572 (Condensed Matter – Superconductivity), Cornell University, Ithaca, NY, July 2018.
5. M. Somayazulu, M. Ahart, A. K. Mishra, Z. M. Geballe, M. Baldini, Y. Meng, V. V. Struzhkin, R. J. Hemley, Evidence for superconductivity above 260 K in lanthanum superhydride at megabar pressures, *Phys. Rev. Lett.*, 122(027001):027001-1–027001-6, 2019.
6. A. Goyal. *Second-Generation HTS Conductors*, Kluwer Academic Publishers, Dordrecht, The Netherlands, 2005.
7. J. Bardeen, L. N. Cooper, J. R. Schrieer, Microscopic theory of superconductivity, *Phys. Rev.*, 106(1):162–164, 1957.
8. J. Bardeen, L. N. Cooper, J. R. Schrieer, Theory of superconductivity, *Phys. Rev.*, 108(5):1175–1204, 1957.
9. L. N. Cooper, Bound electron pairs in a degenerate fermi gas, *Phys. Rev.*, 104(4):1189, 1956.
10. S. S. Kalsi (ed.), Rotating AC machines, Chapter 4. In *Applications of High Temperature Superconductors to Electric Power Equipment*, John Wiley & Sons, Inc., Hoboken, NJ, 2011.
11. P. J. Masson, D. S. Soban, E. Upton, J. E. Pienkos, C. A. Luongo, HTS motors in aircraft propulsion: Design considerations, *IEEE Trans. Appl. Supercond.*, 15(2):2218–2221, 2005.
12. C. A. Luongo, P. J. Masson, T. Nam, D. Mavris, H. D. Kim, G. V. Brown, M. Waters, D. Hall, Next generation more-electric aircraft: A potential application for HTS superconductors, *IEEE Trans. Appl. Supercond.*, 19(3):1055–1068, 2009.
13. M. Armstrong, C. Ross, D. Phillips, M. Blackwelder. *Stability, Transient Response, Control, and Safety of a High-Power Electric Grid for Turboelectric Propulsion of Aircraft*, NASA/CR-2013-217865, 2013.
14. K. L. Kovalev, D. S. Dezhin, L. K. Kovalev, V. N. Poltavets, R. I. Ilyasov, D. S. Golovanov, B. Oswald, K.-J. Best, W. Gawalek, HTS high-dynamic electrical motors, Proceedings of the IEEE/CSC & ESAS European Superconductivity News Forum (ESNF), January 2010.
15. B. Oswald, K.-J. Best, M. Soell, E. Duffner, W. Gawalek, L. K. Kovalev, G. Krabbes, W. Prusseit, HTS motor program at OSWALD, present status, *IEEE Trans. Appl. Supercond.*, 17(2):1583–1586, 2007.
16. B. Oswald, A. T. A. M. de Waele, M. Söll, T. Reis, T. Maier, J. Oswald, J. Teigelkötter, T. Kowalski, Project Sutor: Superconducting speed-controlled torque motor for 25.000 Nm, *Phys. Procedia*, 36:765–770, 2012.
17. O. Tsukamoto, Present status and future trends of R&D for HTS rotational machines in Japan, *Phys. C Supercond.*, 504:106–110, 2014.

18. T. Okazaki, H. Sugimoto, T. Takeda, Liquid nitrogen cooled HTS motor for ship propulsion, Proceedings of the Power Engineering Society General Meeting, Montreal, QC, Canada, June 18–22, 2006.

19. H. J. Chad, F. S. Rich, Design and fabrication of high temperature superconducting field coils for a demonstration DC motor, *IEEE Trans. Appl. Supercond.*, 3(1):373–376, March 1993.

20. P. W. Eckels, G. Snitchler, 5 MW high temperature superconductor ship propulsion motor design and test results, *Nav. Eng. J.*, 117(4):31–36, 2005.

21. M. Frank, J. Frauenhofer, P. Van Hasselt, W. Nick, H. W. Neumueller, G. Nerowski, Long-term operational experience with first Siemens 400 kW HTS machine in diverse configurations, *IEEE Trans. Appl. Supercond.*, 13(2):2120–2123, 2003.

22. H. W. Neumuller, W. Nick, B. Wacker, M. Frank, G. Nerowski, J. Frauenhofer, W. Rzadki, R. Hartig, Advances in and prospects for development of high-temperature superconductor rotating machines at Siemens, *Supercond. Sci. Technol.*, 19(3):S114–S117, 2006.

23. C. H. Joshi, C. B. Prum, R. F. Schiferl, D. I. Driscoll, Demonstration of two synchronous motors using high temperature superconducting field coils, *IEEE Trans. Appl. Supercond.*, 5(2):968–971, 1995.

24. Y. K. Kwon, H. M. Kima, S. K. Baik, E. Y. Lee, J. D. Lee, Y. C. Kim, S. H. Lee, J. P. Hong, Y. S. Jo, K. S. Ryu, Performance test of a 1 MW class HTS synchronous motor for industrial application, *Phys. C Supercond.*, 468(15–20):2081–2086, 2008.

25. S. Ishmael, C. Goodzeit, P. Masson, R. Meinke, R. Sullivan, Flux pump excited double-helix rotor for use in synchronous machines, *IEEE Trans. Appl. Supercond.*, 18(2):693–696, 2008.

26. W. Nick, J. Grundmann, J. Frauenhofer, Test results from Siemens low-speed, high-torque HTS machine and description of further steps towards commercialization of HTS machines, *IEEE/CSC & ESAS Eur. Supercond. News Forum*, 19:1–10, January 2012.

27. D. Hu, M. D. Ainslie, J. Zou, D. A. Cardwell, 3D modelling of all-superconducting synchronous electric machines by the finite element method, Excerpt from the Proceedings of the 2014 COMSOL Conference, Cambridge, UK.

28. R. Bause, M. D. Ainslie, M. Corduan, M. Boll, M. Filipenko, M. Noe, Electromagnetic design of a superconducting electric machine with bulk HTS material, arXiv:1903.08906v1 [physics.app-ph] March 2019.

29. D. Inácio, J. M. Pina, J. Martins, M. V. Neves, A. Álvarez, Lumped parameters equivalent circuit of a superconducting hysteresis motor, superconductivity centennial conference, *Phys. Procedia*, 36:975–979, 2012.

30. R. Vepa, Modeling and Dynamics of HTS Motors for Aircraft Electric Propulsion, *Aerospace* (MDPI), February 23, 2018, Published online, http://www.mdpi.com/2226-4310/5/1/21/pdf.

31. A. González-Parada, M. Guía, O. Ibarra, R. Guzmán, Development of axial flux HTS induction motors, International Meeting of Electrical Engineering Research ENIINVIE-2012, *Procedia Eng.*, 35:4–13, 2012.

32. F. Xiao, Y. Du, Y. Wang, M. Chen, T. W. Ching, X. Liu, Modeling and analysis of a linear stator, permanent-magnet Vernier HTS machine, *IEEE Trans. Appl. Supercond.*, 25(3):5202104, June 2015.

33. A. Chen, X. Liu, F. Xu, J. Cao, L. Li, Design of the Cryogenic System for a 400 kW Experimental HTS Synchronous Motor, *IEEE Trans. Appl. Supercond.*, 20(3):2062–2065, June 2010.

34. J. D. Waltman, J. M. Superczynski, Jr., High temperature superconducting magnet motor demonstration, *IEEE Trans. Appl. Supercond.*, 5(4):3532–3535, December 1995.

35. J. K. Lee, S. H. Park, Y. Kim, S. Lee, H. G. Joo, W. S. Kim, K. Choi, S. Y. Hahn, Test results of a 5 kW fully superconducting homopolar motor, *Prog. Supercond. Cryog.*, 15(1):35–39, 2013.

36. J. E. Pienkos, P. J. Masson, S. V. Pamidi, C. A. Luongo, Conduction cooling of a compact HTS motor for Aeropropulsion, *IEEE Trans. Appl. Supercond.*, 15(2):2150–2153, June 2005.

37. P. C. Vratny, P. Forsbach, A. Seitz, M. Hornung, Investigation of universally electric propulsion systems for transport aircraft, 29th Congress of the International Council of Aeronautical Sciences, St. Petersburg, Russia, pp 1–13, September 7–12, 2014.

38. K. S. Haran, S. Kalsi, T. Arndt, H. Karmaker, R. Badcock, B. Buckley, T. Haugan, M. Izumi, D. Loder, J. W. Bray, P. Masson, E. W. Stautner, High power density superconducting rotating machines—Development status and technology roadmap, *Supercond. Sci. Technol.* (IOP Publ.) 30(123002):1–41, 2017.

8 Aeroacoustics and Low Noise Design

8.1 AEROACOUSTIC ANALOGIES

The basic equations of motions characterizing any fluid flow are the continuity and momentum equations. In the absence of body forces, the equations of continuity and momentum may be expressed, respectively, as

$$\frac{\partial \rho}{\partial t} + \frac{\partial \left(\rho u_j \right)}{\partial x_j} = 0, \tag{8.1}$$

$$\frac{\partial \left(\rho u_i \right)}{\partial t} + \frac{\partial \left(\rho u_i u_j \right)}{\partial x_j} = -\frac{\partial p}{\partial x_i} + \frac{\partial \tau_{ji}}{\partial x_j}, \tag{8.2}$$

where, x_i ($i = 1, 2, 3$) are the Cartesian coordinates corresponding to (x, y, z), u_i is the Cartesian component of the local flow velocity vector, t is the time, p is the local pressure, ρ is the corresponding density and τ_{ij} is the viscous stress tensor.

Although these equations form the evolution for any fluid flow, in the case of turbulent flows, when it is customary to decompose the flow components into mean flow and fluctuating flow components, more variables arise. Thus, there is need for further physically relevant equations, to solve the flow field explicitly. The additional equations are generally obtained from considerations of the kinetic energy, dissipation and shear stress transport, to resolve the turbulence closure problem.

An alternative approach for solving for the aeroacoustically generated noise distributions is to consider the generation of sound by comparing the fluid flow equations in terms of the unsteady pressure components to the acoustic equations for sound generation. With aim of studying the noise generated by jet engines, Lighthill [1] introduced one of the fundamental analogies used in studying the generation of noise that is known as the acoustic analogy. The basic idea of the acoustic analogy approach is to replace the regions of unsteady fluid flow by an equivalent distribution of fundamental solutions to the acoustic propagation dynamics and thus derive linear perturbations to the base flow that generate noise. The formation of the analogy is based on manipulating the momentum and mass continuity equation to obtain a linear wave equation with non-linear forcing terms that are independent of the far-field radiation. In order to derive the acoustic analogy, consider a jet of air streaming into a quiescent medium with density ρ and at the speed of sound a. Away from that jet, the perturbation pressure distribution defining the acoustic flow field $\tilde{p}$ is known to satisfy the wave equation of the form

$$\frac{1}{a^2}\frac{d^2\tilde{p}}{dt^2} - \left(\frac{\partial^2}{\partial x_1^2} + \frac{\partial^2}{\partial x_2^2} + \frac{\partial^2}{\partial x_3^2}\right)\tilde{p} \equiv \frac{1}{a^2}\frac{d^2\tilde{p}}{dt^2} - \Delta^2\tilde{p} = S\left(x_1, x_2, x_3\right). \qquad (8.3)$$

In Equation (8.3), $S\left(x_1, x_2, x_3\right)$ is a distribution, responsible for generating the perturbation noise pressure distribution, $\tilde{p}$, $a = d\tilde{p}/d\tilde{\rho}$, where $\tilde{\rho}$ is the corresponding density distribution of the flow field. Manipulating the continuity and momentum Equations (8.1) and (8.2), one obtains the equation

$$\frac{\partial}{\partial t}\left(\frac{\partial \rho}{\partial t} + \nabla \cdot (\rho \mathbf{V})\right) = \frac{\partial}{\partial t}\left(\frac{\partial \rho}{\partial t} + \frac{\partial(\rho u_i)}{\partial x_i}\right) = 0, \qquad (8.4)$$

and

$$-\frac{\partial}{\partial x_i}\left(\frac{\partial(\rho u_i)}{\partial t} + \nabla \cdot (\rho u_i \mathbf{V})\right) = -\frac{\partial}{\partial x_i}\left(\frac{\partial(\rho u_i)}{\partial t} + \frac{\partial(\rho u_i u_j)}{\partial x_j}\right) = \frac{\partial^2 p}{\partial x_i \partial x_j}. \qquad (8.5)$$

Adding Equations (8.4) and (8.5), and replacing the density and pressure by the corresponding perturbation density and pressure,

$$\frac{\partial}{\partial t}\frac{\partial \tilde{\rho}}{\partial t} - \frac{\partial}{\partial x_i}\frac{\partial(\tilde{\rho} u_i u_j)}{\partial x_j} = \frac{\partial^2 \tilde{p}}{\partial x_i \partial x_j}. \qquad (8.6)$$

Thus Lighthill [1] was able to show that

$$\frac{1}{a^2}\frac{d^2\tilde{p}}{dt^2} - \Delta^2\tilde{p} = \frac{\partial^2}{\partial x_i \partial x_j}\left(\tilde{\rho} u_i u_j + \tilde{p} - a^2\tilde{\rho}\delta_{ij}\right) \equiv \frac{\partial^2 T_{ij}}{\partial x_i \partial x_j}, \qquad (8.7)$$

where $T_{ij} = \tilde{\rho} u_i u_j + \tilde{p} - a^2\tilde{\rho}\delta_{ij}$ is known as Lighthill's turbulence stress tensor.

As Lighthill [2] was primarily interested in jet noise, he did not include other possible distributions to the right-hand side of Equation (8.7). Lighthill conceptually used integral methods based on the Green's function (or fundamental solution) of the classical wave equation and the convolution or Duhamel's superposition integral. The solution involves the superposition of fundamental solutions. The most basic fundamental solutions to Equation (8.7) which are analogous to sources in potential flow are monopoles. By differentiating these solutions along one direction, one obtains a dipole while differentiating again a second orthogonal direction results in a quadrupole. Thus, a dipole is constructed by placing two monopoles in a line, and reducing the distance to zero and increasing their strength to maintain the strength of the dipole at unity. A quadrupole may be constructed in a similar way by using two dipoles. In order to replicate the noise generated by the turbulence stress tensor, Lighthill found it necessary to use a distribution of quadrupoles.

Lighthill [1] derived his acoustic analogy by rearranging the Navier–Stokes equations into a single equation for perturbation in the density variable. The analogy is derived by assuming the medium is at rest, with the left-hand side having the wave operator acting on the density perturbation and the right-hand side with all the other terms such as pressure and velocity. Lighthill's analogy facilitates the application of classical acoustics to predict the noise generated by an aerodynamic flow. Three distinct source terms may be identified in Lighthill's analogy which include changes in the flow velocity, entropy and viscous stresses. Thus, the effects of turbulence, temperature fluctuations due to combustion, as well as viscous friction stresses could be accounted for and modeled as sources of noise. Yet it is required that the right-hand side of Lighthill's analogy is explicitly modeled or simulated. Furthermore, several sound propagation effects such as refraction at shear and boundary layers appear on the right-hand side as equivalent sources dependent on the acoustic field, although these are kinematic effects. Thus, the interpretation of the right-hand side as source terms is questionable in cases where the flow includes any form of feedback or resonances. There are several flows which include such phenomena. To account for such situations, Lighthill's analogy has been extended in a number of ways. These include considering moving surfaces in the flow, including the effects of the motion of the base flow or considering moving media and by including different variables on the left-hand side. A key requirement in defining an analogy is to ensure that the left-hand side of the equation admits wave-like solutions in the primary variables, even in the absence of all source terms on the right-hand side. Some of the extensions proposed to Lighthill's analogy have clear physical interpretations that could be applied to engineering systems such as airfoils, wings, lifting surfaces, rotor blades, propellers and jet engines. Curle [3] and Ffowcs Williams and Hawkings [4] were one of the first to proposed extensions to Lighthill's analogy. Curle considered solid static surfaces, while Ffowcs Williams and Hawkings generalized the analogy to account for arbitrary, moving impermeable or permeable surfaces. The approach by Ffowcs Williams and Hawkings [4] uses the theory of generalized functions, and allows for a precise definition of the source terms as monopole, dipole or quadrupole sources. Moreover, all of the above analogies are based on using the perturbed density as the primary propagating variable.

The Ffowcs Williams–Hawkings (FW–H) analogy is a generalization of Lighthill's acoustic analogy. To generalize Lighthill's analogy, a volume source is added to the continuity equation, as well as an external force is added to the momentum equation on the right-hand sides. This results in an inhomogeneous form of the wave equation, given by,

$$\frac{\partial^2 \rho}{\partial t^2} - a_0^2 \frac{\partial^2 \rho}{\partial x_j^2} = \rho_0 \frac{\partial q}{\partial t} - \nabla \cdot \mathbf{F} + \frac{\partial^2 T}{\partial x_i \partial x_j}. \tag{8.8}$$

To understand the FW–H formulation, consider a moving, impenetrable body described by $(f(\mathbf{x},t) = 0)$, such that a point satisfying $f(\mathbf{x},t) > 0$ is outside the body and $\nabla f = \hat{\mathbf{n}}$ (outward normal to $f(\mathbf{x},t) = 0$). Inside the body, the fluid is at rest and with the same conditions as the formulation of Lighthill's acoustic analogy. Based on this setup, there is an artificial discontinuity at the body $(f(\mathbf{x},t) = 0)$. To take into account the jump present at the surface, the mathematical concept of derivatives of generalized functions is used to make the required corrections to the conservation laws. The generalized conservation laws of mass continuity and momentum are respectively expressed as,

$$\frac{\overline{\partial}\rho}{\partial t} + \frac{\overline{\partial}(\rho u_j)}{\partial x_j} = \rho_0 v_n \delta(f),$$

(8.9)

$$\frac{\overline{\partial}(\rho u_i)}{\partial t} + \frac{\overline{\partial}(\rho u_i u_j + P_{ij})}{\partial x_j} = l_i \delta(f),$$

(8.10)

where $v_n = -\partial f/\partial t$ is the local normal velocity at the surface of the body, $l_i = P_{ij} n_j$ is the local force intensity that acts on the fluid and $\delta(f)$ is the Dirac delta function. The bars over the derivatives denote generalized differentiation. Combining Equations (8.9) and (8.10),

$$\frac{\overline{\partial}}{\partial t}\frac{\overline{\partial}\tilde{\rho}}{\partial t} - \frac{\overline{\partial}}{\partial x_i}\frac{\overline{\partial}(\tilde{\rho}u_i u_j + \tilde{p})}{\partial x_j} = \frac{\overline{\partial}}{\partial t}\left(\rho_0 v_n \delta(f)\right) - \frac{\overline{\partial}}{\partial x_i}\left(l_i \delta(f)\right).$$

(8.11)

Introducing P_{ij} the compressive stress tensor with the constant $p_0 \delta_{ij}$ subtracted,

$$\frac{1}{a^2}\frac{d^2\tilde{p}}{dt^2} - \Delta^2 \tilde{p} = \frac{\overline{\partial}}{\partial t}\left(\rho_0 v_n \delta(f)\right) - \frac{\overline{\partial}}{\partial x_i}\left(l_i \delta(f)\right) + \frac{\partial^2\left(T_{ij}H(f)\right)}{\partial x_i \partial x_j},$$

(8.12)

where $H(f)$ is the Heaviside function.

While Ffowcs Williams and Hawkings (FW–H) generalized analogy was applicable to moving surfaces (or bodies), Curle's generalized analogy, which required the use of dipoles, and included only the second of the three terms on the right-hand side of Equation (8.12), was able to model the sound obstructed by or arising from solid boundaries. Thus, by using monopoles, dipoles and quadrupoles, one can explain sound generated from airfoils, fans and jets. Lowson's [5] frequency domain formulation in terms of rotating sources and Kirchhoff's time domain formulation of moving and rotating sources are particularly applicable to propellers and motors.

In Kirchhoff's time domain formulation, which was originally proposed by Gustav Kirchhoff [6], one defines a generalized pressure distribution on the surface of the body, which is zero inside, and uses generalized derivatives to define appropriate monopoles and dipoles. Thus, one defines a generalized inhomogeneous wave equation in a moving coordinate frame to obtain Kirchhoff's governing equation, which is given by,

$$\left(\frac{1}{a^2}\frac{\overline{\partial}^2}{\partial t^2} - \sum_i \frac{\overline{\partial}^2}{\partial x_i^2}\right)\tilde{p}(\overline{x},t) = -\left(\frac{\partial\tilde{p}}{\partial x_i}\cdot \mathbf{n}_i + \frac{1}{a}M_n\dot{\tilde{p}}\right)\delta(f)$$

$$-\frac{1}{a}\frac{\partial}{\partial t}\left[M_n\tilde{p}\delta(f)\right] - \frac{\partial}{\partial x_i}\left[\frac{\partial\tilde{p}}{\partial x_i}\cdot\mathbf{n}_i\delta(f)\right].$$

(8.13)

For propellers, Gutin [7] developed an analytical model to describe the noise generated by the steady thrust and torque of a propeller, given certain approximations. By considering the far field noise radiated by a distribution of unsteady sources over a volume, Lowson [5] was able to show that a force accelerating along the line connecting the source and observer will create a sound disturbance. In the time domain, the most popular noise prediction methods are due to Farassat [8, 9] who reformulated the FW–H equations, to convert the spatial derivative in the loading term into time derivative, thus making the numerical computation of the unsteady noise fields feasible. Farassat [10] developed four different time domain formulations, of which two involving moving observers in subsonic and supersonic stationary mediums, are used as standard techniques for unsteady noise prediction. Alternate formulations involving stationary observers in moving media have also been developed. Farassat and Myers [11] and Farassat and Posey [12] have extended Kirchhoff formulae to moving surfaces as well as presenting an alternate and useful derivation.

8.1.1 Sound Pressure Level

The directivity of a sound source refers to the manner in which the measured or predicted sound pressure, at a fixed distance r from the source, varies with angular position θ. For all plots, sound pressures are converted to sound pressure levels,

$$L_p\left(r,\ \theta\right) = 10\log\frac{\left\langle\left|p^2\left(r,\theta,t\right)\right|\right\rangle_t}{p_{ref}^2}, \tag{8.14}$$

where $\langle\ \rangle_t$ indicates a time average at a fixed (r,θ) and p_{ref} is a standard reference pressure taken to be 20 µPa. The units of $L_p(r,\theta)$, or the sound pressure level (SRP) are in decibels (dB). In addition, all sound pressure level values are normalized to the value at $\theta = 0°$, which is the accepted practice for directivity plots.

8.2 INTEGRAL METHODS OF LIGHTHILL, FFOWCS WILLIAMS AND HAWKINGS, AND KIRCHHOFF

Sources of noise may be modeled as fundamental solutions to the classical wave equation for the unsteady pressure distribution. Thus, the sound propagates from a combination of several fundamental sources, which include monopolar noise (from non-lifting thickness vibrations), dipolar noise (from unsteady lifting pressure loading, vortex shedding) and quadrupolar noise (from vortices, turbulence, jets and boundary layers). The general solution for the modified FW–H equation on a moving surface is written as the sum of 3 terms:

$$p'\left(\overline{\mathbf{x}},t\right) = p'_Q\left(\overline{\mathbf{x}},t\right) + p'_L\left(\overline{\mathbf{x}},t\right) + p'_T\left(\overline{\mathbf{x}},t\right), \tag{8.15}$$

where $p'_Q\left(\overline{\mathbf{x}},t\right)$, is the quadrupole noise, $p'_L\left(\overline{\mathbf{x}},t\right)$, is the loading or dipole noise and $p'_T\left(\overline{\mathbf{x}},t\right)$ is the thickness or monopole noise. A similar three-term decomposition for the solution of Kirchhoff's equation also exists.

In the Green's function approach, given the speed of sound in the undisturbed medium as a_0, one may define a Green's function as follows given an observer at $\bar{\mathbf{x}}$ and a source at $\bar{\mathbf{y}}$ as,

$$\left(\frac{1}{a_0^2}\frac{\partial^2}{\partial t^2}-\frac{\partial^2}{\partial x_i^2}\right)G\left(\bar{\mathbf{x}},t\right)=\delta\left(\bar{\mathbf{x}}-\bar{\mathbf{y}}\right)\delta\left(t-\tau\right),\tag{8.16}$$

where $G\left(\bar{\mathbf{x}},t\right)=0$ for $t<\tau$ and

$$G\left(\bar{\mathbf{x}},t-\tau\right)=\frac{1}{4\pi\left|\bar{\mathbf{x}}-\bar{\mathbf{y}}\right|}\delta\left(t-\tau-\frac{\left|\bar{\mathbf{x}}-\bar{\mathbf{y}}\right|}{a_0}\right),\tag{8.17}$$

otherwise. The Green's function is also known as the retarded potential, due to the presence of the time delay. Hence, given $Q\left(\bar{\mathbf{x}},t\right)$ and the equation,

$$\left(\frac{1}{a_0^2}\frac{\partial^2}{\partial t^2}-\frac{\partial^2}{\partial x_i^2}\right)P\left(\bar{\mathbf{x}},t\right)=Q\left(\bar{\mathbf{x}},t\right),\tag{8.18}$$

after integration with time,

$$P\left(\bar{\mathbf{x}},t\right)=\frac{1}{4\pi}\int_{-\infty}^{\infty}\frac{1}{\left|\bar{\mathbf{x}}-\bar{\mathbf{y}}\right|}Q\left(\bar{\mathbf{y}},t-\frac{\left|\bar{\mathbf{x}}-\bar{\mathbf{y}}\right|}{a_0}\right)d^3\bar{\mathbf{y}}.\tag{8.19}$$

To apply the method to the FW–H equation, first dealing with monopole and dipole. Thus, given,

$$\frac{\partial^2\rho}{\partial t^2}-a_0^2\frac{\partial^2\rho}{\partial x_j^2}=\rho_0\frac{\partial q}{\partial t}-\nabla\cdot\mathbf{F},\tag{8.20}$$

the general solution is,

$$\rho\left(\bar{\mathbf{x}},t\right)=\frac{1}{4\pi}\int_{-\infty}^{\infty}\frac{\rho_0}{\left|\bar{\mathbf{x}}-\bar{\mathbf{y}}\right|}\frac{\partial}{\partial t}q\left(\bar{\mathbf{y}},t-\frac{\left|\bar{\mathbf{x}}-\bar{\mathbf{y}}\right|}{a_0}\right)d^3\bar{\mathbf{y}}$$

$$-\frac{1}{4\pi}\int_{-\infty}^{\infty}\frac{1}{\left|\bar{\mathbf{x}}-\bar{\mathbf{y}}\right|}\frac{\partial}{\partial y_i}F_i\left(\bar{\mathbf{y}},t-\frac{\left|\bar{\mathbf{x}}-\bar{\mathbf{y}}\right|}{a_0}\right)d^3\bar{\mathbf{y}}.\tag{8.21}$$

Integrating by parts, the integral in the second term is,

$$\int_{-\infty}^{\infty}\frac{1}{\left|\bar{\mathbf{x}}-\bar{\mathbf{y}}\right|}\frac{\partial}{\partial y_i}F_i\left(\bar{\mathbf{y}},t-\frac{\left|\bar{\mathbf{x}}-\bar{\mathbf{y}}\right|}{a_0}\right)d^3\bar{\mathbf{y}}=\frac{\partial}{\partial x_i}\int_{-\infty}^{\infty}\frac{1}{\left|\bar{\mathbf{x}}-\bar{\mathbf{y}}\right|}F_i\left(\bar{\mathbf{y}},t-\frac{\left|\bar{\mathbf{x}}-\bar{\mathbf{y}}\right|}{a_0}\right)d^3\bar{\mathbf{y}}.\tag{8.22}$$

Thus, the complete solution is

$$
\rho(\bar{\mathbf{x}},t) = \frac{1}{4\pi} \int_{-\infty}^{\infty} \frac{\rho_0}{|\bar{\mathbf{x}}-\bar{\mathbf{y}}|} \frac{\partial}{\partial t} q\left(\bar{\mathbf{y}},t - \frac{|\bar{\mathbf{x}}-\bar{\mathbf{y}}|}{a_0}\right) d^3\bar{\mathbf{y}}
$$

$$
- \frac{1}{4\pi} \frac{\partial}{\partial x_i} \int_{-\infty}^{\infty} \frac{1}{|\bar{\mathbf{x}}-\bar{\mathbf{y}}|} F_i\left(\bar{\mathbf{y}},t - \frac{|\bar{\mathbf{x}}-\bar{\mathbf{y}}|}{a_0}\right) d^3\bar{\mathbf{y}}.
$$

(8.23)

Now let us consider the more general equation, including the Lighthill stress tensor, in the form of a volume distribution of acoustic stress

$$
\frac{\partial^2 \rho}{\partial t^2} - a_0^2 \frac{\partial^2 \rho}{\partial x_j^2} = \rho_0 \frac{\partial q}{\partial t} - \nabla \cdot \mathbf{F} + \frac{\partial^2 T_{ij}}{\partial x_i \partial x_j}.
$$

(8.24)

First observe that integral in the second term of the general solution may also be expressed as

$$
\int_{-\infty}^{\infty} \frac{1}{|\bar{\mathbf{x}}-\bar{\mathbf{y}}|} \frac{\partial}{\partial y_i} F_i\left(\bar{\mathbf{y}},t - \frac{|\bar{\mathbf{x}}-\bar{\mathbf{y}}|}{a_0}\right) d^3\bar{\mathbf{y}} = \int_{-\infty}^{\infty} \frac{1}{a_0 |\bar{\mathbf{x}}-\bar{\mathbf{y}}|} \frac{\partial}{\partial t} F_i\left(\bar{\mathbf{y}},t - \frac{|\bar{\mathbf{x}}-\bar{\mathbf{y}}|}{a_0}\right) d^3\bar{\mathbf{y}}.
$$

(8.25)

Consequently, the general solution is

$$
\rho(\bar{\mathbf{x}},t) = \frac{1}{4\pi} \int_{-\infty}^{\infty} \frac{\rho_0}{|\bar{\mathbf{x}}-\bar{\mathbf{y}}|} \frac{\partial}{\partial t} q\left(\bar{\mathbf{y}},t - \frac{|\bar{\mathbf{x}}-\bar{\mathbf{y}}|}{a_0}\right) d^3\bar{\mathbf{y}}
$$

$$
- \frac{1}{4\pi} \int_{-\infty}^{\infty} \frac{1}{|\bar{\mathbf{x}}-\bar{\mathbf{y}}|} \left(\frac{\partial}{a_0 \partial t} F_i\left(\bar{\mathbf{y}},t - \frac{|\bar{\mathbf{x}}-\bar{\mathbf{y}}|}{a_0}\right) - \frac{\partial^2}{a_0^2 \partial t^2} T_{ij}\left(\bar{\mathbf{y}},t - \frac{|\bar{\mathbf{x}}-\bar{\mathbf{y}}|}{a_0}\right) \right) d^3\bar{\mathbf{y}}.
$$

(8.26)

The extension of the general solution to the case of a moving media (with a stationary body) or the case of a moving body in a stationary media is quite straightforward. In this case, the general equation for sound generation reduces to

$$
\frac{D^2 \rho}{Dt^2} - a_0^2 \frac{\partial^2 \rho}{\partial x_j^2} = \rho_0 \frac{Dq}{Dt} - \nabla \cdot \mathbf{F} + \frac{\partial^2 T_{ij}}{\partial x_i \partial x_j}.
$$

(8.27)

In the above equation

$$
\frac{D}{Dt} = \frac{\partial}{\partial t} + U_i \frac{\partial}{\partial x_i},
$$

(8.28)

is the convective or advective derivative. It is also known as the material derivative. It can be obtained when the total derivative with respect to time is expanded by using the chain rule and the path follows the fluid current described by the fluid's undisturbed velocity vector U. Equation (8.27) is applicable to a stationary body immersed in a moving medium. For a stationary medium, $U \equiv 0$. If one assumes that $U = \begin{bmatrix} 0 & 0 & U_\infty \end{bmatrix}$, the frequency domain Green's function defined by Garrick and Watkins (1953) is given by,

$$G(S,\omega) = \frac{1}{4\pi S} \exp(jk\sigma - j\omega t), \quad j = \sqrt{-1} \tag{8.29}$$

where $k = \omega/a$. The phase and source magnitude, σ and S are defined differently depending on whether the source is moving or is stationary in a uniform flow. In the case of a moving source in a quiescent fluid, with a stationary observer,

$$S = \sqrt{\beta^2 \left(r^2 + r_1^2 - 2rr_1 \cos(\theta - \theta_1)\right) + (z - U_\infty t)^2}, \quad \sigma = (S + M_\infty z)/\beta^2, \tag{8.30}$$

with $M_\infty = U_\infty/a$, $\beta = \sqrt{1 - M_\infty^2}$, $z = x_3 - y_3$. When the propeller and observer are both stationary in a uniformly moving fluid,

$$S = \sqrt{\beta^2 \left(r^2 + r_1^2 - 2rr_1 \cos(\theta - \theta_1)\right) + z^2}, \quad \sigma = (S + M_\infty z)/\beta^2. \tag{8.31}$$

However, in the case of the moving propeller, the reference frame is chosen such that the propellers axial motion is given by, $z = U_\infty t$.

In the case of the stationary propeller in a uniform flow, given $\bar{\theta}$ is the direction of the flow, $S_1 = S/\beta^2$, $S_2 = S_1 - M_\infty z/\beta^2$ and the observer at time $\tau = t - S_2/a_0$ with $a_\theta = a_0 (1 - M_\infty \cos \bar{\theta})$, the time domain Green's function and the general solution are respectively given by

$$G(\bar{x}, t - \tau) = \frac{1}{4\pi S_1} \delta \left(t - \tau - \frac{S_2}{a_0} \right), \tag{8.32}$$

$$\rho(\bar{x}, t) = \frac{1}{4\pi} \int_{-\infty}^{\infty} \frac{\rho_0}{S_1 (1 - M_\infty \cos \bar{\theta})} \frac{\partial}{\partial t} q \left(\bar{y}, t - \frac{S_2}{a_0} \right) d^3 \bar{y} \tag{8.33}$$

$$- \frac{1}{4\pi} \int_{-\infty}^{\infty} \frac{1}{S_1 (1 - M_\infty \cos \bar{\theta})} \left(\frac{\partial}{a_\theta \partial t} F_i \left(\bar{y}, t - \frac{S_2}{a_0} \right) - \frac{\partial^2}{a_\theta^2 \partial t^2} T_{ij} \left(\bar{y}, t - \frac{S_2}{a_0} \right) \right) d^3 \bar{y}.$$

For the three-term solution of the FW–H equation,

$$p'(\bar{x}, t) = p'_Q(\bar{x}, t) + p'_L(\bar{x}, t) + p'_T(\bar{x}, t), \tag{8.34}$$

in the frequency domain, based on the Fourier series expansions, the nth harmonics of the fundamental with rotational frequency Ω in terms of the Bessel functions of the first kind are

$$\tilde{p}'_Q(\bar{x}) = \frac{Q'}{2\pi S_1} J_n \left(nM_\infty \sin \bar{\theta} \right), \quad \tilde{p}'_L(\bar{x}) = \frac{n\Omega F \cos \bar{\theta}}{2\pi a_0 S_1} J_n \left(nM_\infty \sin \bar{\theta} \right),$$

$$\tilde{p}_T'(\overline{\mathbf{x}}) = \frac{n^2\Omega^2 T \cos^2\overline{\theta}}{2\pi a_0^2 S_1} J_n(nM_\infty \sin\overline{\theta}). \tag{8.35}$$

The above equations are useful in evaluating the sound radiated by the loading on the propeller blades.

8.3 MONOPOLES, DIPOLES AND QUADRUPOLES

The characterization of noise generated by monopoles, dipoles and quadrupoles in terms of the typical sound pressure levels generated and in terms of the spatial distribution of the relative intensity of propagation or directivity are fundamental to the description of aeroacoustically generated noise.

In the first instance, consider the pressure field generated by an impulsive source in the frequency domain, defining the Green's function in the frequency domain

$$\left(\frac{\partial^2}{\partial x_i^2} + \frac{\omega^2}{a_0^2}\right)\hat{G}(\overline{\mathbf{x}},\omega) \equiv \left(\frac{\partial^2}{\partial x_i^2} + k^2\right)\hat{G}(\overline{\mathbf{x}},\omega) = \delta(\overline{\mathbf{x}} - \overline{\mathbf{y}}), \tag{8.36}$$

with, the condition,

$$\frac{\partial \hat{G}}{\partial y_n} = 0, \text{ on } S, \tag{8.37}$$

to ensure that $\hat{G}(\overline{\mathbf{x}},\omega)$ is causal. A particular solution for $\hat{G}(\overline{\mathbf{x}},\omega)$ in a fluid medium with density ρ, sound speed a and source strength Q is,

$$\hat{G}(\overline{\mathbf{x}},\omega) = \frac{AQ\rho a e^{jk|\mathbf{x}-\mathbf{y}|}}{4\pi|\mathbf{x}-\mathbf{y}|}, \quad j = \sqrt{-1}. \tag{8.38}$$

Constructing a far field expansion of $\hat{G}(\overline{\mathbf{x}},\omega)$, it may be expressed as,

$$\hat{G}(\overline{\mathbf{x}},\omega) = \frac{jkQ\rho a e^{-jkr}}{4\pi r}. \tag{8.39}$$

The corresponding pressure is

$$p(r,\theta,t) = \frac{jkQ\rho a e^{-j(kr-\omega t)}}{4\pi r}. \tag{8.40}$$

The pressure amplitude of this source is

$$p_s = |p| = \frac{kQ\rho a}{4\pi r}, \tag{8.41}$$

which is uniform in all directions. It represents the sound radiated by a typical monopole. An acoustic monopole radiates sound equally in all directions. A typical example of an acoustic monopole is a small sphere whose volume alternately expands and contracts, while maintaining its spherical shape. In practice, a sound source will act as a monopole radiating sound equally in all directions when its dimensions are much smaller than the wavelength of the sound being radiated.

Two monopoles of equal source strength, but opposite phase, and separated by a small distance d, such that $kd \ll 1$, comprise an acoustic dipole. When compared to a single monopole, there is no net flow of fluid mass by a dipole. As one source "generates" a flow, the other source of opposite phase or sink "absorbs" it and the fluid surrounding the dipole simply flows back and forth between the source and the sink. There is a net force on the fluid which causes energy to be radiated in the form of soundwaves. The far-field expression for the pressure radiated by an acoustic dipole may be written as,

$$p(r,\theta,t) = -j\frac{k^2 Q\rho a d e^{-j(kr-\omega t)}}{4\pi r}\cos\theta. \tag{8.42}$$

The amplitude of these spherically diverging pressure waves is,

$$\left| p(r,\theta,t) \right| = \frac{kQ\rho a}{4\pi r} kd\cos\theta = p_s \times kd\cos\theta. \tag{8.43}$$

The pressure radiated by a dipole may be interpreted as the product of the pressure amplitude radiated by a monopole, a term kd which relates the radiated wavelength to the source separation, and a directivity function which depends on the angle θ. A dipole does not radiate equally in all directions.

A quadrupole source consists of two identical dipoles, with opposite phase and separated by small distance e. In the case of the quadrupole, there is not only no net flux of fluid, but also no net force on the fluid. There is a net fluctuating stress on the fluid which generates the sound waves. However, since only viscous fluids support shear stresses well and quadrupoles in inviscid fluids are poor radiators of sound. For a lateral quadrupole source, the dipole axes do not lie along the same line, while for a longitudinal quadrupole source, the dipole axes do lie along the same line.

The far-field sound pressure amplitude produced by a lateral quadrupole may be written as

$$\left| p(r,\theta,t) \right| = \frac{kQ\rho a}{4\pi r} 2kd\cos\theta\ 2ke\sin\theta = p_s \times 2kd\cos\theta \times 2ke\sin\theta. \tag{8.44}$$

The pressure radiated by a lateral quadrupole may be interpreted as the product of a simple source, a term $4k^2 de$ which relates the radiated wavelength to the quadrupole source separations, and a directivity function which depends on the angle θ. There are four directions where sound is radiated very well, and four directions in which destructive interference occurs and no sound is radiated.

The far-field sound pressure amplitude produced by a longitudinal quadrupole may be written as

$$\left| p(r,\theta,t) \right| = \frac{kQ\rho a}{4\pi r} 2kd\cos\theta\ 2ke\sin\theta = p_s \times 2kd\cos\theta \times 2ke\cos\theta. \tag{8.45}$$

The pressure radiated by a longitudinal quadrupole may also be interpreted as the product of a simple source, a term $4k^2de$ which relates the radiated wavelength to the quadrupole source separations, and a directivity function which depends on the angle θ. The directivity pattern, is different from a lateral quadrupole and looks similar to that of a dipole. There are two directions in which sound is radiated extremely well, and two directions in which no sound is radiated. However, the width of the lobes is narrower than for the dipole. The directivity patterns of a monopole, dipole, lateral quadrupole and a longitudinal quadrupole are compared in Figure 8.1.

8.3.1 Tonal Characterization of Aeroacoustically Generated Noise

Aeroacoustic tones provide an alternate method of characterizing typical discrete tones generated by flow fields are generally modeled by dominant distribution of monopoles, dipoles and quadrupoles. As air passes around an object, it generates noise by the physical process of vortex shedding. These types of tones are referred

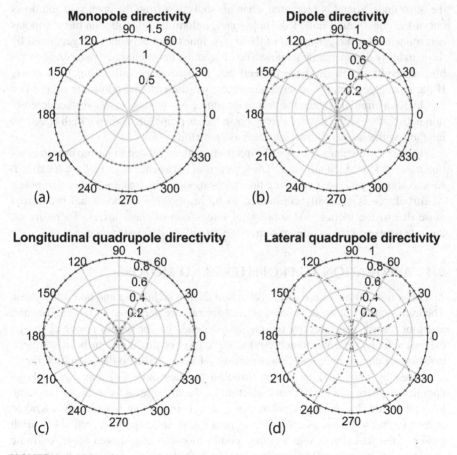

FIGURE 8.1 Directivity patters from a) Monopole b) Dipole c) Longitudinal quadrupole and d) Lateral quadrupole.

to as Aeolian tones and the dominant source of the noise is a distribution of dipoles. Such noise is generated by the Aeolian harp and by swinging objects. As air flows over an open cavity, there is a resulting hydrodynamic feedback loop due to multiple reflections of waves at the cavity boundaries which in turn results in sustained pressure oscillations. The dominant source that is responsible for such cavity generated tones is a distribution of dipoles for shallow cavities, and a distribution of monopoles for deep cavities. When a planar jet strikes an edge or a wedge, vortices are shed and there is a resulting hydrodynamic feedback loop. Consequently, there are sustained pressure oscillations and the dominant distribution of dipole is responsible for the generation of these edge tones. When a circular jet strikes a plate with a circular hole of similar dimensions to the jet, a hole-related tone is caused by the pressure pulses generated at the hole, with the jet causing the hydrodynamic feedback. Consequently, there are sustained pressure oscillations and the dominant distribution of monopoles are responsible for the generation of these hole tones which are similar to edge tones and may be considered a limiting case of cavity tones. Pipe tones are caused by an open-ended cylindrical cavity (a pipe that is open at one end and has a square hole at the other end). Sound is generated when air sucked in from the open side and flows out of the constricting hole. Like hole tones, a dominant distribution of monopoles is responsible for the generation of these pipe tones. Screech tones are generated by the instability of supersonic jet flow. The interaction between shock pulses and instabilities results in a hydrodynamic feedback loop resulting in high intensity noise. If the nozzle generating the jet is square or two-dimensional dominant source is a dipole; for a circular nozzle, the dominant source is a monopole. Jet-surface interaction noise refers to the low-frequency noise that is propagated when a high-speed jet interacts with an external surface, such as an airfoil.

Figure 8.2 illustrates the typical directivity pattern related to jet–surface interaction noise at low Mach numbers. The interaction between shocks that are localized, such as shocks generated at a wing tip, can be modeled by longitudinal quadrupoles.

Turbulence is generally considered to be responsible for non-tonal broadband noise due to the motion and subsequent interaction of shear layers. There are no discrete tones and the dominant source is a lateral quadrupole source.

8.4 APPLICATION TO PROPELLERS AND MOTORS

The peak noise from an aircraft is generated during its landing and take-off phases. The main sources of noise in conventional aircraft are generated by the engines and airframe. Noise generated by the engines depends on the eighth power of the jet exhaust velocity. In order to achieve the same level of thrust as a turbojet engine, a turbofan jet engine requires a lower exhaust jet velocity due to the additional thrust generated by the fan. This makes turbofan engines more efficient than turbojet engines, as there is a significant reduction of the noise level as well, for the same level of thrust. Moreover, turbofan engines with increasing bypass ratios tend to reduce the noise levels, as cold ducts generate less noise (proportional to the sixth power of the jet exhaust velocity) than combustion chamber-driven ducts.Airframe noise, which is comparable to engine noise, includes noise generated by high lift devices and landing gear.

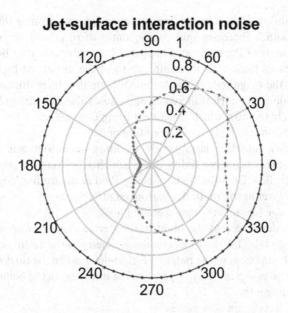

FIGURE 8.2   Directivity pattern of jet-surface interaction noise.

Sounds generated by a moving body such as a propeller may be modeled as a series of narrow band noise pulses that are summed with the time between the pulses varied inversely proportional to the speed of rotation. The Kirchhoff integral is used to produce a multi-pole expansion replicating the aeroacoustic sound of a propeller. A review of the noise generated by propellers by Marte and Kurtze [13] indicated loading sounds, thrust and torque, are one of the main sources of propeller noise. The interaction and distortion effects result in amplitude and frequency modulation of the noise and "blade slap", which are characteristic of helicopter rotor noise. Thickness noise is important at transonic tip speeds and even then, it is quite low in intensity.

Wideband noise generated by propellers is known to be analogous to Aeolian tones generated by the rotating large aspect ratio blades in the air. The mechanism behind this kind of noise is that eddies shed from the blade induce oscillatory pressures on the blade surface which in turn radiate noise which is akin to vortex noise. Vortex shedding sources of sound are the dominant sources of broadband noise. In propellers, the frequency of vortex shedding depends on the rotational speed of the blade.

8.4.1   Sources of Airfoil and Propeller Noise

There are several sources of airfoil noise [14]. These can be briefly classified as:

i) Turbulent boundary layer trailing edge noise
    The turbulent boundary layer trailing edge noise source is the most common source of noise from an airfoil, particularly at high Reynolds numbers.

ii) Separated flow noise

The possibility of flow separation begins to increase in magnitude as the angle of attack increases and consequently there is also an increase in the magnitude of the noise as it happens. Unsteady flow after the onset of stall produces noise over the entire chord of the airfoil. At higher angles of attack, the magnitude of the turbulent boundary layer thickness on the suction side of the airfoil increases and causes the formation of unsteady structures in the flow which generate trailing edge noise.

iii) Laminar boundary layer vortex shedding noise

The laminar boundary layer vortex shedding noise source is due to the Tollmien–Schlichting instability waves which arise from the positive feedback of the flow from the vortices being shed at the trailing edge into the boundary layer upstream of the trailing edge.

iv) Trailing-edge bluntness vortex shedding noise

Vortex shedding due to the blunt trailing edge can generate noise. This noise source is highly dependent on the geometry of the trailing edge and is the main component of the radiated airfoil noise when the thickness of the trailing edge is significantly larger than the thickness of the boundary layer at the trailing edge.

v) Turbulent freestream flow noise

An important source of noise is due to the interaction of the turbulent freestream flow interaction with the leading edge of airfoil.

While the far field sound power radiations due to a monopole and dipole in two dimensions depend on the third and fifth power of the distance from the source, and the corresponding radiation in three dimensions on the fourth and sixth powers, turbulent high Mach number flows depend primarily on the third power and low Mach number flows can depend on the sixth, seventh or eighth powers provided there are significant density variations, the flow is two-dimensional or uniform.

vi) Tip vortex formation noise

The vortices shed by the wingtips are another source of noise. All vortex-generated sounds may be generated from combinations of pairs of dipoles. The shock surfaces, particularly at the tip, can produce quadrupole noise from rotating high-speed propellers.

The primary periodic sources of noise in a propeller are loading noise (due to thrust and torque which can be individually modeled by dipole distributions) and thickness noise (which can be modeled by moving dipoles). A special case of unsteady loading noise is caused by blade vortex interactions in propellers. In Figure 8.3, the directivity patterns of vortex noise, torque loading noise, thrust loading and the noise due to a combination of both torque and thrust are compared. The other periodic noise is due to interactions and distortion. Broadband noise is generated by distributed vortices and turbulence. With propellers the trailing tip vortices that cause blade vortex interaction noise can also cause high levels of turbulence that generate broadband noise and this is referred to as blade wake interaction noise. Rotating propellers thus emit two distinctly different types of acoustic signatures. These and other sources are identified in Figure 8.4. The first

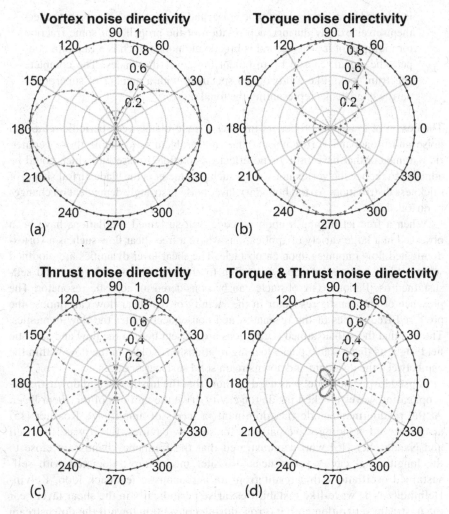

FIGURE 8.3 Directivity patters from a) Vortex b) Torque loading c) Thrust loading and d) Combined loading due to torque and thrust.

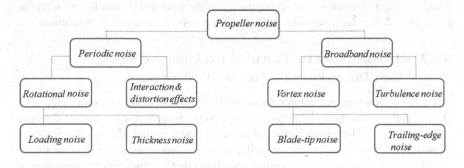

FIGURE 8.4 Propeller noise sources (from Marte and Kurtz [13]).

is referred to as tone or harmonic noise, and is caused by sources that repeat themselves exactly during each rotation of the propeller or some fraction (or multiple) of it. The second is broadband noise which is a random, non-periodic signal, caused by turbulent flow over the blades. The complete spectrum looks like a broadband spectrum with a number of superposed narrowband peaks representing the tonal noise.

The design of low-noise propellers involves a trade-off between periodic rotational noise and broadband vortex noise. Thus, the problem is to define those geometric parameters that have little or no effect on rotational noise but which could be adjusted to reduce the vortex noise. One such parameter is the blade airfoil shape, or thickness distribution. Airfoil boundary layer noise is strongly dependent on changes of airfoil shape.

When a free jet impinges upon an edge, self-sustained oscillations have been observed in a large variety of applications where a free shear flow such as a vortex-dominated flow impinges upon an obstacle. The shear layer dynamics are modified when a bluff body is introduced, leading to an enhanced organization, and self-sustained oscillations. The obstacle can be considered an acoustic resonator. The presence of an acoustic resonator in the vicinity of the shear flow can impose the preferred frequencies of the resonator, and consequently alter the flow dynamics. The fact that the self-sustained oscillations are present both in the flow and acoustic field suggests that a stable phase-locking relationship evolves in the flow within the gap between the upstream and downstream solid surfaces or edges.

A pseudo-piston model was used to show that the upstream- and downstream-propagating waves blocked inside the cavity drive the motions of the shear layer. At the predominant mode (i.e. dominant in terms of amplitude), Rossiter [15] also observed standing-wave patterns for shallow cavities. But it was Rockwell and Naudascher [16] who demonstrated that the frequency bands are close to the longitudinal cavity resonances. Rossiter modes are associated with self-sustained oscillations that result from an aeroacoustic feedback loop. Kelvin-Helmholtz-type wave-like instabilities, arise periodically in the shear layer near the upstream separation corner, grow during convection toward the downstream edge.

It can be noted that a hysteresis cycle is often observed for jet-wedge interaction, whether the jet-wedge distance, or equivalently the freestream velocity, is increased or decreased. Similar hysteresis cycles are not observed for cavity flow interactions.

8.4.2   HAMILTON-STANDARD PROCEDURE FOR ESTIMATING THE NOISE DUE TO PROPELLER AERODYNAMIC LOADING

The noise estimation method developed by Hamilton Standard facilitates [17], [18], [19], [20] the estimation of the maximum observed noise at a prescribed distance from the source. The parameters used in estimating the noise SPL are the propeller diameter, the number of blades, propeller speed, engine shaft power in horse power (1 HP = 746 W), airspeed and speed of sound. While the atmospheric attenuation of sound is not considered in the original procedure, the method has been modified by

Selfridge, Moffat and Reiss [21] to incorporate atmospheric attenuation, directionality and spherical attenuation effects.

The procedure followed is based on Selfridge, Moffat and Reiss [21] and is conveniently defined in eight distinct steps.

Step 1: The first step is to obtain a reference SPL, L_α, in dBs, which relates to the power input to the propeller:

$$L_\alpha[n] = 15.11 \log(P[n]) + 83.57, \tag{8.46}$$

where $P[n]$ is the engine power given in units of horse power (HP)

Step 2: The next step is to calculate a correction factor L_β for the number of blades and the blade diameter.

$$L_\beta = -20 \log \frac{B}{4} - 40 \log \frac{D}{4.72}, \tag{8.47}$$

where B is number of blades and D is the blade diameter.

Step 3: A correction factor L_γ, based on the rotation speed of the propeller and the distance between the propeller to what is labelled as a point of interest, is then estimated. According to the review by Marte and Kurtz [13], using the conversion factor 1 ft = 0.305 m, the correction factor is given as

$$L_\gamma = (25.12 M_T[n] - 33.40) \log \frac{0.305}{D} + 33.37 M_T[n] - 36.88, \tag{8.48}$$

where $M_T[n]$ is the blade tip Mach number.

Step 4: Following Selfridge, Moffat and Reiss [21], a directional correction may be incorporated which accounts for the directional characteristics of the loading sound sources in the propeller. This is consistent with Gutin's [7] theory of propeller aeroacoustics modeling the propeller sound as a combination of several ideal sources. The greatest acoustic intensity is in a direction of 120° and does not reduce to zero as in an ideal case. The direction correction is:

$$L_\delta[n] = 1.19 \theta[n] - 0.0053 \theta^2[n] - 62.32, \tag{8.49}$$

where $\theta[n]$ is the azimuth angle between source and observer.

Step 5: A correction factor for spherical attenuation between the source and the observer is estimated as:

$$L_\varepsilon[n] = -20 \log(3.375 r[n] - 1), \tag{8.50}$$

where r is the distance between the observer and source in meters.

Step 6: The correction factors are summed and added to the reference SPL:

$$L_\zeta[n] = L_\alpha[n] + L_\beta[n] + L_\gamma[n] + L_\delta[n] + L_\varepsilon[n]. \tag{8.51}$$

Step 7: The gain for the fundamental and the nine harmonics is calculated using:

$$H_N[n] = 26\exp\left(-\left(0.79 - 0.7M_T[n]\right)N\right) - 22, \tag{8.52}$$

where $H_N[n]$ is the gain for each harmonic and N is the harmonic number, $1 \le N \le 10$.

Step 8: The gains obtained are then individually subtracted from the total SPL in Step 6 for each of the ten individual harmonics.

$$L_{N_t}[n] = L_\zeta[n] - H_N[n]. \tag{8.53}$$

Step 9: The attenuation due to atmospheric absorption is subtracted from each of the SPL values for each harmonic:

$$L_{N_\psi}[n] = L_{N_t}[n] - A_N[n]. \tag{8.54}$$

Selfridge, Moffat and Reiss [21], have provided values for $A_N[n]$ ranging from 0 dB at low frequencies to 20 dB beyond 11.2 kHz. The SPL correction factor is converted to a gain value:

$$G_N[n] = 0.00002 \times 10^{L_{N_t}[n]/20}. \tag{8.55}$$

The frequencies of each harmonic are then calculated:

$$f_{L_N}[n] = N \times B \times RPM[n]/60. \tag{8.56}$$

This information is used to construct ten band pass filters which together are used by Selfridge, Moffat and Reiss [21] to simulate the noise due to the aerodynamic loading.

8.5 THEORETICAL MODELING OF THE NOISE FIELDS

8.5.1 THEORETICAL MODELING OF THE PROPELLER NOISE FIELDS

The theory of propeller noise is closely related to the problem of determining the steady and unsteady loads on the propeller blade. The principal difference is that in propeller noise theory, the focus is directed towards the far field. However, for the surface loading problem the focus is directed at the near field. In the near field case certain boundary conditions must also be satisfied, particularly that the blade surface must be considered to be impenetrable.

Typical propeller noise is characterized by tonal noise involving a superposition of several tones that occur at the blade passing frequency (BPF) and its harmonics. The blade passing frequency is the frequency at which a blade passes a fixed point in the propeller's plane of rotation and is a function of the number of blades in the propeller and rotational speed of the propeller, Ω. Discrete tones or rotational tonal noise is caused by the periodic disturbances generated as the blades cut through the air. The primary source of this noise is the steady loading on the propeller blades due to the thrust and torque distributions along each blade. As seen in Chapter 6, each element of the propeller blade has a pressure distribution across the blade cross-section as the blade rotates through the air. The pressure distribution can be resolved into the axial thrust force F_T and a circumferential component. The resultant of the circumferential components is the net torque Q, on the rotor.

Consider the far field at a distance r_0 from the propeller disc center and position in space defined by the azimuth angle δ, and the polar coordinate θ for the nth harmonic of the BPF for a single propeller with N blades. In terms of the three components of forces, Y_1, Y_2 and Y_3 acting at a point, the velocity potential due to a stationary source of noise in fact is modeled as a set of three dipoles of differing strengths acting in three directions and is,

$$\phi = \frac{j\exp(-jkr_0)}{4\pi k\rho a_0 r_0}\left(Y_1\frac{\partial}{\partial y_1} + Y_2\frac{\partial}{\partial y_2} + Y_3\frac{\partial}{\partial y_3}\right), \quad j = \sqrt{-1}. \tag{8.57}$$

The location of a field point relative to a small element on the blade can be expressed as $r_0 - r\cos\theta\sin\delta$. Thus, assuming the blades are thin, the total far-field velocity potential due increments in the thrust force dF_T and the net torque dQ is obtained by the principle of superposition and is given by

$$\phi_{nN} = \frac{\exp\left(jk(a_0 t - r_0)\right)}{4\pi^2\rho a_0 r_0}\iint L\exp\left(j(kr\sin\delta\cos\theta - nN\theta)\right)dr d\theta, \tag{8.58}$$

with the loading L given by

$$L = \frac{dF_T}{dr}\cos\delta + \frac{1}{r}\frac{dQ}{dr}\sin\delta\sin\theta. \tag{8.59}$$

After integration with respect to θ, the corresponding far-field pressure amplitude due to the nth harmonic is given by

$$P_{nN} = \left|\rho\frac{\partial\phi_{nN}}{\partial t}\right| = \frac{nN\Omega}{2\sqrt{2}\pi a_0 r_0}\int_{r_h}^{R}\left\{-\frac{dF_T}{dr}\cos\delta + \frac{a_0}{\Omega r^2}\frac{dQ}{dr}\right\}J_{nN}(kr\sin\delta)dr. \tag{8.60}$$

Similarly, for the far-field noise due to the pressure distribution generated by the blade thickness is

$$P_{nN} = \frac{\rho N\Omega^2}{2\sqrt{2}\pi r_0}\int_{r_h}^{R}k_c t_b c J_{nN}(kr\sin\delta)dr, \tag{8.61}$$

where k_c is a correction factor for the finite solidity of the rotor, t_b is the blade thickness and c is the blade chord.

Consider the forward motion of the propeller disc and reconsider the expression for the velocity potential given by equation (8.57). Given the normal velocity on the airfoil surface v_n, and force vector applied to the fluid F, the radiated thickness and loading noise or pressure disturbances, are respectively given by

$$p_T = \frac{1}{4\pi} \frac{D}{Dt} \int\limits_{-\infty}^{\infty} \int\limits_{A} \frac{\rho v_n}{S_0} \delta\left(t - \tau - \frac{\sigma_0}{a}\right) dA d\tau, \tag{8.62}$$

$$p_L = -\frac{1}{4\pi} \nabla \cdot \int\limits_{-\infty}^{\infty} \int\limits_{A} \frac{F}{S_0} \delta\left(t - \tau - \frac{\sigma_0}{a}\right) dA d\tau, \tag{8.63}$$

where

$$S_0 = \sqrt{\beta^2\left((x_1 - y_1)^2 + (x_2 - y_2)^2 + (x_3 - y_3)^2\right)}, \tag{8.64}$$

and

$$\sigma_0 = \left(S_0 + M_\infty(x_3 - y_3)\right)/\beta^2. \tag{8.65}$$

In the above Equations (8.62), the convective or advective derivative is defined by Equation (8.28).

In the frequency domain, when the propeller and observer are both stationary relative to a uniformly moving fluid

$$\phi = \frac{j\exp(-jk\sigma_0)}{4\pi k\rho a_0 S_0}\left(Y_1 \frac{\partial}{\partial y_1} + Y_2 \frac{\partial}{\partial y_2} + Y_3 \frac{\partial}{\partial y_3}\right). \tag{8.66}$$

The pressures due to the nth harmonics in thickness and loading noise are

$$p_T = -jn\Omega\rho \frac{1}{4\pi} \frac{D}{Dt} \int\limits_{r_h}^{R} \int\limits_{0}^{2\pi} \frac{\exp(j(k\sigma_0 - n\Omega t - n\theta))}{S_0} h_n(r) d\theta dr,$$

$$p_L = -\frac{1}{4\pi} \nabla \cdot \int\limits_{r_h}^{R} \int\limits_{0}^{2\pi} \frac{\exp(j(k\sigma_0 - n\Omega t - n\theta))}{S_0} F_n(r) r d\theta dr, \tag{8.67}$$

where $h_n(r)$ is the nth harmonic corresponding to the propeller blade normal velocity which is equal to

$$v_n = -jn\Omega h_n(r)\exp(j(-n\Omega t)). \tag{8.68}$$

Using the non-dimensional pressure, length and velocity after dividing by ρa_0^2, R and a_0 respectively, the pressures due to the nth harmonics in thickness and loading noise due to the propeller blade may be expressed as

$$p_T = -n^2 M_{tip}^2 e^{jn\theta} I\left(r, \theta, z, n, M_{tip}, M_\infty\right), \qquad (8.69)$$

$$p_L = e^{jn\theta} L\left(r, \theta, z, n, M_{tip}, M_\infty\right) - jne^{jn\theta} Q\left(r, \theta, z, n, M_{tip}, M_\infty\right), \qquad (8.70)$$

where, following Carley [22],

$$I = \int_0^{2\pi} \int_{\bar{r}_h}^1 \frac{\exp\left(j\left(k\sigma_0 - n\theta\right)\right)}{4\pi S_0^3} \left(S_0 \sigma_0 + j\frac{M_\infty z}{M_{tip} n}\right) h_n(r) r dr d\theta, \qquad (8.71a)$$

$$L = \int_0^{2\pi} \int_{\bar{r}_h}^1 \frac{\exp\left(j\left(k\sigma_0 - n\theta\right)\right)}{4\pi S_0^3} \left(z - j\frac{kS_0}{\beta^2}\left(z + M_\infty S_0\right)\right) g_n(r) r dr d\theta, \qquad (8.71b)$$

and

$$Q = \frac{1}{r} \int_0^{2\pi} \int_{\bar{r}_h}^1 \frac{\exp\left(j\left(k\sigma_0 - n\theta\right)\right)}{4\pi S_0} q_n(r) r dr d\theta. \qquad (8.71c)$$

In the above Equations (8.71a–c), $\bar{r}_h$ is the non-dimensional hub radius, M_{tip} is the blade tip rotational Mach number, g_n and q_n are the thrust and circumferential drag components of F_n and the acoustic wave number is $k = M_{tip} n$. Carley [22] has discussed the evaluation of I, L and Q. Carley [22] has also presented formulas for evaluating I, L and Q.

Based on the work of Garrick and Watkins [23], Runyan [24] presents the approximate formula for the loading noise on an N bladed propeller as,

$$p = -\frac{1}{2\pi S_0} j^{nN} e^{j\omega t - \frac{jk}{\beta^2}(S_0 + M_\infty x)} J_{nN}\left(\frac{kR_e z}{S_0}\right)\left(F_T \frac{jk}{\beta^2}\left(M_\infty + \frac{x}{S_0}\right) - \frac{jnN}{R_e^2} Q\right), \qquad (8.72)$$

where R_e is about 0.8 of the blade tip radius, J_{nN} is a Bessel function of the first kind, x and y are the lateral and vertical coordinates in the propeller disc plane and z is the coordinate along the forward axis.

In addition to the tonal noise, there is also a significant amount of broadband noise present which is specified by a spectral density function. Typical examples of such spectral density functions are the spectra associated with the trailing and leading edge noise generated by the blades and presented in the original contributions of Amiet [25, 26], Amiet and his co-workers [27–30] and Roger and Moreau [31–33]. In the case of airfoil noise, as with the trailing and leading-edge noise generated by the blades, the unsteady blade section flow field is decomposed into the sum of a convective gust part and an acoustic response flow field characterized by a velocity distribution which can be derived from a velocity potential. The velocity potential ϕ is harmonic in time and spatially satisfies the convected Helmholtz equation,

$$\left\{\left(1-M^2\right)\frac{\partial^2}{\partial x^2}+\frac{\partial^2}{\partial y^2}+\frac{\partial^2}{\partial z^2}+2jM\frac{\omega}{a_0}\frac{\partial}{\partial x}+\frac{\omega^2}{a_0^2}\right\}\phi=0. \tag{8.73}$$

The boundary conditions are the zero normal velocity boundary condition on the airfoil surface as well as conditions that must be imposed far upstream of the flow. By applying a transformation of the velocity potential, $\phi=\varphi\exp\left(j\omega x M\big/\left(a_0\beta^2\right)\right)$, the preceding equation is reduced to

$$\left\{\left(1-M^2\right)\frac{\partial^2}{\partial x^2}+\frac{\partial^2}{\partial y^2}+\frac{\partial^2}{\partial z^2}+\frac{\omega^2}{a_0^2\left(1-M^2\right)}\right\}\varphi=0. \tag{8.74}$$

The solution is obtained by reducing it to a convolution integral in terms of the Kussner indicial response function or by the application of Fourier transforms.

8.5.2 FARASSAT'S FORMULATION OF THE FW–H EQUATION

An alternate method of estimating the far field noise is by Farassat's [10, 34] retarded-time formulation of the FW–H equation, known as formulation 1A, could be employed, to estimate the loading and thickness noise generated by the propeller. The quadrupole terms, which are essentially non-linear, are generally neglected, and hence the contribution of the quadrupole source term has been dropped from the FW–H equation. The linear terms in Farassat's formulation are the thickness and loading noise.

i) Thickness noise can be expressed in terms of the components of outward normal velocity of the blade surface as

$$4\pi p_T'\left(\vec{x},t\right)=\iint_S\left[\frac{\rho_0\left(\dot{V}_n+V_{\dot{n}}\right)}{r\left(1-M_r\right)^2}\right]_{ret}dS_y$$

$$+\iint_S\left[\frac{\rho_0 V_n\left(r\dot{M}_r+a_0\left(M_r-M^2\right)\right)}{r^2\left(1-M_r\right)^3}\right]_{ret}dS_y, \tag{8.75}$$

where $r=\left|\mathbf{x}-\mathbf{y}\left(\tau\right)\right|$ is the source-to-observer distance and the quantity $\vec{M}$ of magnitude M is the vector Mach number of a source point on the blade surface S, which moves with an outward normal velocity V_n. The blade surface is chosen as integration surface. The dotted quantities denote time derivative with respect to the source (retarded) time $\tau=t-\left|\mathbf{x}-\mathbf{y}\left(\tau\right)\right|\big/a_0$. The quantity M_r is the projection of the vector Mach number $\vec{M}$ in the direction of the observer which is the direction in which the sound radiates. Quantities in the square brackets are evaluated at the retarded time. The subscripts a, n and r denote, respectively, quiescent fluid quantities,

quantities projected along the direction of the surface normal $\hat{n}$, and quantities projected along the direction of the source-to-observer.

ii) Loading noise can be expressed in terms of the pressure forces acting on the blade surface as

$$4\pi p'_L(\vec{x},t) = \frac{1}{a_0} \iint_S \left[\frac{\dot{F}_r}{r(1-M_r)^2} \right]_{ret} dS_y + \iint_S \left[\frac{F_r - F_M}{r^2(1-M_r)^2} \right]_{ret} dS_y$$

$$+ \frac{1}{a_0} \iint_S \left[\frac{F_r(r\dot{M}_r + a_0(M_r - M^2))}{r^2(1-M_r)^3} \right]_{ret} dS_y,$$

(8.76)

where $\vec{F} = (p - p_0)\hat{n}$ is the pressure force acting on the blade surface S, while F_r is the projection of the pressure force in the sound radiation direction and $F_M = (p - p_0)\mathbf{M} \cdot \hat{n}$ is the pressure force projected along the direction of the surface velocity, normal to the surface.

8.5.3 FORMULATION OF THE FAR-FIELD NOISE BASED ON A ROTATING SOURCE

An alternate approach for obtaining the noise distribution radiated by an open rotor is to use a distribution of rotating sources. The "point source" model uses a distribution of rotating monopole and dipole sources over the surface of each blade of an open rotor and can be applied to the problem of estimating both the independent rotor and the interaction tones caused by the effect of scattering of open rotor tones by an adjacent cylindrical fuselage.

One is interested in the problem of predicting the sound field that arises from an open rotor that is attached either to the fuselage or the wing, using a large pylon. The pylon is assumed to be attached to the open rotor center body which is modeled as a cylinder. The problem is formulated as follows. First, the incident pressure field due to a distribution of rotating, single-frequency, monopole or dipole point sources is determined using Fourier transform methods. This is the pressure field that would be present in the absence of the open rotor center body cylinder. It is convenient to determine the incident field in terms of a moving reference frame which is expressed in terms of cylindrical polar coordinates as $\begin{bmatrix} x & y & z \end{bmatrix} = \begin{bmatrix} r\cos\varphi & r\sin\varphi & z \end{bmatrix}$, centered on the source's axis of rotation. In this moving frame, it is assumed that the source is located in the plane $z=0$, and there is a mean axial flow with velocity $-U$ in the positive z-direction. The incident field impinging on a solid object produces scattered waves. The complete solution for the pressure distribution is given by the sum of the incident and the scattered waves. The method of solution is based on McAlpine and Kingan [35]. In this section, only a brief outline of the methodology is presented.

As a consequence of the combination of the forward and circular motion, a rotating point source moves along a helical path, with its center on an axis of rotation which lies parallel to the axis of the cylinder. The cylinder is of infinite length and is rigid. Along the axis of rotation, the axis of the cylinder is at a horizontal distance of $b\cos\beta$ and a vertical distance of $b\sin\beta$. The radius of the helical path is r_a, and the radius of the cylinder is r_0, where $r_a + r_0 < b$. The angular velocity of the point source is Ω and its axial velocity is U. In a uniform stationary medium with constant sound speed a_0 and density ρ_0, from Equation (8.20), the acoustic pressure field $p'(\mathbf{x},t)$ due to a monopole or dipole source satisfies the inhomogeneous wave equation,

$$\frac{\partial^2 p'}{\partial t^2} - a_0^2 \frac{\partial^2 p'}{\partial^2 x_j} = \rho_0 \frac{\partial q}{\partial t} - \nabla \cdot \mathbf{F}, \tag{8.77}$$

where, for a monopole source, $q = q(\mathbf{x},t)$ is the volume flow rate per unit volume, or for a dipole source $\mathbf{F} = \mathbf{F}(\mathbf{x},t)$ is the force per unit volume. Only single-frequency subsonic sources are considered. Since the sources are moving,

$$q(\mathbf{x},t) = Q_0(\mathbf{x})e^{j\omega_0 t}\delta(\mathbf{x} - \mathbf{X}(t)), \quad \mathbf{F}(\mathbf{x},t) = \mathbf{F}_0(\mathbf{x})e^{j\omega_0 t}\delta(\mathbf{x} - \mathbf{X}(t)), \tag{8.78}$$

where $X(t)$ is the position vector of the source at time t, $Q_0(x)$ is the magnitude of the initial volume flow rate of the monopole source and $F_0(x)$ is the initial force vector at time $t = 0$. The incident field for a rotating single-frequency monopole source is found by solving for the pressure in cylindrical coordinates, where the time derivatives are replaced by the convective derivative given by Equation (8.28). Thus, the pressure field due to the incident wave satisfies

$$\frac{D^2 p_i'}{Dt^2} - a_0^2 \left(\frac{\partial^2 p_i'}{\partial r^2} + \frac{1}{r}\frac{\partial p_i'}{\partial r} + \frac{1}{r}\frac{\partial^2 p_i'}{\partial \varphi^2} + \frac{\partial^2 p_i'}{\partial z^2} \right) = \rho_0 \frac{Dq}{Dt}, \tag{8.79}$$

where, from Equation (8.28),

$$\frac{D}{Dt} = \frac{\partial}{\partial t} - U\frac{\partial}{\partial z}. \tag{8.80}$$

Furthermore, given the position of the point source at $r = r_a$, $\varphi = \Omega t$ and $z = 0$, the volume source is given by

$$q(r,\varphi,z,t) = Q_0 \frac{e^{j\omega_0 t}}{r}\delta(r - r_a)\delta(z)\sum_{n=-\infty}^{\infty}\delta(\varphi - \Omega t - 2\pi n). \tag{8.81}$$

The solution of Equation (8.79) is found using Fourier transform methods.

Introduce the Fourier transform in z and t, and the Fourier series in φ and express the solution for the incident field by the pair of equations:

$$\tilde{p}_m'(r,k_z,\omega) = \int_{-\infty}^{\infty}\int_{-\pi}^{\pi}\int_{-\infty}^{\infty} p'(r,\varphi,z,t)\exp\big(i \times (k_z z + m\varphi - \omega t)\big)dz d\varphi dt, \tag{8.82a}$$

$$p'(r,\varphi,z,t) = \frac{1}{(2\pi)^3}\sum_{m=-\infty}^{\infty}\tilde{P}_m(r)\exp(-im\varphi), \tag{8.82b}$$

with,

$$\tilde{P}_m(r) = \int_{-\infty}^{\infty}\int_{-\infty}^{\infty} \tilde{p}'_m(r,k_z,\omega)\exp(-i\times(k_z z - \omega t))dk_z d\omega. \qquad (8.82c)$$

Thus, derivatives are transformed as

$$\frac{\partial}{\partial z} = -jk_z, \quad \frac{\partial}{\partial \varphi} = -jm \text{ and } \frac{\partial}{\partial t} = j\omega. \qquad (8.83)$$

With the application of the Fourier transform, the Equation (8.79) for the incident field transforms to

$$\frac{\partial^2 \tilde{p}_{im}}{\partial r^2} + \frac{1}{r}\frac{\partial \tilde{p}_{im}}{\partial r} + \left\{\left(\frac{\omega}{a_0} + k_z\frac{U}{a_0}\right)^2 - k_z^2 - \left(\frac{m}{r}\right)^2\right\}\tilde{p}'_{im} = Q_m, \qquad (8.84)$$

$$Q_m = -jQ_0\rho_0 a_0\left(\frac{\omega}{a_0} + k_z\frac{U}{a_0}\right)\frac{2\Omega(-1)^{m+1}\sin\pi\left\{\dfrac{\omega-\omega_0}{\Omega}\right\}}{m\Omega - [\omega - \omega_0]}$$

$$\times \frac{\delta(r - r_a)}{r}\sum_{n=-\infty}^{\infty}\delta(\omega - [\omega_0 + n\Omega]). \qquad (8.85)$$

The solution of Equation (8.85) can be obtained using the method of variation of parameters. Taking $Q_m = 0$, then Equation (8.85) is in the form of Bessel's differential equation i.e.

$$\frac{\partial^2 \tilde{p}_{im}}{\partial r^2} + \frac{1}{r}\frac{\partial \tilde{p}_{im}}{\partial r} + \left\{\Gamma^2 - \left(\frac{m}{r}\right)^2\right\}\tilde{p}'_{im} = Q_m. \qquad (8.86)$$

where the radial wave number Γ is defined as

$$\Gamma^2 = \left(\frac{\omega}{a_0} + k_z\frac{U}{a_0}\right)^2 - k_z^2. \qquad (8.87)$$

The equation encountered here is similar in form to the equation for the perturbation velocity potential encountered in Chapter 6, although Equation (6.134) has additional terms. The solution for Equation (8.86) may be expressed as a superposition of the Bessel and Hankel functions, $J_m(\Gamma r)$ and $H_m^{(2)}(\Gamma r)$ respectively. Thus, the solution is expressed in terms of a convolution integral as

$$\tilde{p}'_{im} = J_m(\Gamma r)\left\{A_m(k_z,\omega) - j\frac{\pi}{2}\int_0^r Q_m(s)H_m^{(2)}(\Gamma s)sds\right\} \qquad (8.88)$$

$$+ H_m^{(2)}(\Gamma r)\left\{B_m(k_z,\omega) + j\frac{\pi}{2}\int_0^r Q_m(s)J_m(\Gamma s)sds\right\},$$

where A_m and B_m are only functions of k_z and of ω.

Introducing the inverse Fourier time transform, $\vec{p}'_{im}$ defined by

$$\vec{p}'_{im} = \frac{1}{2\pi} \int\limits_{-\infty}^{\infty} \tilde{p}'_{im}\left(r,k_z,\omega\right)\exp\left(j\omega t\right)d\omega, \qquad (8.89)$$

and following the procedure given by McAlpine and Kingan [35] and transforming the solution to alternate cylindrical polar coordinates defined as $\begin{bmatrix}\bar{x} & \bar{y} & \bar{z}\end{bmatrix} = \begin{bmatrix}\bar{r}\cos\bar{\varphi} & \bar{r}\sin\bar{\varphi} & \bar{z}\end{bmatrix}$ and fixed in the pylon, the solution for the incident field $\vec{p}'_{im}\left(\bar{r},k_z,t\right)$ may be expressed in terms of Bessel and Hankel functions. The reference frames defining the location of the rotating point source, the field point and the geometry of the pylon, which is modeled as a cylinder, are shown in Figure 8.5.

The pressure field due to the scattered waves satisfies the convected wave equation

$$\frac{D^2 p'_s}{Dt^2} - a_0^2\left(\frac{\partial^2 p'_s}{\partial \bar{r}^2} + \frac{1}{\bar{r}}\frac{\partial p'_s}{\partial \bar{r}} + \frac{1}{\bar{r}}\frac{\partial^2 p'_s}{\partial \bar{\varphi}^2} + \frac{\partial^2 p'_s}{\partial \bar{z}^2}\right) = 0. \qquad (8.90)$$

It is the homogeneous version of Equation (8.79) expressed in the cylinder's (the pylon's) polar coordinate system $\begin{bmatrix}\bar{r} & \bar{\varphi} & \bar{z}\end{bmatrix}$. Applying a procedure similar to that applied in the case of the incident field and following McAlpine and Kingan [35], the solution for the scattered field $\vec{p}'_{sm}\left(\bar{r},k_z,t\right)$ is also found. The total field is obtained by superposition which is,

$$\vec{p}'_{tm}\left(\bar{r},k_z,t\right) = \vec{p}'_{im}\left(\bar{r},k_z,t\right) + \vec{p}'_{sm}\left(\bar{r},k_z,t\right). \qquad (8.91)$$

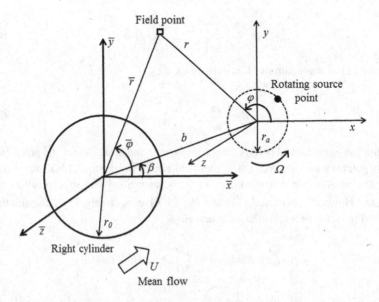

FIGURE 8.5   The reference frames defining the location of the rotating point source, the field point and the geometry of the pylon cylinder.

The incident pressure field was shown by McAlpine and Kingan [35] to take the form,

$$\bar{P}'_{im}\left(\bar{r},k_z,t\right) = \sum_{n=-\infty}^{\infty} S_n\left(k_z\right)\exp\left(j\beta m\right)J_n\left(\Gamma_{0n}a\right)G\exp\left(j\left(\omega_0 + n\Omega\right)t\right), \qquad (8.92)$$

where,

$$G = \begin{cases} \exp\left(-j\beta n\right)H^{(2)}_{m-n}\left(\Gamma_{0n}b\right)J_m\left(\Gamma_{0n}\bar{r}\right), & r > a, \bar{r} < b, \\ \exp\left(-j\bar{\phi}n\right)H^{(2)}_{m+n}\left(\Gamma_{0n}\bar{r}\right)J_m\left(\Gamma_{0n}b\right), & r \gg b, \bar{r} \gg b. \end{cases}$$

The scattered pressure field is transformed similar to the incident field as

$$\bar{p}'_s\left(\bar{r},\bar{\phi},\bar{z},t\right) = \frac{1}{\left(2\pi\right)^2}\sum_{m=-\infty}^{\infty}\exp\left(-jm\bar{\phi}\right)\left(\int_{-\infty}^{\infty}\bar{P}'_{sm}\left(\bar{r},k_z,t\right)\exp\left(-jk_z\bar{z}\right)dk_z\right). \qquad (8.93)$$

The scattered pressure field is defined by

$$\bar{p}'_{sm}\left(\bar{r},k_z,t\right) = \frac{1}{2\pi}\int_{-\infty}^{\infty} C_m\left(k_z,\omega\right)H^{(2)}_m\left(\Gamma\bar{r}\right)\exp\left(j\omega t\right)d\omega. \qquad (8.94)$$

After applying the appropriate boundary conditions, to evaluate $C_m(k_z,\omega)$, the solution for the case of rotating ring of sources may be found. For several other examples and applications, the reader is referred to the original paper by McAlpine and Kingan [35] as well as the original papers by Hanson [36–38] and by Gaffney, McAlpine and Kingan [39] and McAlpine, Gaffney and Kingan [40].

8.5.4 LILLEY'S ANALOGY AND ITS APPLICATION TO DUCTS

By considering non-zero base flows, one could extend the acoustic analogy to moving media. Lilley [41] extended Lighthill's analogy to account for transversely sheared mean effects in the base flow. Lilley derived his analogy by retaining on the left-hand side of the analogy a third order, non-linear operator acting on the logarithm of the pressure. This non-linear operator could be approximated by the linear Pridmore-Brown operator [42]. The source terms corresponding to the linear Pridmore-Brown equation [43] appearing on the right-hand side could also be derived. The exact source terms on the right-hand side of Lilley's analogy with the Pridmore-Brown operator are given in Colonius et al. [44], which also gives a simpler approximate form of the source terms, based on the work in Goldstein [45]. Moreover, Goldstein [46] also showed that for small fluctuations in the flow, the logarithm of pressure could be replaced by the pressure itself. Powell [47] formulated an aeroacoustic analogy which highlights the significance of vorticity as an acoustic source. In this formulation, the Lamb vector, which is the cross-product of the vorticity vector and the velocity vector, acts as the source. Powell's analogy is an approximate version of Lighthill's analogy where the independent variable is expressed in terms of the pressure and the right-hand side in terms of the vorticity vector. It can also be considered to be an approximation of Lilley's analogy and naturally leads to the vortex theory of sound. To obtain Powell's equation, it is further assumed that the fluid is incompressible inside the source region. The vortex sound theory reduces the source term only to the region where vorticity is not negligible which is typically smaller than the source region described by Lighthill's analogy. Other extensions consider the acoustic analogy acting on different variables. These include Howe's [48] and Doak's analogies [49, 50] that are applicable to highly compressible flows as well as other extensions which are only applicable to specific flows. Finally, Morfey and Wright [51] introduced the idea of developing non-linear acoustic analogies that can be used with methods of computational aero acoustics. Based on an aeroacoustic analogy proposed by Posson and Peake [52], Mathews [53] has modeled the pressure modes of a turbofan engine and obtained them numerically by approximating the modes using Chebyshev polynomials.

The dynamics of a compressible fluid is described by the Navier–Stokes equations which consist of the triple, the conservation of mass or mass balance equation, the conservation of momentum or momentum balance equation and the energy balance equation and are respectively given by,

$$\frac{\partial \rho}{\partial t} + \nabla \cdot (\rho \mathbf{v}) = \dot{m}, \tag{8.95a}$$

$$\frac{\partial \rho \mathbf{v}}{\partial t} + \nabla \cdot (\rho \mathbf{v} \mathbf{v}) + \nabla p = \nabla \cdot \tau + \mathbf{f} + \dot{m} \mathbf{v}, \tag{8.95b}$$

$$\frac{\partial \rho e_t}{\partial t} + \nabla \cdot (\rho e_t \mathbf{v}) + \nabla \cdot (p \mathbf{v}) = -\nabla \cdot \mathbf{q} + \nabla \cdot (\tau \mathbf{v}) + \mathbf{f} \cdot \mathbf{v} + \dot{m} e_t, \tag{8.95c}$$

where ρ represents the fluid density, $\dot{m}$ is source of mass, v its velocity vector, p the pressure, $e_t = e + (1/2) \mathbf{v} \cdot \mathbf{v}$ the specific total energy, made up of specific internal energy e and the specific kinetic energy, τ denotes the friction-related stress tensor, f is a source of force and the heat flux vector is denoted by q. It has been implicitly assumed that there are no heat sources.

An alternate set of equations are obtained by multiply the mass balance equation by the velocity vector v and subtracting it from the momentum balance equation and by taking the dot product of the momentum balance equation with v as well as the product of the mass balance equation with the specific internal energy e and subtracting them both from the energy balance equation. Thus, one has,

$$\frac{D\rho}{Dt} + \rho \nabla \cdot \mathbf{v} = \dot{m}, \tag{8.96a}$$

$$\rho \frac{D\mathbf{v}}{Dt} + \nabla p = \nabla \cdot \tau + \mathbf{f}, \tag{8.96b}$$

$$\rho \frac{De}{Dt} + p\nabla \cdot \mathbf{v} = \Phi_D - \nabla \cdot \mathbf{q}, \tag{8.96c}$$

where Φ_D is a viscous dissipation function and henceforth

$$D/Dt = \partial/\partial t + \mathbf{v} \cdot \nabla \tag{8.97}$$

denotes the substantial or material derivative. Recognizing the relationship between the specific internal energy, pressure, density, temperature and entropy, s may be expressed as,

$$\frac{De}{Dt} = \frac{p}{\rho^2} \frac{D\rho}{Dt} + T \frac{Ds}{Dt}, \tag{8.98}$$

and using the alternate mass balance equation, one may write the last equation in the alternate set as

$$\rho T \frac{Ds}{Dt} = \Phi_D - \nabla \cdot \mathbf{q} - \frac{p}{\rho} \dot{m}. \tag{8.99}$$

Finally using the fact that the density is a function of the pressure and entropy, one may write,

$$\frac{D\rho}{Dt} = \frac{1}{a^2} \frac{Dp}{Dt} - \frac{\rho}{c_p} \frac{Ds}{Dt}, \tag{8.100}$$

where a is the local speed of sound. The preceding equation is expressed as

$$\rho \frac{Ds}{Dt} = -c_p \frac{D\rho}{Dt} + \frac{c_p}{a^2} \frac{Dp}{Dt} = -c_p \dot{m} + c_p \rho \nabla \cdot \mathbf{v} + \frac{c_p}{a^2} \frac{Dp}{Dt}. \tag{8.101}$$

Hence one obtains for the substantial derivative of the pressure, the equation

$$\frac{1}{a^2} \frac{Dp}{Dt} + \rho \nabla \cdot \mathbf{v} = \frac{1}{c_p T} (\Phi_D - \nabla \cdot \mathbf{q}) + (1 - p/\rho c_p T) \dot{m}. \tag{8.102}$$

With $\dot{m} \equiv 0$, and using Equation (8.101), the equation for the substantial derivative of the pressure is

$$\frac{1}{\rho a^2} \frac{Dp}{Dt} + \nabla \cdot \mathbf{v} = \frac{1}{\rho c_p T} (\Phi_D - \nabla \cdot \mathbf{q}) = \frac{1}{c_p} \frac{Ds}{Dt}. \tag{8.103}$$

Using the fact that the speed of sound may be expressed as, $a = \sqrt{\gamma p/\rho}$, where γ is the ratio of specific heats, one obtains

$$\frac{1}{\gamma p} \frac{Dp}{Dt} + \nabla \cdot \mathbf{v} = \frac{1}{c_p} \frac{Ds}{Dt}. \tag{8.104}$$

One now defines a new variable as

$$\Pi = \frac{1}{\gamma} \ln \frac{p}{p_\infty}. \tag{8.105}$$

Hence,

$$\frac{D\Pi}{Dt} = \frac{1}{\gamma p}\frac{Dp}{Dt} = -\nabla \cdot \mathbf{v} + \frac{1}{c_p}\frac{Ds}{Dt}. \tag{8.106}$$

However, with $f=0$,

$$\frac{D\mathbf{v}}{Dt} = \frac{\partial \mathbf{v}}{\partial t} + \mathbf{v}\cdot\nabla\mathbf{v} = -\frac{\nabla p}{\rho} + \frac{1}{\rho}\nabla\cdot\tau = -a^2\nabla\Pi + \frac{1}{\rho}\nabla\cdot\tau. \tag{8.107}$$

Hence,

$$\nabla\cdot\frac{D\mathbf{v}}{Dt} = \frac{\partial(\nabla\cdot\mathbf{v})}{\partial t} + \nabla\cdot(\mathbf{v}\cdot\nabla\mathbf{v}) = -\nabla\cdot(a^2\nabla\Pi) + \nabla\cdot\left(\frac{1}{\rho}\nabla\cdot\tau\right), \tag{8.108}$$

$$\nabla\cdot\frac{D\mathbf{v}}{Dt} = \frac{D(\nabla\cdot\mathbf{v})}{Dt} + \nabla\cdot(\mathbf{v}\cdot\nabla\mathbf{v}) - \mathbf{v}\cdot\nabla(\nabla\cdot\mathbf{v})$$

$$= -\nabla\cdot(a^2\nabla\Pi) + \nabla\cdot\left(\frac{1}{\rho}\nabla\cdot\tau\right). \tag{8.109}$$

Moreover,

$$\nabla\cdot(\mathbf{v}\cdot\nabla\mathbf{v}) - \mathbf{v}\cdot\nabla(\nabla\cdot\mathbf{v}) = \text{trace}\left\{(\nabla\mathbf{v})^T(\nabla\mathbf{v})\right\} \equiv (\nabla\mathbf{v})^T : (\nabla\mathbf{v}). \tag{8.110}$$

Thus,

$$\frac{D(\nabla\cdot\mathbf{v})}{Dt} = \nabla\cdot\frac{D\mathbf{v}}{Dt} - (\nabla\mathbf{v})^T : (\nabla\mathbf{v})$$

$$= -(\nabla\mathbf{v})^T : (\nabla\mathbf{v}) - \nabla\cdot(a^2\nabla\Pi) + \nabla\cdot\left(\frac{1}{\rho}\nabla\cdot\tau\right). \tag{8.111}$$

Taking the substantial derivative of Equation (8.106),

$$\frac{D^2\Pi}{Dt^2} = -\frac{D}{Dt}(\nabla\cdot\mathbf{v}) + \frac{D}{Dt}\left(\frac{1}{c_p}\frac{Ds}{Dt}\right)$$

$$= (\nabla\mathbf{v})^T : (\nabla\mathbf{v}) + \nabla\cdot(a^2\nabla\Pi) - \nabla\cdot\left(\frac{1}{\rho}\nabla\cdot\tau\right) + \frac{D}{Dt}\left(\frac{1}{c_p}\frac{Ds}{Dt}\right). \tag{8.112}$$

Upon re-arranging the preceding equation, one obtains the Phillips analogy [54],

$$\frac{D^2\Pi}{Dt^2} - \nabla \cdot \left(a^2 \nabla \Pi\right) = \left(\nabla \mathbf{v}\right)^T : \left(\nabla \mathbf{v}\right) - \nabla \cdot \left(\frac{1}{\rho} \nabla \cdot \tau\right) + \frac{D}{Dt}\left(\frac{1}{c_p}\frac{Ds}{Dt}\right). \tag{8.113}$$

Again, taking the substantial derivative of the first term on the right-hand side of the Phillips analogy,

$$\frac{D}{Dt}\left(\left(\nabla \mathbf{v}\right)^T : \left(\nabla \mathbf{v}\right)\right) = 2\left(\nabla \mathbf{v}\right)^T : \nabla\left(\frac{D\mathbf{v}}{Dt}\right) - 2\left(\nabla \mathbf{v}\right)^T : \left(\nabla \mathbf{v} \nabla \mathbf{v}\right). \tag{8.114}$$

Hence,

$$\frac{D}{Dt}\left(\left(\nabla \mathbf{v}\right)^T : \left(\nabla \mathbf{v}\right)\right) = -2\left(\nabla \mathbf{v}\right)^T : \nabla\left(a^2 \nabla \Pi - \frac{1}{\rho} \nabla \cdot \tau_1\right) - 2\left(\nabla \mathbf{v}\right)^T : \left(\nabla \mathbf{v} \nabla \mathbf{v}\right). \tag{8.115}$$

Taking the substantial derivative of the Phillips analogy,

$$\frac{D}{Dt}\left(\frac{D^2\Pi}{Dt^2} - \nabla \cdot \left(a^2 \nabla \Pi\right)\right) + 2\left(\nabla \mathbf{v}\right)^T : \nabla\left(a^2 \nabla \Pi\right) = -2\left(\nabla \mathbf{v}\right)^T : \left(\nabla \mathbf{v} \nabla \mathbf{v}\right) + \Psi, \tag{8.116}$$

with,

$$\Psi = 2\left(\nabla \mathbf{v}\right)^T : \nabla\left(\frac{1}{\rho} \nabla \cdot \tau\right) - \frac{D}{Dt}\left\{\nabla \cdot \left(\frac{1}{\rho} \nabla \cdot \tau\right)\right\} + \frac{D^2}{Dt^2}\left(\frac{1}{c_p}\frac{Ds}{Dt}\right). \tag{8.117}$$

The above equation is called "Lilley's analogy". Just as in the case of Lighthill's analogy, Lilley's analogy is nothing but a re-arrangement of the general conservation equations of the flow. Lilley's equation is fundamentally non-linear and also not self-adjoint. The principle of superposition does not hold and a common method adopted to find approximate solutions to it is to linearize it around a time independent base flow, subtracting the base flow balance equation and adding all the terms removed from the linearized left-hand side to the source terms at the right-hand side. The resulting linear equation is thus still exact, with all of the non-linear terms being interpreted as source terms. Alternately one could linearize the basic equations for mass, momentum and energy balance and derive a linear equation for the pressure perturbation to the base flow.

Thus defining the base flow substantial or material derivative as, $D_0/D_0 t = \partial/\partial t + \mathbf{V}_0 \cdot \nabla$, where $\mathbf{V}_0 = U_0(y,z)\mathbf{e}_x$, the mass balance, momentum balance and pressure equations, in the absence of viscous stresses and the source terms, respectively are

$$\frac{D_0\rho}{D_0 t} + \rho \nabla \cdot \mathbf{V}_0 + \nabla \cdot \left(\rho_0 \mathbf{v}\right) = 0, \tag{8.118a}$$

$$\rho_0 \frac{D_0\mathbf{v}}{D_0 t} + \rho_0 \left(\mathbf{v} \cdot \nabla\right)\mathbf{V}_0 + \rho\left(\mathbf{V}_0 \cdot \nabla\right)\mathbf{V}_0 + \nabla p = 0, \tag{8.118b}$$

$$\frac{1}{a_0^2}\frac{D_0 p}{D_0 t} = \frac{D_0 \rho}{D_0 t} + \mathbf{v}\cdot\nabla\rho_0. \tag{8.118c}$$

For the assumed mean base flow distribution, the equation for the pressure reduces to

$$D_0^3 p + 2a_0^2 \frac{\partial}{\partial x}\left(\nabla U_0 \cdot \nabla p\right) - D_0 \nabla\cdot\left(a_0^2 \nabla p\right) = 0. \tag{8.119}$$

The preceding equation is the linearized equivalent of the Lilley equation. Assuming a wave-like solution for the pressure p, so,

$$p(x,y,z,t) = P_{0b} + \exp\left(j\omega t - jkx\right)P(y,z), \quad \Omega = \omega - kU_0, \tag{8.120}$$

where P_{0b} is the pressure in the base flow, one obtains the generalized Pridmore-Brown equation given by

$$\frac{\Omega^2}{a_0^2}\nabla\cdot\left(\frac{a_0^2}{\Omega^2}\nabla P\right) + \left(\frac{\Omega^2}{a_0^2} - k^2\right)P = 0. \tag{8.121}$$

Assuming cylindrical symmetry and using cylindrical polar coordinates, with

$$P(y,z) = P_m(r)\exp\left(-jm\theta\right), \tag{8.122}$$

$$\frac{\Omega^2}{ra_0^2}\left(\frac{ra_0^2}{\Omega^2}P_m'\right)' + \left(\frac{\Omega^2}{a_0^2} - k^2 - \frac{m^2}{r^2}\right)P_m = 0. \tag{8.123}$$

Thus, expressing the pressure and density respectively as,

$$p(x,y,z,t) = P_{0b} + P_m(r)\exp\left(j\omega t - jm\theta - jkx\right), \tag{8.124}$$

$$\rho(x,y,z,t) = \rho_{0b} + \rho_m(r)\exp\left(j\omega t - jm\theta - jkx\right), \tag{8.125}$$

the Pridmore-Brown equation, for the pressure modes $P_m(r)$, reduces to

$$P_m'' + \left(\frac{1}{r} + 2\frac{a_0'}{a_0} + 2\frac{kU_0'}{\Omega}\right)P_m' + \left(\frac{\Omega^2}{a_0^2} - k^2 - \frac{m^2}{r^2}\right)P_m = 0. \tag{8.126}$$

The density can then be obtained from the solution for the pressure.

It is now important to specify the boundary conditions. In general, one may observe that the sound field at large distances from the wave centers i.e. $r \to \infty$, so in terms of the velocity potential ϕ is governed by the same wave operator as the pressure. Moreover, in the far field the pressure and acoustic normal particle velocity are proportional. Thus, one may consider the proportionality constant as an acoustic impedance which is defined as the product of the density and the velocity of sound and is, $Z \equiv p/v_n$. The acoustic impedance provides a frequency-dependent ratio of pressure and normal velocity at the bounding surface. An acoustically passive boundary exhibits a non-negative resistance $\text{Re}(Z) \geq 0$ and a boundary that absorbs sound has a strictly positive impedance $\text{Re}(Z) > 0$. The two special cases of an ideal opening and a hard wall correspond to impedances $Z_{\text{open}} = 0$ and $|Z_{\text{wall}}| = \infty$, receptively. The impedance perceived by an acoustic wave traveling within a fluid is called specific impedance and is given by $Z \equiv p/v_n = \rho_0 a_0$. The impedance is often normalized by the specific value and the normalized impedance $z = Z/\rho_0 a_0$ is denoted by the lower-case variable, z. The boundary condition for a wall with impedance Z follows from matching fluid and solid normal displacement over an infinitely thin shear layer in the vicinity of the boundary, rather than normal velocity, and is known as the Ingard–Myers boundary condition [55, 56]. Rienstra [57, 58] has adopted such an approach and shown that the boundary condition at the outer wall of a cylindrical duct of radius $r = r_a$ can be expressed as

$$j\omega Z p'_m\big|_{r=r_a} = \rho_{0b} a_0^2 \Omega^2 p_m\big|_{r=r_a}. \tag{8.127}$$

At $r=0$, the solution is assumed to be regular.

Reinstra [57, 58] has studied the prediction and stability of duct modes and presented results for non-uniform mean flow profiles, described by Pridmore-Brown equations for an axi-symmetric configuration of a circular cylindrical lined duct. Numerical solutions were constructed based on the Galerkin method with Chebyshev basis functions, leading to a non-linear eigenvalue problem that was solved by a Newton-type approach. For further details, the reader is referred to Reinstra's papers [57, 58].

In the absence of a mean flow and with a_0 a constant, Equation (8.109) reduces to,

$$D_0^2 p - a_0^2 \nabla^2 p = 0. \tag{8.128}$$

The general solution for the scattered pressure modes then takes the form

$$p(r,x,\theta,t) = \exp(j\omega t + jm\theta) J_m(\alpha_{m,n} r/r_a)$$
$$\times (A_{m,n} \exp(-jk_{m,n}x) + B_{m,n} \exp(jk_{m,n}x)), \tag{8.129}$$

where (m,n) are integers representing the azimuthal and radial modal numbers respectively, J_m are the Bessel functions of order m, $\alpha_{m,n}$ are constants obtained by solving an eigenvalue problem for the pressure modes of the cylinder, $A_{m,n}$ and $B_{m,n}$ are the wave amplitudes travelling in the positive and negative x directions and the axial wave numbers are given by

$$k_{m,n} = \sqrt{\frac{\omega^2}{a_0^2} - \frac{\alpha_{m,n}^2}{r_a^2}}. \tag{8.130}$$

In addition, there is the noise due to a moving dipole which contributes both the near-field and to the far-field noise. The far-field noise is composed of the noise due to the fluctuating forces and the steady forces (Gutin noise). The former consists of a periodic part and a random part.

An application of the above formulation is to the propagation of pressure modes in a duct, where the source of the noise is a propeller at one end of the duct. This application is particularly relevant to ducted propellers used for the propulsion of electric aircraft. In this case the work of Tyler and Sofrin [59] related to the sound generated by rotors in ducts is applicable. The representations of pressure modes in cylindrical ducts of finite length, have also been used to study the acoustic wave stability and control within combustion chambers (see, for example, Yazar, Caliskan and Vepa [60]).

In the case of ducted propellers, the sound field is generated both by the basic flow as well the distribution of forces; the propeller applies to the flow. The resulting sound field depends also on the acoustic response due to the moving dipoles ("Green's functions") representing the blade forces, which depends on the surrounding system where the propeller operates. The periodic part of the noise from a ducted propeller due to the fluctuating forces is repeated at each blade passage. This creates a tonal spectrum, with multiple tones at several harmonics of the blade passing frequency. In applications to ducted propellers, tuned liners (noise suppressors) may be used within the duct to damp the propagating tonal noise before it radiates to the surroundings. The methodology of Mathews [53] based on the analogy of Posson and Peake [52] may also be applied, in principle, to a ducted propeller and the Wentzel–Kramers–Brillouin (WKB) method for solving a differential equation with a small parameter may be used to obtain the acoustic field related to the tonal noise, at the high frequencies.

The random part of the noise from a ducted propeller due to the fluctuating forces is related to unsteady turbulent inflow and the flow separation. This creates a broad-band spectrum. Considering the physics of the problem, the broadband noise produced by a ducted propeller is influenced by several factors and is not easy to predict. On one hand, the propeller itself generates noise by its very presence in the turbulent mean flow, while on the other, it can also generate flow instabilities from the upstream flow or the boundary layers separated from the duct walls, which contribute to the turbulence and overall noise generation. Based on the work of Hlaváček [61], the spectral density of the broadband noise can be determined from the integral spectral relation

$$\Phi_{p,R} = \frac{\pi z_0 a_0}{\omega^2} \int_0^R U^2(r)\{\Phi_{p,p}(r,\omega) + \Phi_{p,s}(r,\omega)\}\, dr. \qquad (8.131)$$

The functions denoted $\Phi_{p,p}(r,\omega)$ and $\Phi_{p,s}(r,\omega)$ are the acoustic pressure-related power spectral densities on the pressure and suction side of the blades respectively which could be determined either experimentally or computationally. When measurements of the spectral response are used, the data must be curve fitted to obtain a continuous representation of the spectral densities. This methodology has often been adopted in predicting the broadband noise generated by ducted fans. Thus, the complete acoustic field generated by a ducted propeller may be estimated, in principle.

CHAPTER SUMMARY

Propeller induced noise is a key feature of propellers that are expected to be used for the propulsion of future electric aircraft. For this reason, the underlying theories of propeller induced noise prediction and their application to low noise design of propellers, are discussed at some length in this chapter.

REFERENCES

1. M. J. Lighthill, On sound generated aerodynamically. I. General theory, *Proc. R. Soc.*, 211A:564–587, 1952.
2. M. J. Lighthill, On sound generated aerodynamically. II. Turbulence as a source of sound, *Proc. R. Soc.*, 222A:1–32, 1954.
3. N. Curle, The influence of solid boundaries upon aerodynamic sound, *Proc. R. Soc.*, 231A:505–514, 1955.
4. J. E. Ffowcs Williams, D. L. Hawkings, Sound generation by turbulence and surfaces in arbitrary motion, *Phil. Trans. R. Soc. Lond. A*, 264(1151):321–342, 1969.
5. M. V. Lowson, Theoretical study of compressor noise. NASA CR 1287, 1969.
6. G. R. Kirchhoff, Zur Theorie der Lichtstrahlen. *Ann. Phys. Chem.*, 18(4):663–695, 1883.
7. L. Gutin, On the sound field of a rotating propeller, NACA TM 1195, 1936.
8. F. Farassat, Theory of noise generation from moving bodies with an application to helicopter rotors. NASA Technical Report TR–451, Washington, DC, 59 p, 1975.
9. F. Farassat, Linear acoustic formulas for calculation of rotating blade noise, *AIAA J.*, 19(9):1122–1130, 1981.
10. F. Farassat, Theoretical analysis of linearized acoustics and aerodynamics of advanced supersonic propellers, AGARD-CP-366, (10), pp 1–15, 1985.
11. F. Farassat, M. K. Myers, Extension of Kirchhoff's formula to radiation from moving surfaces, *J. Sound Vib.*, 123(3):451–460, 1988.
12. F. Farassat, J. Posey, A fast method of deriving the Kirchhoff formula for moving surfaces, *J. Acoust. Soc. Am.*, 122(5):2965, 2007.
13. J. E. Marte, D. W. Kurtz, A review of aerodynamic noise from propellers, rotors, and lift fans, Technical Report 32-1462, Jet Propulsion Laboratory, NASA, 1970.
14. B. Magliozzi, D. B. Hanson, R. K. Amiet, Propeller and propfan noise, In Hubbard H. H. (Ed.). *Aeroacoustics of Flight Vehicles: Theory and Practice*, Vol. 1: Noise Sources, NASA Ref. Publ. 1258, 1991.
15. J. E. Rossiter, Wind-tunnel experiments on the flow over rectangular cavities at subsonic and transonic speeds, Aeronautical Research Council Reports and Memoranda, Technical Report 3438, 1964.
16. D. Rockwell, E. Naudascher, Review self-sustaining oscillations of flow past cavities, *J. Fluids Eng.*, 100(2):152, 1968.
17. R. Worobel, M. G. Mayo, Advanced general aviation propeller study, NASA CR 114399, Hamilton Standard, Windsor Locks, CT, December 1971.
18. F. W. Barry, B. Magliozzi, Noise detectability prediction method for low tip speed propellers, Hamilton Standard, Technical Report AFAPL-TR-71-37, June 1971.
19. D. Brown, J. B. Ollerhead, Propeller noise at low tip speeds, AFAPL TR 71 55, Wyle Laboratories, Hampton, VA, September 1971.
20. F. B. Metzger, B. Magliozzi, *New Directions in Aircraft Propulsor Noise Research*, No. 750515, Society of Automotive Engineers, New York, 1975.
21. R. Selfridge, D. Moffat, J. D. Reiss, Physically derived sound synthesis model of a propeller, Proceedings of the AM '17, London, UK, August 23–26, 2017, 8 pages, doi:10.1145/3123514.3123524.

22. M. Carley, Sound radiation from propellers in forward flight, *J. Sound Vib.*, 225(2):353–374, 1999.

23. I. E. Garrick, C. E. Watkins, A theoretical study of the effect of forward speed on the free-space sound pressure field around propellers, NACA Report 1198, 1953.

24. H. L. Runyan, Unsteady lifting surface theory applied to a propeller and helicopter rotor, Doctoral Thesis, Submitted in partial fulfilment of the requirements for the award of Doctor of Philosophy by Loughborough University, Loughborough, UK, 1973.

25. R. K. Amiet, Acoustic radiation from an airfoil in a turbulent stream, *J. Sound Vib.*, 41(4):407–420, 1975.

26. R. K. Amiet, High frequency thin-airfoil theory for subsonic flow, *AIAA J.*, 14(8):1076–1082, 1976.

27. R. W. Paterson, R. K. Amiet, Acoustic radiation and surface pressure characteristics of an airfoil to incident turbulence. NASA Contractor Report CR-2733, 1976.

28. R. W. Paterson, R. K. Amiet, Noise of a model helicopter rotor due to ingestion of turbulence, NASA CR-3213, 1979.

29. R. H. Schlinker, R. K. Amiet, Helicopter trailing edge noise, NASA CR-3470, 1981.

30. R. K. Amiet, J. C. Simonich, R. H. Schlinker, Rotor noise due to atmospheric turbulence ingestion. Part II: Aeroacoustic results, *J. Aircr.*, 27(1):15–22, 1990.

31. M. Roger, S. Moreau, Broadband self-noise from loaded fan blades, *AIAA J.*, 42(3):536–544, 2004.

32. M. Roger, S. Moreau, Back-scattering correction and further extensions of Amiet's trailing-edge noise model, *J. Sound Vib.*, 286(3):477–506, 2005.

33. S. Moreau, M. Roger, Competing broadband noise mechanisms in low-speed axial fans, *AIAA J.*, 45(1):48–57, 2007.

34. F. Farassat, Prediction of advanced propeller noise in the time domain, *AIAA J.*, 24(4):578–584, 1986.

35. A. McAlpine, M. J. Kingan, Far-field sound radiation due to an installed open rotor, *Int. J. Aeroacoust.*, 11(2):213–245, 2012.

36. D. B. Hanson, Compressible helicoidal surface theory for propeller aerodynamics and noise, *AIAA J.*, 21(6):881–889, 1983.

37. D. B. Hanson, Near-field frequency-domain theory for propeller noise, *AIAA J.*, 23(4):499–504, 1985.

38. D. B. Hanson, Sound from a propeller at angle of attack: A new theoretical viewpoint, *Proc. Math. Phys. Sci.*, 449(1936):315–328, 1995.

39. J. Gaffney, A. McAlpine, M. J. Kingan, Sound radiation of fan tones from an installed turbofan aero-engine: Fuselage boundary-layer refraction effects, Proceedings of the 22nd AIAA/CEAS Aeroacoustics Conference, Paper no. AIAA-2016-2878, Lyon, France, 30 May–1 June 2016.

40. A. McAlpine, J. Gaffney, M. Kingan, Near-field sound radiation of fan tones from an installed turbofan aero-engine, *J. Acoust. Soc. Am.*, 138(3):131–1324, 2015.

41. G. M. Lilley, On the noise from jets. Noise mechanism, AGARD-CP-131, pp. 13.1–13.12, 1974.

42. M. E. Goldstein, An exact form of Lilley's equation with a velocity quadrupole/temperature dipole source term, *J. Fluid Mech.*, 443:231–236, 2001.

43. D. Pridmore-Brown, Sound propagation in a fluid flowing through an attenuating duct, *J. Fluid Mech.*, 4(4):393–406, 1958.

44. T. Colonius, S. K. Lele, P. Moin, Sound generation in a mixing layer, *J. Fluid Mech.*, 330:375–409, 1997.

45. M. E. Goldstein, Aeroacoustics of turbulent shear flows, *Annu. Rev. Fluid Mech.*, 16(1):263–285, 1984.

46. M. E. Goldstein. *Aeroacoustics*, McGraw–Hill, New York, 293 pages, 1976.

47. A. Powell, Theory of vortex sound, *J. Acoust. Soc. Am.*, 36(1):177–195, 1964.

48. M. S. Howe, Contributions to the theory of aerodynamic sound, with application to excess jet noise and the theory of the flute, *J. Fluid Mech.*, 71(4):625–673, 1975.

49. P. E. Doak, Fluctuating total enthalpy as a generalized acoustic field, *Acoust. Phys.*, 41(5):677–685, 1995.

50. P. E. Doak, Fluctuating total enthalpy as the basic generalized acoustic field, *Theoret. Comput, Fluid Dyn.*, 10:115–133, 1998.

51. C. L. Morfey, M. C. M. Wright, Extensions of Lighthill's acoustic analogy with application to computational aeroacoustics, *Proc. Math. Phys. Eng. Sci.*, 463(2085):2101–2127, September 8, 2007.

52. H. Posson, N. Peake, The acoustic analogy in an annular duct with swirling mean flow, *J. Fluid Mech.*, 726:439–475, 2013.

53. J. R. Mathews. Mathematical modelling of noise generation in turbofan aero-engines using Green's functions, PhD Dissertation, Darwin College, University of Cambridge, Cambridge, July 2016.

54. O. M. Phillips, On the generation of sound by supersonic turbulent shear layers, *J. Fluid Mech.*, 9(1):1–28, 1960.

55. K. U. Ingard, Influence of fluid motion past a plane boundary on sound reflection, absorption, and transmission, *J. Acoust. Soc. Am.*, 31(7):1035–1036, 1959.

56. M. K. Myers, On the acoustic boundary condition in the presence of flow, *J. Sound Vib.*, 71(3):429–434, 1980.

57. S. W. Rienstra, Solutions and properties of the Pridmore-Brown equation, AIAA-2019-2594, 25th AIAA/CEAS Aeroacoustics Conference, May 20–24, 2019, Delft, The Netherlands, pp 1–38.

58. S. W. Rienstra, Fundamentals of duct acoustics, VKI Lecture Series 2016-02. Notes of course "Progress in Simulation, Control and Reduction of Ventilation Noise", November 16–18, 2015. ISBN-13 978-2-87516-098-0, 2016.

59. J. M. Tyler, T. G. Sofrin, Axial flow compressor noise studies, *Trans. Soc. Automot. Eng.*, 70:309–332, 1962.

60. I. Yazar, F. Caliskan, R. Vepa, Influence of flame dynamics on the optimal control of combustion with uncertainties, *Combust. Sci. Technol.*, 190(6):983–1006, 2018, doi:10 .1080/00102202.2017.1423293.

61. D. Hlaváček, Methods of ducted fan aircraft propulsion unit noise prediction, Technical Transactions, Mechanics, Czech Technical University, Prague, pp 155–165, 2013.

9 Principles and Applications of Plasma Actuators

9.1 FLOW CONTROL AND PLASMA ACTUATION

Flow control can be defined as modifying the flow field around the airfoil primarily to decrease drag. This could be achieved by using different flow control techniques, such as plasma actuators, blowing and suction, morphing wing, changing the shape of the airfoil and other techniques.

Plasma is state of matter such as ionized gases or ionized air where positively charged ions of the gas (or air), negatively charged electrons and neutral particles carrying no charge can co-exist in an equilibrium state, but are not bonded together. In such a state they can easily be accelerated by the application of an electric or magnetic field. Plasma actuators have been built and used as thrusters for spacecraft in deep space. Under certain circumstances, air may be ionized by the application of high voltages and maintained as a plasma. Once air is reduced to an ionized state or is a plasma, a momentum can be imparted to it by the application of an electric field which in turn can be used to alter the velocity profile within a boundary layer. Thus, it can be used in principle to alter the velocity profile within a boundary layer and consequently can be used to modify the skin friction or drag coefficient of a body in a flow field.

Plasma actuators are electrical systems driven by a high voltage source that produce a plasma discharge. Thus, plasma actuators function as ion generators that produce a plasma discharge when subjected to a large supply voltage. The plasma discharge ionizes the surrounding air, thus creating a thrust that energizes the flow locally within the boundary layer and prevents the separation of the flow from the airfoil surface. A dielectric-barrier-discharge (DBD) plasma actuator is a device that has been developed for flow control applications and appears to be most promising for energizing a boundary layer and inhibiting flow separation. It is an all-electric device with a layered construction, consisting only of a pair of planar electrodes separated by a layer of dielectric material. The size of the electrodes is typically of the order of a few centimeters for flow actuation applications, and the voltage used is in the range of 1–10 kV at an alternative frequency of 1–20 kHz, which helps to sustain the plasma and reduce the power consumption, as well as the erosion of the electrodes. When a high AC voltage is applied, the flow in the vicinity of the upper electrode is ionized, and the velocity of the flow field in the vicinity of the upper electrode functions as almost like a classical wall jet [1], [2]. The solution for the flow field around the wall jets may be obtained by an approximate method [3] which can be used to represent a plasma actuator's influence. A typical configuration of

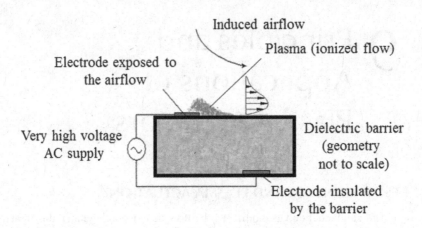

FIGURE 9.1   A typical schematic diagram of a DBD plasma actuator.

an asymmetric DBD plasma actuator is shown in Figure 9.1, and also shows the region of the plasma, the direction of the induced flow and the velocity profile just beyond the plasma region. The performance and suitability of DBD plasma actuators for flow control has been experimentally verified by Hoskinson, Hershkowitz and Ashpis [4] and Little et al. [5].

Another method of controlling drag is to introduce continuous suction from within the boundary layer region. Continuous suction, using carefully designed and positioned slots lying in a region below the laminar boundary layer, could be used to keep the boundary layer laminar. However, once the boundary layer becomes turbulent, it is almost impossible to return it to the laminar state by suction. A reduction in the form drag is possible by sucking away the boundary layer in the region of the trailing edge, while a reduction in the turbulent skin-friction drag is possible by blowing away the slowly moving air. In the region behind a suction slot the thinning of the boundary layer, when it is not in a laminar state, always increases the skin-friction drag and decreases the form drag. It has been observed experimentally that the nearer the suction slot is to the trailing edge, the less is the disadvantage of the increased skin-friction drag.

Passive flow control devices can prevent flow separation by facilitating the mixing between the boundary layer and the free stream, so as to ensure that the higher momentum fluid in the outer flow is redirected into the boundary layer region. Thus, the flow within the boundary layer is energized so the adverse pressure gradients do not cause the flow to separate. There are a number of passive devices that have been introduced which have the tremendous advantage that they do not need to use any source of external energy. Thus, the operation of several classes of passive flow control devices are briefly reviewed first.

9.2   PASSIVE METHODS OF FLOW CONTROL

The occurrence of flow separation is related to the wall shear stress, which is obtained from the product of the viscosity and the transverse velocity gradient. Flow separation occurs if the velocity gradient at the wall is zero i.e. when the wall shear stress is zero and a separation bubble develops at the rear of the airfoil. Trailing edge stall occurs in thick airfoils at high Reynolds numbers when the laminar to turbulent flow transition occurs at a point along the airfoil's surface towards the trailing edge. In this case, separation is determined by the ability of the flow to cope with the suction pressure at the trailing edge, which leads to trailing edge separation at a high angle of attack. Passive flow separation control is based on either directly increasing the momentum in the boundary layer, or by creating flow structures such as vortices for transporting higher momentum in the free stream flow to within the boundary layer, but without the use of any external source of energy to aid the flow. Increasing the momentum of a boundary layer will generally increase the ability to overcome the adverse pressure gradient. In practice one would like to achieve this by using a minimum external energy input.

A number of passive devices have been proposed in the literature and most cases have been experimentally proven to reduce drag or serve as useful flow control devices.

9.2.1 RIBLETS

Riblets are a biomimetic concept arising from shark skins which are micro-grooves on the surface and aligned to the freestream direction, as illustrated in Figure 9.2, that influence the turbulent vortices, resulting in a decrease of the momentum transfer and the wall shear stress. They are deemed to generate turbulent structures that interfere with development of the near wall structures in the turbulent boundary layer. They are manufactured as symmetric V-grooves (height equal to spacing) on adhesive-backed polymeric film so the film can be bonded to the surface of the airfoil. The boundary layer is influenced in both the streamwise and spanwise directions by the pressure gradients and for this reason they must be properly located and the spacing optimized for minimizing drag. When properly located and designed so the spacing is optimum, the maximum skin friction drag reduction is 5–10% for a range of heights h.

Generally the height to spacing ratio h/s is chosen to be 0.5, while the optimum spacing depends on the kinematic viscosity and the friction velocity ($\sqrt{\tau_w/\rho}$) [6]. When improperly located without optimally choosing the position or the spacing, they could increase the skin friction drag by 10%! Riblets with other geometries

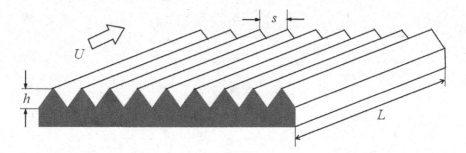

FIGURE 9.2 Geometry and alignment of saw tooth riblets in a flow (Robert [5]).

for the grooves, including trapezoidal grooves, rectangular grooves and scalloped grooves have also been experimentally investigated.

9.2.2 DIMPLES

"Dimples" are semi-spherical protrusions on the surface of a wing. Outward oriented and uniformly spaced multiple "dimple"-like protrusions on the airfoils surface can produce vortices of prescribed strength and duration for the real-time control of aerodynamic flows. Multiple dimple sets can control transition or fully turbulent, attached or separating flows. A typical airfoil section with dimples is shown in Figure 9.3. Ideally dimples should be located in the region 30–60% of the chord from the leading edge and certainly not close to the trailing edge. Furthermore, the dimple depth should be less than 1.5% of its diameter. Drag reduction could be increased by increasing the dimple depth from 1.5% to 5% of its diameter. However, augmenting the dimple depth can result in increased flow separation, causing additional drag.

9.2.3 FENCES

Spanwise fences are used on an airfoil surface for trapping vortices. When a fence type device is attached to a wing, part of the bound circulation of the wing is carried over to the fence. Upper surface fences have a favorable effect on longitudinal flow characteristics due to the retardation of spanwise boundary layer flow near the trailing edge, along with a reduction in separation over the outer wing. Wing fences act as barriers to the tipward flow on swept-back wings. They also generate powerful streamwise vortices. This results in a net reduction of the induced drag, due to a reduction in the strength of the tip vortex.

For the example of a thin swept wing shown in Figure 9.4, where the breakdown of flow begins at the leading edge and at low speeds, the main effect of the fence is to act as a partial-reflector, thus modifying pressure distributions and chordwise loadings on either side of it, and also affecting the spanwise loading and the downwash distribution.

9.2.4 VORTEX GENERATORS (VGS) AND MICRO-VGS

Vortex generators (VGs) are the most commonly used passive control devices that are used on the wings of several aircraft to reduce drag. A vortex generator is built

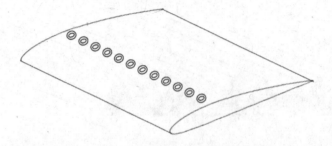

FIGURE 9.3 Typical geometry of dimples on an airfoil in a flow.

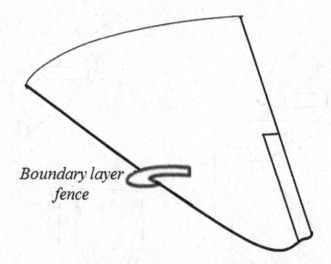

FIGURE 9.4   Typical example of a boundary layer fence.

out of a pair of small shaped vertical plates positioned at an angle with respect to the local freestream flow, and either parallel to each other (Figure 9.5(a)) or at a symmetric angle to each other, as shown in Figure 9.5(b).

The plates could be rectangular as shown in Figure 9.5(a) and (b) or shaped with linearly increasing height as shown in Figure 9.5(c) and (d). In a parallel configuration they generate co-rotating vortices while they generate counter-rotating vortices in symmetrically angled cases. With appropriate sizing and positioning of these vortex generators, streamwise vortices are created which together can be used to control the flow. On the upper surface of the wing, a VG produces vortices in such a way that it then acts as an aerodynamic barrier to the spanwise boundary layer flow. Typically, a VG transports energy into the boundary layer and can be used to control the separated flow even after separation has been initiated. Thus, streamwise vortices generated by VGs can entrain the high-momentum fluid within the boundary layer not only to eliminate or delay separation, but also to induce reattachment. Micro-VGs can additionally sweep away uncontrolled separation of the airflow over the airplane's flaps, while increasing lift marginally and reducing drag significantly. Several applications of VGs have been discussed in the literature by Casper, Lin and Yao [8], Mai et al. [9], Babinsky, Loth and Lee [10] and Gao et al. [11]. Yang, Zhang and Xu [12] considered the general aerodynamic performance of airfoils with blunt trailing edges, with and without VGs. Goddard and Stanislas [13] have discussed the application of VGs to the control of a boundary layer while Lin [14] and Lu et al. [15] have discussed the application of micro-VGs to the control of the turbulent boundary layer and high speed flows.

9.2.5 VORTILONS

Vortilons, which were first introduced by Taylor [16], were invented to control flow separation at transonic speeds by generating vorticity and consequently delaying or

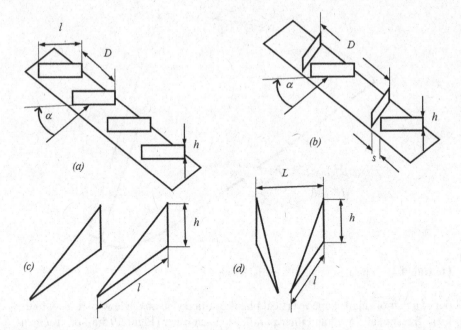

FIGURE 9.5 Examples of vortex generators a) parallel configuration of rectangular VGs, b) rectangular VGs placed at an angle to each other, c) parallel configuration of triangular VGs, d) triangular VGs placed at an angle to each other.

eliminating flow separation. Vortilons are under-wing fences which were introduced in the design of the DC9 by Shevell, Schaufele and Roensch [17] to control stall. Vortilons are known not only to improve stall performance, but are also effective beyond stall according to Shevell, Schaufele and Roensch [17].

The design of the vortilons does not affect the maximum lift capacity, but reduces the drag. As in the case of VGs, vortilons can be configured in counter-rotational or co-rotational arrangements. In both cases, the inclination angles are generally very small. Vortilons are placed under the surface of an aircraft wing and its leading edge so it intersects the wing slightly ahead of the stagnation point at stall. Vortilons which are also fence-like devices generate vortices of relatively high strength as the aircraft approaches stall. Vortilons avoid the blocking of the lateral air movement on the upper wing surface, while generating the vortex in the lower wing surface which flows opposite to the direction of the lateral air movement in the boundary layer. A typical example of a vortilon fitted to a wing along with a leading-edge droop is redrawn based on an original drawing and illustrated in Figure 9.6.

9.2.6 WINGLETS

The basic idea of wing tip modification was inspired by the nature of bird wings. Winglets are a bioinspired concept, where the spanwise wing tip of a wing is modified primarily to reduce induced drag. Winglet types include a raked tip, a continuously varying dihedral winglet where the wingtip is finally pointing up, a canted

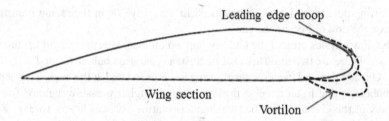

FIGURE 9.6 Vortilons as well as other passive flow control devices fitted to a wing.

winglet, MD 11, 12 style up-down winglets, a spiroid, a tip fence, a scimitar winglet, a "sharklet" Airbus A350 type winglet, blended winglets and feathers to emulate the birds. It was found several decades ago that a non-planar lifting system was required in order to reduce induced drag. It was also found that it is possible to increase the efficiency of wings and reduce induced drag by imitating bird feathers in splitting the wing-tip. The optimal performance could be achieved by imitating birds and spreading the wing tip as is done by birds when they spread their feathers over the wing tip. Thus, the development of winglets has evolved and most current aircraft employ a winglet to reduce induced drag. A negative aspect of winglets, in particular for cruise flight conditions, is that they introduce additional friction drag through the added wetted area. But the benefits seem to far outweigh the losses.

9.2.7 CAVITIES

The use of cavities, particularly to trap vortices that develop over the upper surface of a wing and eventually cause separation, can grossly influence the flow over the airfoil. While a number of airfoil configurations with small cavities have been tested for reducing drag, a successful concept that has emerged is the idea of using a cavity with suction to trap vortices that develop over the upper surface of a wing.

9.2.8 GURNEY FLAPS

A Gurney flap, illustrated in Figure 9.7, is small rectangular flap-like device located on the pressure side of the airfoil, towards the trailing edge, and having a width of around 1–3% of the chord. The Gurney flap increases lift produced by the air-foil, with only a small increase in drag, making it an effective, low-maintenance,

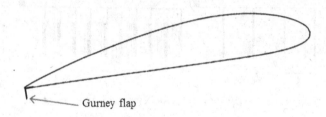

FIGURE 9.7 The Gurney flap.

high-lifting device, which was introduced in the early 70s in the racing industry to increase the downforce on the cars.

The flow physics around the Gurney flap which increases the overall lift are very intricate. There are two main areas of interest: a separation bubble located at the front of the Gurney flap and two counter-rotating vortices located at the back. This separation bubble is due to an adverse pressure region (higher pressure region). Towards the back of the Gurney flap, the two counter-rotating vortices act as a wake, which means that it is a region of low pressure. The presence of these two regions gives rise to the increase in pressure drag due to the presence of the Gurney flap. The increase in lift however, is due to multiple reasons, one example being that the Gurney flap delays separation on the suction surface (the upper surface) of the airfoil, therefore increasing the pressure difference across the chord, thus increasing lift. The Gurney flap can delay separation and consequently it can be used at higher angles of attack to produce lift.

9.3 PASSIVE METHODS COUPLED WITH PLASMA ACTUATION

The idea of coupling passive flow control methods with plasma actuators was pioneered by K-S Choi [7] and his co-workers (Feng et al. [18], Jukes and Choi [19], Choi et al. [20]) and by others (Zhang, Liu and Wang [21] and Jukes, Segawa and Furutani [22]). Plasma actuation can be combined with any of the preceding passive techniques to enhance the role of the passive device. For example, it can be combined with a vortex generator to enhance the processes of streamwise vortex generation and energize the boundary layer, as was demonstrated by Jukes and Choi [19]. The velocity profiles obtained by them were very similar to an analytical wall jet proposed by Tetervin [23].

The direction of actuation which determines the direction of the wall jet is a key parameter in determining the effectiveness of the DBD-VG. DBD-VGs can be easily configured into VG arrays. Figure 9.8 (redrawn from Jukes, Segawa and Furutani [22]) illustrates the counter-rotating (left) and co-rotating (right) DBD plasma VG arrays on an airfoil. Figure 9.9 illustrates a plasma actuator on the downstream surface of a Gurney flap, which is essentially enhancing the action of the Gurney flap.

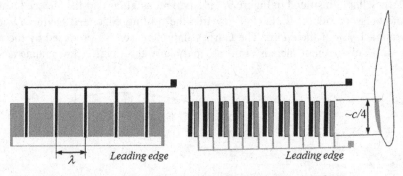

Array of counter-rotating DBD-VGs   Array of counter-rotating DBD-VGs

FIGURE 9.8 Plasma actuators configured into vortex generators.

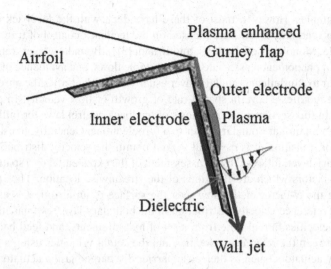

FIGURE 9.9   A plasma actuator assisted Gurney flap.

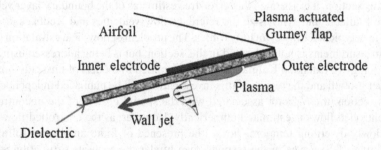

FIGURE 9.10   Alternate plasma assisted Gurney flap proposed by Feng et al. [18].

The experiments by Zhang, Liu and Wang [21] (reported in Choi et al. [20]) suggest that a 1% increase in the momentum coefficient of the plasma forcing corresponds to an effective Gurney flap height increment of 1%. The report and figure (redrawn) are from Choi et al. [20]. Alternate plasma Gurney flap configurations have also been extensively researched and Figure 9.10 illustrates an alternate arrangement of the plasma actuator and the airfoil which could increase the lift coefficient by 5% for all angles of attack.

9.4   REDUCTION OF SKIN-FRICTION DRAG BY FEEDBACK

The reduction of the skin-friction drag of an airfoil using low energy actuation of the flow within a boundary layer is a technique that is receiving considerable attention in recent times. The papers by Bewley and Liu [24], Bewley [25], Bewley, Moin and Temam [26], Chevalier, Hœpffner, Åkervikb and Henningson [27], Ahuja [28], Ahuja and Rowley [29], Dadfar, Semeraro, Hanifi and Henningson [30], Monokrousos, Lundell and Brandt [31] and Semeraro, Bagheri, Brandt and Henningson [32] are typical examples. However, most of these have dealt with the feedback control of instabilities, and only a few have focused on the feedback control of transition (see, for example, Monokrousos, Lundell, and Brandt [31]). Boundary layer transition is a post-critical phenomenon associated with unstable flows, and avoidance of transition of a laminar to a turbulent boundary layer usually takes place after the onset of instability, as it requires a certain spatial rate of growth of flow velocity. In particular, of interest in this section is the synthesis of feedback control laws for inhibiting the transition of a laminar boundary layer to a turbulent one. Generally, to establish the criterion for transition, it is essential to first obtain the velocity distribution within the boundary layer, followed by an assessment of the exponential (e^N) spatial growth rates or N-factors which are functions of the streamwise location. Thus to be able to regulate the N-factor of the flow over the surface of an airfoil, it is essential to establish a reference velocity distribution in the boundary layer, estimate the boundary layer velocities in real time from a set of measurements, and feed back the difference between the reference velocities and the actual velocities using a set of low power flow actuators (such as dielectric barrier discharge plasma actuators) located at certain finite locations on the airfoil surface, to boost the flow within the boundary layer.

In this section, it is assumed that error-free estimates of the boundary layer velocities are available and synthesize the reference flow velocities and feedback control inputs for a typical family of airfoil profiles. The problem of flow-field estimation from noisy measurements is not addressed in this section, but is being addressed using the methodology of extended Kalman filtering as applied to models of unsteady boundary layer growth and the results of this investigation will be published independently. In this section, the N-factor associated with the spatial growth of the uncontrolled and controlled flows is estimated numerically. However, as the controlled flows are only slowly diverging temporal flows, the presence of finite amplitude Tollmien-Schlichting (T-S) waves in the response can hinder the accurate estimation of the N-factor, and consequently the simulated flow response or measurements in practice must be post-processed to isolate the spatially growing signals from other wave-like components. This problem is also addressed and discussed in a later section. In this section, following the synthesis of a family of reference velocity profiles for the flow within the boundary layer for inhibiting transition, a typical noisy simulation of a fluid flow signal is de-noised by applying wavelet decomposition. It is shown with examples that the technique is capable of decomposing and de-noising a real measured fluid flow signal while also isolating the elements contributing to the spatial growth of wave. The reference velocity profiles and measured velocity profiles are used to construct feedback control laws for inhibiting transition. The designed feedback control laws are then applied to the flow simulation to demonstrate the feasibility of regulating the N-factor to maintain it below 9 over the entire upper surface of the airfoil.

9.4.1 FEEDBACK CONTROL OF TRANSITION

To understand the principles underpinning the feedback control of transition, it is important to recognize that the fundament parameter, albeit a distributed parameter, is the velocity distribution of the flow within the boundary layer. The dynamics of the flow within the boundary layer is governed by the continuity and Navier–Stokes (NS) equations, while the linear stability equations may be derived by perturbing the equilibrium steady flow within the boundary layer. For our part the velocity-vorticity decomposition of the flow variables will be employed, as these facilitate the numerical simulation of both the uncontrolled and controlled flows.

The scalar vorticity equation in two-dimensional flow is

$$\left(\partial/\partial t\right)\omega + \mathbf{u}\cdot\nabla\omega - \nu\nabla^2\omega = 0. \tag{9.1}$$

The velocity components are assumed to be obtained from a stream function, ψ. Hence if,

$$\mathbf{u} = \left(u_1, u_2\right) = \left(\partial/\partial y, -\partial/\partial x\right)\psi, \tag{9.2}$$

the stream function, ψ must satisfy $\nabla^2\psi = \omega$. The base flow velocity is defined when $\mathbf{u} = \mathbf{u}_B\left(x, y\right) = \left(U_B\left(\text{x,y}\right), 0\right)$ which is assumed to satisfy the steady NS equations.

Consider a perturbation to the base flow. The velocity and vorticity may be expressed as the sum of the base flow quantities and the perturbation quantities. Hence, dropping the subscripts "p", $\mathbf{u} \to \mathbf{u}_p + \mathbf{u}_B \equiv \mathbf{u} + \mathbf{u}_B$, $\omega \to \omega_p + \omega_B \equiv \omega + \omega_B$. Thus, from here on, $\mathbf{u} = \mathbf{u}_p$ is the perturbation velocity, $\omega = \omega_p$ is the perturbation vorticity. The perturbation vorticity satisfies,

$$\frac{\partial}{\partial t}\omega + \left(\mathbf{u} + \mathbf{u}_B\right)\cdot\nabla\left(\omega + \frac{\partial}{\partial y}U_B\right) - \nu\nabla^2\left(\omega + \frac{\partial}{\partial y}U_B\right) = 0, \tag{9.3}$$

where

$$\omega = \left(\partial/\partial y, -\partial/\partial x\right)\cdot\left(u_1, u_2\right) \tag{9.4}$$

and the perturbation stream function ψ is defined as,

$$\mathbf{u} = \left(u_1, u_2\right) = \left(\partial/\partial y, -\partial/\partial x\right)\psi \tag{9.5}$$

and satisfies,

$$\nabla^2\psi = \omega. \tag{9.6}$$

If the base flow satisfies NS equations, the linearized perturbation vorticity equation is

$$\frac{\partial}{\partial t}\omega + \left(u_1\frac{\partial}{\partial x} + u_2\frac{\partial}{\partial y}\right)\frac{\partial}{\partial y}U_B + U_B\frac{\partial}{\partial x}\omega - \nu\nabla^2\omega = 0, \tag{9.7}$$

where ω satisfies,

$$\omega = \frac{\partial}{\partial y} u_1 - \frac{\partial}{\partial x} u_2 \text{ and } \nabla^2 \psi = \omega, \tag{9.8}$$

ψ is the perturbation stream function. Making the further assumption that the base flow and its normal derivative are parallel to first order, the term, $\left(u_1 \partial/\partial x\right)\left(\partial/\partial y\right)U_B$ may be assumed to be a second order term and ignored. Hence to first order

$$\frac{\partial}{\partial t}\omega + u_2 \frac{\partial^2}{\partial y^2} U_B + U_B \frac{\partial}{\partial x}\omega - \nu\nabla^2\omega = 0, \tag{9.9}$$

where u_2 satisfies,

$$\nabla^2 u_2 = v = -\nabla^2 \frac{\partial}{\partial x}\psi = -\frac{\partial}{\partial x}\omega. \tag{9.10}$$

Now consider the perturbation control flow which is assumed to satisfy the NS equations. The velocity and vorticity associated with control flow are respectively given by, $\mathbf{u}_C = \left(U_C, 0\right)$, $\omega_C = \partial U_C / \partial y$.

The control flow, generated by a distribution of flow actuators such dielectric barrier discharge (DBD) plasma actuators, is assumed to be a perturbation in addition to the perturbation flow. Thus, it can be shown that the vorticity equation is modified to

$$\frac{\partial}{\partial t}\omega + u_2 \frac{\partial^2}{\partial y^2} U_{BC} + U_{BC} \frac{\partial}{\partial x}\omega - \nu\nabla^2\omega = 0, \tag{9.11}$$

where $U_{BC} = U_B + U_C$, and u_2 satisfies

$$\nabla^2 u_2 = v = -\nabla^2 \frac{\partial}{\partial x}\psi = -\frac{\partial}{\partial x}\omega. \tag{9.12}$$

Because it was assumed that the base flow satisfies the NS equations, there are no source terms in the right-hand side of Equation (9.11), as in the case of Davies and Carpenter [33]. The Equations (9.11) and (9.12) are solved by the method proposed by Davies and Carpenter [33]. These equations are integrated twice in the wall normal direction and the vorticity ω and the wall normal velocity component, $u_2 \equiv v$ are expressed as weighted linear combinations of Chebyshev polynomials in "y". The integral operator along the wall normal direction is approximated using a pseudo-spectral method (Davies and Carpenter [10]). Furthermore, the streamwise derivatives $\partial/\partial x$ and $\partial^2/\partial x^2$ are approximated by central difference formulae.

The control flow generates a set or distribution of points on the surface of the airfoil. This is done by defining an appropriate reference velocity distribution within the normalized boundary layer for the particular conditions of flow, $U\left(x, y\right) = U_{ref}\left(x, y\right)$. The feedback control is then defined as

$$U_C\left(x, y\right) = U_{ref}\left(x, y\right) - U_{Be}\left(x, y\right) \tag{9.13}$$

where $U_{Be}(x,y)$ is an estimate of the base flow velocity distribution. In this section it will be assumed, for simplicity, that $U_{Be}(x,y)=U_B(x,y)$. The velocity distribution $U_{Be}(x,y)$ could be estimated, in principle, from velocity-related measurements within the boundary layer, based on extended Kalman filtering and the dynamics of unsteady boundary layers. In practice the control flow velocity distribution is applied only over a restricted domain over the surface of the airfoil S_a, which will be denoted as, D_c. Thus,

$$U_C(x,y) = U_{ref}(x,y) - U_{Be}(x,y), \quad S_a \cup D_c, \tag{9.14a}$$

$$U_C(x,y) = 0 \times \left(U_{ref}(x,y) - U_{Be}(x,y) \right), \quad \text{elsewhere.} \tag{9.14b}$$

The methodology adopted to validate the feedback control approach is based on estimating the N-factor distribution from the time domain vorticity–velocity response. These responses are in turn used to estimate the receptivity distribution and the N-factor distribution is obtained from it. Transition is assumed to occur at point where $N \geq 9$ for the first time. The aim of the feedback controller is to ensure laminar flow over the entire upper surface of the airfoil. Consequently, our objective is to regulate the N-factor distribution and ensure that it is $N < 9$ over the entire upper surface of the airfoil.

9.4.2 Modeling the Flow Due to DBD Plasma Actuators

To ensure that the actual control flow distribution closely approximates the desired control flow, consider the flow distribution generated by a distribution of DBD plasma actuators. The modeling of the dielectric barrier discharge (DBD) plasma actuators is based on Glauert's [1] solution for the velocity distribution in a wall jet. The Glauert wall jet is essentially a plane wall jet formed by a thin jet of fluid that flows tangentially to an impermeable wall resting horizontally and surrounded by a fluid of the same type as that in the jet which is in a quiescent ambient state. The jet consists of an inner region wherein the flow resembles a boundary layer, and an outer region where the flow is more like a free shear layer.

Consider a self-similar plane wall jet formed over an impermeable resting wall governed by the equation of a steady boundary layer over a flat plate. Consider the two-dimensional boundary layer equations for a viscous incompressible fluid of kinematic viscosity ν and formulate the problem in terms of a dimensional stream function $\bar{\psi}(x,y)$.

Let (x,y) in a Cartesian coordinate system and t denote the dimensional time. The coordinates are made non-dimensional by defining $y \to yh$, $x \to 2(Uh/\nu)xh = 2R_h xh$ and $t \to t/\Omega$ where the steady, freestream, streamwise base velocity is U, h is a transverse length scale, R_h is the free stream Reynolds number based on the scale length h, which in practice would be the displacement thickness of the boundary layer δ^*, and Ω is a reference unsteady frequency and defines a dimensionless stream function such that $\psi = Uh\bar{\psi}(x,y)$. The stream function, $\psi = \psi(x,y)$ defines the velocity components

$$u = \partial\psi/\partial y, \quad v = -\partial\psi/\partial x. \tag{9.15}$$

Starting with the boundary layer equations for continuity and momentum in two dimensions, in dimensional coordinates and dependent variables

$$\frac{\partial}{\partial x}u + \frac{\partial}{\partial y}v = 0, \tag{9.16a}$$

$$\frac{\partial u}{\partial t} + u\frac{\partial u}{\partial x} + v\frac{\partial u}{\partial y} = v\frac{\partial^2 u}{\partial y^2}, \tag{9.16b}$$

making the transformations to the dimensionless coordinates leads to the dimensionless governing momentum equation for the stream function $\psi = \psi(x,y)$ in terms of $S = \Omega h^2 v$ as

$$S\frac{\partial^2 \psi}{\partial t \partial y} + \frac{\partial \psi}{\partial y}\frac{\partial^2 \psi}{\partial x \partial y} - \frac{\partial \psi}{\partial x}\frac{\partial^2 \psi}{\partial^2 y} = \frac{\partial^3 \psi}{\partial^3 y}. \tag{9.17}$$

The continuity Equation (9.16a) is satisfied by the stream function. The momentum equation reduces in a steady state to

$$\frac{\partial \psi}{\partial y}\frac{\partial^2 \psi}{\partial x \partial y} - \frac{\partial \psi}{\partial x}\frac{\partial^2 \psi}{\partial^2 y} = \frac{\partial^3 \psi}{\partial^3 y}. \tag{9.18}$$

The dimensionless stream function may be expressed as

$$\psi = \psi(x,y) = 4x^{1/4}f(p), \quad p = x^{-3/4}y. \tag{9.19}$$

The impermeability, no-slip and asymptotic boundary conditions are:

$$\psi(x,0) = 0, \quad \frac{\partial \psi(x,0)}{\partial y} = 0, \quad \frac{\partial \psi(x,y)}{\partial y} \to 0, \text{ as } y \to \infty. \tag{9.20}$$

The self-similar part of the stream function, $f(p)$ satisfies the ordinary differential equation

$$f'''(p) + f(p)f''(p) + 2(f'(p))^2 = 0, \tag{9.21}$$

along with the boundary conditions,

$$f(0) = 0, \quad f'(0) = 0, \quad f'(p) \to 0 \text{ as } p \to \infty. \tag{9.22}$$

The Glauert-jet solution corresponds to the normalization $f(\infty) = 1$ of the stream function. It leads to the well-known implicit form of the analytical solution of the problem found by Glauert which is expressed in the form

$$2p = \ln\left(\frac{\left(1 + f + f^{1/2}\right)}{\left(1 - f^{1/2}\right)^2}\right) + 2 \times 3^{1/2}\arctan\left(\frac{(3f)^{1/2}}{2 + f^{1/2}}\right). \tag{9.23}$$

In dimensional terms

$$\bar{\psi}(x, y) = (40\nu E)^{1/4} x^{1/4} f(p),$$ (9.24)

where the parameter E defined by Opaits et al. [2] and Opaits [34] that fully characterizes the jet and is conserved at any cross-section downstream is a product of the volumetric flux rate $\int_0^\infty u\, dy$ and the specific momentum flux $\int_0^\infty u^2\, dy$ and is

$$E = 0.45 \int_0^\infty u\, dy \int_0^\infty u^2\, dy.$$ (9.25)

The equation for p is,

$$2p = \ln\left(\frac{(1 + f + f^{1/2})}{(1 - f^{1/2})^2}\right) + 2 \times 3^{1/2} \arctan\left(\frac{(3f)^{1/2}}{2 + f^{1/2}}\right) = \frac{(40\nu E)^{1/4}}{\nu} x^{-3/4} y.$$ (9.26)

From the governing Equation (9.21) for $f(p)$, it can be shown by successive integration that the non-dimensional downstream velocity profile of the wall-jet solution is

$$u = 4x^{-1/2} f'(p)$$ (9.27)

where $f'(p)$ is given by

$$f'(p) = \frac{2}{3}\sqrt{f}\left(1 - \left(\sqrt{f}\right)^3\right).$$ (9.28)

The skin friction coefficient may be shown to be $f''(0)=2/9$.

In a recent paper, Shahmohamadi and Rashidi [3] have obtained an approximate numerical solution for $f(p)$.

$$f(p) = 2184 p^2 \left(\sum_{i=0}^{3} n_p(i+1) p^{3i} \Big/ \sum_{i=0}^{4} d_p(i+1) p^{3i} \right).$$ (9.29)

In the above equation the arrays n_p and d_p [3] are respectively given by

$$n_p = (78767923259224584,\ 1725658791362352,$$

$$99910515826755,\ 135721670638),$$

$$d_p = (15482622995833184231040,\ 625910732693761394880,$$

$$7676359653888591120,\ 29381362861683924,\ 19389517508143).$$ (9.30)

The velocity profile of a typical wall jet in uniformly and geometrically scaled coordinates (assuming a typical Reynolds number R_h to be 1000) in shown in Figure 9.11. It is quite clear that the velocity subsides rapidly in the streamwise direction.

Moreover, Opaits et al. [2] and Opaits [34] have shown that the jet is completely characterized by the parameter E. Along with the fluid kinematic viscosity ν, it determines the entire flow field. The principal flow characteristics as a function of x the distance along the wall measured from a hypothetical point of origin, where the jet's mean height is zero, and not from the plasma where the jet is actually created, is found analytically by Opaits et al. [2] and Opaits [34]. In particular,

$$u_{max}(x) = 0.4980(E/\nu x)^{1/2}. \tag{9.31}$$

In their papers, Hoskinson, Hershkowitz and Ashpis [4] and Little et al. [5] used an empirical fit to the self-similar dielectric barrier discharge (DBD) induced flow velocity profile:

$$u_{fit}(x,y) = u_0(x)\sqrt{\frac{y}{y_0}}\exp\left\{-\left(\frac{y-y_0}{L}\right)^2\right\}, \tag{9.32}$$

which is a maximum for large L when, $y=y_0$ and $u_{fit}(x,y_0)=u_{max}(x)$. Assuming that the maximum of Equation (9.32) with respect to y, without any loss of generality, is

$$u_{fit}(x,y)\big|_{max\ wrt\ y} = u_0(x)/g_m. \tag{9.33}$$

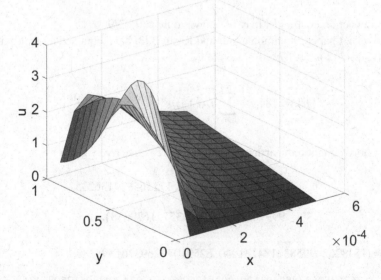

FIGURE 9.11 Two-dimensional velocity profile of a typical Glauert wall jet.

Adopting the model,

$$u_{fit}(x,y) = u_0(x)\sqrt{\frac{y}{y_0}} \exp\left\{-\left(\frac{y-y_0}{L}\right)^2\right\}, \quad (9.34)$$

$$u_0(x) = 0.4980 g_m \left(E(x)/vx\right)^{1/2}. \quad (9.35)$$

In Equation (9.35), $E(x)$ is assumed to be unknown and is defined such that

$$u_{fit}(x,y_0) = u_0(x) = u_{max} g_m. \quad (9.36)$$

Changing the reference point to where the plasma jet is created,

$$u_0(x) = 0.4980 g_m \left(E(x)/v(x+x_{pc})\right)^{1/2}, \quad (9.37)$$

where x_{pc} is the *unknown* distance of the point where the plasma jet is created from a point where the jet's mean height is zero. Moreover, considering that the velocity subsides in the streamwise direction, as observed in Figure 9.11, one may average the velocity over a step from $x=0$ to $x=\Delta x$ where Δx, in practice, is of the order of 0.5% to 5% of the chord of the airfoil or greater. Assuming that $E(x)$ is a constant over each step,

$$\bar{u}_0 = \frac{0.4980 g_m}{\Delta x}\sqrt{\frac{E}{v}} \int_0^{\Delta x} \left(\frac{1}{x+x_{pc}}\right)^{1/2} dx$$

$$= \frac{0.4980 g}{\Delta x}\sqrt{\frac{Ex_{pc}}{v}} \times 2\left\{\left(\frac{\Delta x}{x_{pc}}+1\right)^{1/2} - 1\right\}. \quad (9.38)$$

$$\bar{u}_0 \approx 0.4980 g_m \sqrt{E/v x_{pc}}. \quad (9.39)$$

Hence, if the jet is created at point x_J from a reference point, a model of the jet's streamwise mean velocity distribution is:

$$\bar{u}_{mJ} = 0.4980 g_m \left(\frac{E}{v(x_{pc}-x_J)}\right)^{1/2} H(x+x_{pc}-x_J)\sqrt{\frac{y}{y_0}} \exp\left\{-\left(\frac{y-y_0}{L}\right)^2\right\}, \quad (9.40)$$

$$0 \le x \le \Delta x,$$

where $H(x)$ is a unit step function. Assuming a distribution of N plasma actuators in the streamwise direction,

$$u_m(x,y) = \sum_{J=1}^{N} 0.4980 g_m \left(\frac{E_J}{v(x_{pc,J} - x_J)} \right)^{1/2}$$

$$\times H(x + x_{pc,J} - x_J) \sqrt{\frac{y}{y_{0,J}}} \exp\left\{ -\left(\frac{y - y_{0,J}}{L_J} \right)^2 \right\}, \quad 0 \le x \le \Delta x.$$

(9.41)

Since both E_J and $x_{pc,J}$ are unknown, a new constant is defined as,

$$\kappa_J = 0.4980 g_m \left(\frac{E_J}{v(x_{pc,J} - x_J)} \right)^{1/2}.$$

(9.42)

Equation (9.42) can be used to relate κ_J which is defined from the desired control flow distribution to the parameters specific to a particular DBD plasma actuator, E_J and $x_{pc,J}$. Equation (9.41) reduces to,

$$u_m(x,y) = \sum_{J=1}^{N} \kappa_J H(x + x_{pc,J} - x_J) \sqrt{\frac{y}{y_{0,J}}} \exp\left\{ -\left(\frac{y - y_{0,J}}{L_J} \right)^2 \right\},$$

(9.43)

$$0 \le x \le \Delta x.$$

The parameters κ_J, which relates to the magnitude of the control flow distribution at a particular location, and $y_{0,J}$ and L_J, which define the distribution in the y direction, are found by minimizing the square of the error between the desired control signal distribution and the distribution $u_m(x,y)$ generated by the plasma actuators which are located at the n streamwise locations $x = x_J$, $J = 1, 2, \ldots n$. The parameters for the actuators located closest to the leading edge are found first followed by the actuator next to it and so on until the parameters defining all of actuators are found. Even for a sufficiently large step size Δx, the velocity distribution at the current location is affected by the presence of the DBD plasma actuators ahead of it. A similar approach could be used if a different model, for the variation of the streamwise velocity component in the y direction, is adopted. How well the model parameters $y_{0,J}$ and L_J of the DBD plasma actuators can be made to fit the desired velocity distribution is a matter of how well the actuators can be designed to meet the specified distributions at each location. While the above analysis demonstrates the feasibility of selecting a set of DBD plasma actuators to generate the desired control flow, the actual errors would depend largely on the real designs. As this work, at this stage, is primarily theoretical, it will be assumed that the desired control flow distributions could be generated by a set of DBD plasma actuators. Needless to say, the synthesis must be repeated with actual control velocity distributions in a practical implementation.

9.4.3 DECOMPOSITION OF SIMULATED FLOW FEATURES

Frequency analysis methods using measured fluid flow signals are essential for visualizing T-S waves and other flow features. The interesting features of a measured fluid flow signals are T-S waves and other waves-based instabilities such as Klebanoff waves, cross-flow instabilities (secondary), subharmonics (ternary), Eckhaus/Görtler type instability, Kelvin Helmholtz type instability (quaternary) associated with travelling wave solutions. The Fast Fourier Transform (FFT) is the simplest frequency domain analysis method which can be used for their analysis and classification. The combination of the FFT with envelope detection methods such as Hilbert transforms have been used with some success for fluid-flow signal analysis. The method involves filtering the high-frequency components in the signal spectrum in the first instance. Because the signal, when it is just about unstable, has a relatively low energy, it is often overwhelmed by disturbance signals and noise with higher energy. Consequently, it is difficult to filter the noise in the signal spectrum using conventional FFT methods.

In order to overcome problems with use of the FFT, advanced signal decomposition methods such as Short-Time-Fourier-Transform (STFT), Wigner–Ville distribution and wavelet analysis have been developed. While the STFT uses a single analysis window, the wavelet transform uses a short window at high frequency and a long window at low frequencies. Depending on the primary mother wavelet used, the wavelet transform coefficients at the highest-frequency scales provide high time-resolution with a minimum of signal samples. This flexibility, to effectively track a transient signal at an arbitrarily high time resolution, is absent with the STFT. Moreover, the simplicity and elegance of the wavelet approach cannot be matched by other methods.

For real-time applications, wavelet-based methods are unsuitable and non-linear recursive methods are better suited, as they also permit the determination of the optimal bandwidth. Although there are many types of recursive filters such as Butterworth, Fourier series, Kalman, cubic and quintic splines and finite impulse response (FIR) filters, and several methods to determine the "optimal" bandwidth of the measured fluid-flow signal, the non-linear Kalman filter-based methods are best suited for this purpose and real-time de-noising. Signal decomposition permits the removal of the noise components up to any frequency level.

9.4.4 APPLICATION OF WAVELET DECOMPOSITION AND DE-NOISING

Wavelet transforms and wavelet packet decompositions [35] have recently gained popularity for the analysis of fluid-flow signals. Wavelet packet decomposition is particularly ideally suited for de-noising [36] a measured fluid-flow signal. Over the past ten years, wavelet transform has been used as a powerful tool for image data compression, noise reduction and feature extraction of a signal. Wavelets provide efficient localization in both time and frequency (or scale). To analyze any finite energy signal, the continuous wavelet transform (CWT) provides a decomposition of the signal as a combination of a set of basis function, obtained by means of dilation and translation of a single prototype wavelet function called a mother wavelet. There are over 300 different types of mother wavelets including Haar, Daubechies (db), Symlet, Coiflet, Gaussian, Morlet, complex Morlet, Mexican hat, bio-orthogonal, reverse bio-orthogonal, Meyer, discrete approximation of Meyer, complex Gaussian, Shannon and frequency B-spline families. The continuous wavelet transform maps a signal of one independent variable into another function of two independent variables representing a scaling and translation of the independent variable. The scale factor and/or the translation parameter can both be discretized. The discretization process leads to an orthonormal basis for signal representations. The decomposition process can be iterated, with successive approximations being decomposed in turn, so that one signal is broken down into many lower resolution components. For discrete-time signals, the dyadic discrete wavelet transform (DWT) is equivalent, according to Mallat's algorithm [37] to an octave filter bank, and can be implemented as a cascade of identical cells (low-pass and high-pass finite impulse response filters). These filters split the signal's bandwidth to half. Using downsamplers after each filter, the redundancy of the signal representation can be removed. This is called the wavelet decomposition tree and is the basis for wavelet decomposition.

9.4.5 A Review of Wavelet Decomposition Based on the Wavelet Transform

The motivation in developing the wavelet transforms (WTs) is to overcome the disadvantages of the STFT, which provides constant resolution for all frequencies in the signal, as it uses the same time-scale for the analysis of the signal $f(t)$ being analyzed. On the contrary, WTs use multi-resolution, that is, they use different time-scales and time-shifts to analyze different frequency bands of the signal $f(t)$. Different window functions for resolving the signal $\psi(t; a,b)$ can be generated by a standardized process of scaling and shifting the time axis. These are referred to as *"child"* wavelets, as they can be generated by dilation or compression of a mother wavelet $\psi(t; 0,1)$, which is defined in the primary time frame. A time-scale is the inverse of its corresponding frequency. Furthermore, WTs can be continuous or discrete. The continuous wavelet transform of finite-energy signals $(f(t) \in L^2(r))$ with the analyzing wavelet $\psi(t)$, is defined as the convolution of $f(t)$ with a scaled and conjugated wavelet:

$$W_f(a,b) = \int_{-\infty}^{\infty} f(t) \frac{1}{\sqrt{a}} \psi * \left(\frac{t-a}{b} \right) dt \tag{9.44}$$

where $\psi(t)$ is the wavelet function and a and b are the dilation and translation respectively. The factor $1/\sqrt{a}$ is used to preserve the total energy of the signal, $f(t)$. Equation (9.44) indicates that wavelet analysis is a time–frequency analysis, or more properly termed a time-scaled analysis. Since the continuous parameters a and b cause computational difficulties, they are replaced by discrete factors. A useful selection is: $a = 2^{-j}$ and $b = k2^{-j}$ where k and j are integer values. This form of the wavelet transform is called dyadic wavelet transform. Therefore, wavelet transform may be expressed as:

$$W_f(a,b) = \int_{-\infty}^{\infty} f(t) 2^{j/2} \psi * \left(2^j t - k \right) dt. \tag{9.45}$$

A wavelet packet can be defined as $\omega_{j,k}^{m}(t)$ which is a function with three indices, where the integers m, j and k are the modulation, scale and translation parameters, respectively.

$$\omega_{j,k}^{m}(t) = 2^{j/2}\omega^{m}\left(2^{j}t - k\right). \tag{9.46}$$

Thus, the wavelet packet is defined by dilations and translations of the mother function, or analyzing wavelet, and constitutes an orthogonal basis. The variables j and k are integers that scale and dilate the mother function $\omega^{m}(t)$ to generate wavelets, such as a Daubechies wavelet family. The scale index j indicates the wavelet's width, and the location index k gives its position. Thus, the mother functions are rescaled, or dilated by powers of two, and translated by integers. The scales and dilations cause the wavelet bases to be self-similar. The wavelet functions $\omega^{m}(t)$ satisfy recursive relationships that are the basis for multi-resolution analysis. Any function $f(x)$ may be expanded in an orthogonal series in terms of a wavelet packet. Given any scaling function $\varphi(x)$ which can be used as basis for a wavelet transform, any other function $f(x)$ may be expanded in an orthogonal series with the coefficients of the expansion being defined as

$$c_{j}(k) = \left\langle f(x), 2^{-j}\varphi\left(2^{-j}x - k\right)\right\rangle. \tag{9.47}$$

Moreover, with $k=0$ and $j=1$, the function $(1/2)\varphi(x/2)$ can itself be expanded in a series as,

$$\frac{1}{2}\varphi\left(\frac{x}{2}\right) = \sum_{k}h(k)\varphi(x-k). \tag{9.48}$$

When x is the frequency variable, the coefficients $h(k)$ represent a low pass half-band filter. This relation then allows one to compute the coefficients, $c_{j}(k)$, $j>0$, without computing the scalar product. Thus $c_{j}(k)$ satisfies the recursion

$$c_{j+1}(k) = \sum_{n}h(n-2k)c_{j}(n). \tag{9.49}$$

Similarly, the wavelet functions $\omega^{m}(t)$ satisfy the following recursive relationships:

$$\omega^{2m}(t) = \sqrt{2}\sum_{k}h(k)\omega^{m}(2t-k), \tag{9.50}$$

$$\omega^{2m+1}(t) = \sqrt{2}\sum_{k}g(k)\omega^{m}(2t-k), \tag{9.51}$$

where $g(k)$ is given by the inner product,

$$g(k) = \frac{1}{\sqrt{2}}\left\langle \psi(t), \psi(2t-k)\right\rangle = (-1)^{k}h(-k+1), \tag{9.52}$$

and $\psi(t)$ is the wavelet function. The function $h(k)$ may also be expressed as another inner product,

$$h(k) = \frac{1}{\sqrt{2}} \langle \varphi(t), \ \varphi(2t - k) \rangle, \tag{9.53}$$

where $\varphi(t)$ is known as the scaling function.

The discrete wavelet transform at any given index j of f is

$$Wf[k., j] = \sum_{n=0}^{N-1} f(n)\psi_j *[n - k] \tag{9.54}$$

which is a circular convolution between f and $y_1[n] = y[-n]$. These circular convolutions are typically computed with an FFT which requires $O(N \log_2(N))$ operations. Discrete wavelet decomposition can then be implemented by the scaling filter $h(k)$, which is a low-pass filter related to the scaling function $\varphi(t)$, and the wavelet filter $g(k)$, which is a high-pass filter related to the wavelet function $\psi(t)$:

$$\varphi_j(t) = \sqrt{2} \sum_k h(k) \sqrt{2^j} \varphi(2^{j+1}t - k), \tag{9.55}$$

$$\psi_j(t) = \sqrt{2} \sum_k g(k) \sqrt{2^j} \varphi(2^{j+1}t - k). \tag{9.56}$$

The basic step of a wavelet decomposition algorithm [38] is illustrated in Figure 9.12 which can be implemented in two opposite directions: decomposition and reconstruction. In the decomposition step, the discrete signal f is convolved with a low-pass half-band filter H and a high-pass half-band filter G, resulting in two vectors C_{A1} and C_{D1}. The elements of the vector C_{A1} are called approximation coefficients, and the elements of vector C_{D1} are called detailed coefficients. The decomposition is recursive. Each resulting function from the low-pass filtering is decimated by suppression of one sample out of two. The high frequency signal is left as it is, while the low frequency signal (upper part of figure) is used to iterate. In the reconstruction, the samples are restored by inserting a 0 between each sample, then convolved with the conjugate filters $\tilde{H}$ and $\tilde{G}$, adding the resulting functions and multiplying the result by 2. The iteration is continued up to the smallest scale (lower part of figure).

At each of the wavelet decomposition level, the half-band filters produce signals spanning only half the frequency band. This doubles the frequency resolution, as the uncertainty in frequency is reduced by half [39]. In accordance with Nyquist's rule, if the original signal has a highest frequency of ω, which requires a sampling frequency of 2ω radians, then it now has a highest frequency of $\omega/2$ radians. It can now be sampled at a frequency of ω radians, thus discarding half the samples with no loss of information. This decimation by 2 halves the time resolution, as the entire signal is now represented by only half the number of samples. Thus, while the half-band low-pass filtering removes half of the frequencies and thus halves the resolution, the decimation by 2 doubles the scale. The filtering and decimation process is continued

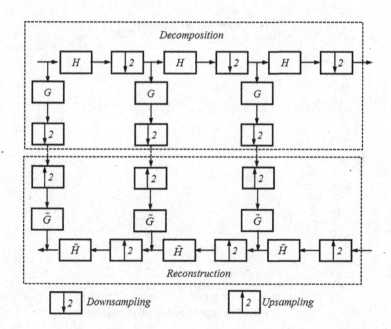

FIGURE 9.12 Wavelet packet decomposition by a tree of low-pass and high-pass half-band filters.

until the desired level is reached. The maximum number of levels depends on the length of the signal. The complete wavelet decomposition of the original signal is then obtained by concatenating all the coefficients, a[n] and d[n], starting from the last level of decomposition.

Wavelet packet decomposition is an ideal technique for de-noising a fluid-flow signal, while retaining the option of reconstructing the original signal and can be easily implemented in MATLAB [40]. Following the de-noising of the fluid-flow signal, the lowest frequency component at the highest level of decomposition is the signal that contains the most significant information.

9.4.6 APPLICATION TO THE REGULATION OF LAMINAR FLOW OVER AN AIRFOIL

Figure 9.13 shows a typical spatial distribution of the N-factor over the upper surface of a NACA 65(2)415 6-digit uncontrolled airfoil section, in a flow field with the free stream Mach number equal to 0.64 at sea level, corresponding to a Reynolds number of about 30 million, at zero angle of attack. The same uncontrolled spatial distribution of the N-factor is shown in Figure 9.14 after the simulated outputs have been decomposed by wavelet packet decomposition and reconstructed after removing the high frequency components. In the uncontrolled case, the spatial growth rates being much higher, the relative contribution of the T-S wave motion is insignificant and this is apparent when Figure 9.14 is compared with Figure 9.13. In this section, the wavelet decomposition is used as a check, to verify that the wave motion components do not have a significant influence on the estimated N-factor distribution.

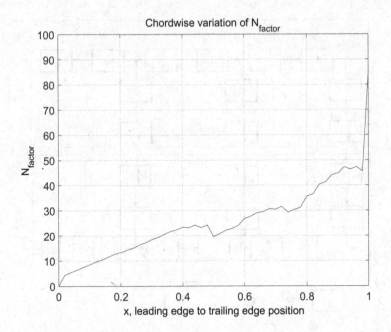

FIGURE 9.13 Spatial distribution of the N-factor over the upper surface of a NACA 65(2)415 airfoil section, with the freestream Mach number = 0.64 at zero angle of attack.

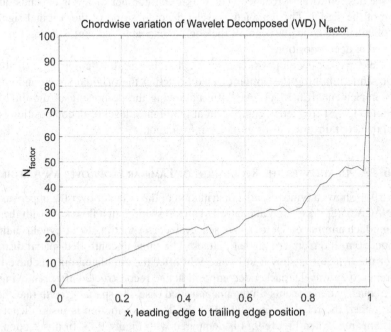

FIGURE 9.14 Spatial distribution of the N-factor for the same case as in Figure 9.3 with wavelet decomposition and reconstruction.

Figure 9.15 shows the uncontrolled base velocity profiles defined for the entire airfoil for the same flight conditions for the same case as in Figure 9.13. The profile close to the leading edge indicates transition in the vicinity, as the velocity ratio just crosses over a magnitude of unity as it approaches the boundary layer edge.

Figure 9.16 shows a typical distribution of the reference base velocity profiles defined for the entire airfoil for the specified flight conditions. These are derived for a family of airfoil sections, for the specified flight conditions, by assuming active control using the methods discussed by Vepa [41]. They are almost self-similar, although there is no requirement for them to be so. It must be emphasized that although they are defined for the complete upper surface of an airfoil section, they could be applied over limited regions of a particular airfoil.

Figure 9.17 shows the closed loop streamwise spatial distribution of the N-factor. The plasma actuators are assumed to be located over 50% of the airfoil's upper surface, evenly distributed starting at the leading edge of the airfoil and up to mid-chord. No change is made to the base velocity distribution aft of mid-chord. Figure 9.18 shows the effect of reducing the extent over which the actuators are distributed from 50% of the airfoil's upper surface to 38.89% and yet starting at the leading edge. Figure 9.19 shows the effect of further reducing the extent over which the actuators are distributed from 38.89% of the airfoil's upper surface to 27.78% and yet again starting at the leading edge. It appears that placing the actuators at the leading edge was fairly crucial for this particular airfoil, as any attempt to reduce the intensity of the control flow at the leading edge led to a dramatic increase in the N-factor. Furthermore, any further decrease of the extent over which the actuators were distributed, below 25%

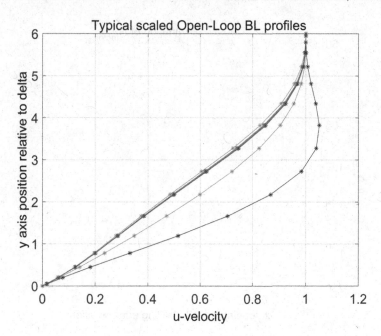

FIGURE 9.15 Typical distribution of the uncontrolled base velocity profiles defined for the entire airfoil for the specified flight conditions.

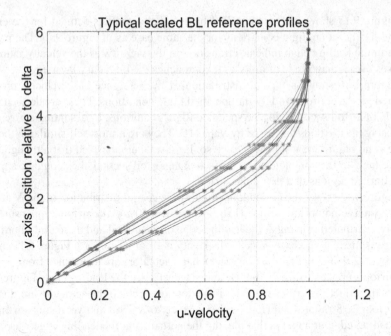

FIGURE 9.16 Typical distribution of the reference base velocity profiles defined for the entire airfoil for the specified flight conditions.

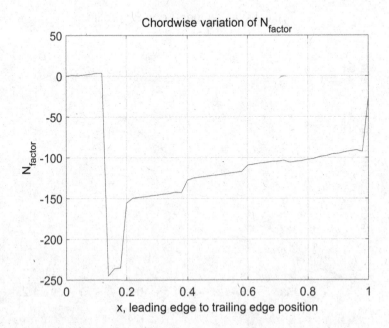

FIGURE 9.17 Closed-loop spatial distribution of the *N*-factor over the upper surface of a NACA 65(2)415 airfoil section, with plasma actuators from the leading edge to mid-chord.

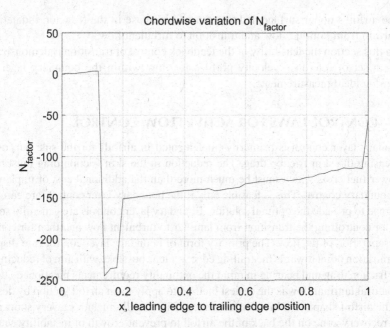

FIGURE 9.18 The effect of reducing the extent of plasma actuation from 50% to 38.89% over the airfoil's upper surface, starting from the leading edge.

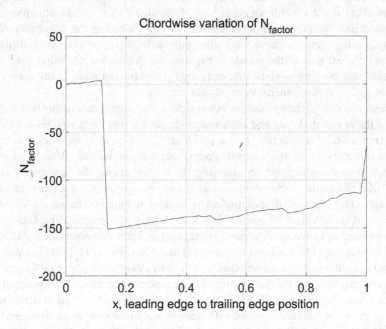

FIGURE 9.19 The effect of further reducing the extent of plasma actuation from 38.89% to 27.78% over the airfoil's upper surface, starting from the leading edge.

of the airfoil's upper surface, led to a marked increase in the N-factor, indicating a departure from laminar flow and transition to turbulent flow.

In this section the feasibility of the feedback control of transition is demonstrated, using a set of reference velocity profiles for flow within the boundary layer and assuming ideal measurements.

9.5   CONTROL LAWS FOR ACTIVE FLOW CONTROL

Boundary layer control is primarily implemented in airfoils for the sole purpose of decreasing the skin friction drag. The reduction in the skin friction drag must result in lower fuel costs which must be much more than the additional cost of implementing boundary control. This is feasible only if the boundary layer controls are carefully designed to provide the optimal solution. Boundary layer controls are generally sought first for controlling the transition from laminar to turbulent flow and then for controlling separation of the flow. The primary form of boundary layer control is to displace the transition point towards the trailing edge as far as possible, with aim of reducing the skin friction drag and bearing in mind the optimality requirements mentioned above.

Natural laminar flow is the easiest method to apply to an airfoil section by designing the airfoil shape to inhibit disturbances. This usually includes a very short pressure recovery zone on the back of the airfoil to prevent growth of instability causing T-S waves. In laminar flow control the boundary layer is prevented from developing and consequently also preventing the instability causing disturbances to grow. One method of accomplishing laminar flow control is to use a porous or perforated surface with a suction system to regulate the boundary layer thickness and shape. One could also use a combination of both of the above-mentioned techniques. Such a method has the additional advantage of not only reducing the possibility of, for example, using suction close to the leading edge to inhibit cross-flow instabilities and tailoring the aft part of the profile to regulate the T-S waves. On other hand other methods have been proposed based on re-energizing the boundary using vortex generators, micro-vortex generators or plasma actuators.

There is a considerable ongoing effort directed at manipulating laminar flows to reduce the associated drag and simultaneously ensure that the flow does not experience transition to turbulence. It is a problem of continuously delaying the transition to turbulence by some controlled adjustments to the laminar flow. Much of the theoretical work is focused on assessing the receptivity of the flow based on the linearized dynamics with realistic disturbances introduced into the laminar boundary layer. The concept of applying active control techniques based on MEMS to problem of stabilization of laminar boundary layers was proposed by Gaster [42]. The application of feedback control strategies has been demonstrated by Högberg and Henningson [43], Kim and Bewley [44] and Chevalier et al. [27]. The practical feasibility of the techniques was demonstrated in a landmark paper by Gaster [45]. More recently Belson et al. [46] have used direct numerical simulations coupled with reduced order modeling and the Eigensystem realization algorithm to demonstrate the feasibility of feedback control with realistic combinations of actuators and sensors. Hanson et al. [47] have demonstrated the application of plasma actuators to the control of bypass transition. It has been suggested by several researchers, led by

Schmid and Henningson [48], that delaying the onset of turbulence in shear flows is a fully non-linear problem and that transition is triggered by finite-amplitude background disturbances and immediately results in a temporally and spatially complex final state. While this substantially true, it cannot be disputed that the process is triggered by infinitesimal disturbances and so can be effectively controlled, provided the instability is adequately compensated for at an early stage of the temporal and spatial evolution. On the other hand, it is physically meaningful to be able to control the boundary layer velocity profiles by suction rather than controlling the disturbance flow field. Such an approach also facilitates a limited amount of feedback of the disturbance flow field, although the main process of stabilization involves changing the boundary layer velocity profiles to ensure a well-behaved receptivity response. The objective of this section is to demonstrate the feasibility of designing laminar flow controllers that alter the boundary layer velocity profiles so as to avoid undesirable spatial growth rates of disturbances. In order to demonstrate the feasibility of implementing the control of the laminar boundary layer, a reasonably accurate model of the dynamics of flow within the laminar boundary layer is used. To perform these simulations a variation of the simulation proposed by Davies and Carpenter [33] based on a vorticity–velocity formulation, is adopted.

In this section, the vorticity–velocity formulation is used to model the flow around an airfoil. The performance output of the stability model is chosen to be the pressure receptivity. The flow controls for maintaining the laminarity are assumed to be the suction holes that are located on the boundary in the pressure recovery region so they directly influence the normal velocity component in that region. Considering the momentum and displacement thicknesses of the boundary layer on the top surface of the airfoil with, and without, suction and the receptivity response in the absence of suction, due to the modeled flow, the N-factor distribution is established. Thus, the spatial growth of receptivity over the airfoil's surface is obtained. To demonstrate the feasibility of laminar flow control, the suction velocities in the recovery region are constructed to ensure a desired velocity profile. This is done by adjusting the suction velocities so a desired set of displacement and momentum thicknesses are obtained. An inverse boundary layer control approach is adopted to synthesize an active control system for regulating the boundary layer. The efficacy of the control system is established by estimating the N-factor of the growth of the disturbances before and after the application of the active controller. By prescribing the shape factor in the trailing edge region, the instabilities in the laminar boundary layer, and consequently the spatial growth rates, are substantially reduced, although this form of control does not seem to influence the point of transition from laminar to turbulent flow.

The active flow controller considered in his section is not based on the feedback. The previously imposed suction is coupled with a flow controller that can modify the base flow velocity profiles in the streamwise direction, over sections of the airfoil, starting at the leading edge. There is dramatic reduction in the spatial growth rates within the region where the control is imposed and a movement of the transition point towards the trailing edge. As the region of control is increased, the transition point moves further downstream, although the relationship is not linear. This is partly due to the fact that the transition from a laminar to a turbulent boundary layer is a post critical phenomenon and not just an issue of inhibiting instabilities.

The use of suction is in fact an active control technique similar in principle to the concept of feed forward. It can be shown that very little power is actually expended while it is deployed. The only power consumption is in the steady suction in the recovery region which is designed to be uniform. The suction velocity is directly related to the first and second order gradients of the boundary layer velocity profile in the recovery region. Thus, it is shown that the synthesis low power suction controller for the control of transition of the laminar boundary layers is a feasible proposition. When this is supplemented with other base flow active controls using actuation based on principles of plasma electrodynamics or electromagnetics, it is possible to successfully regulate the boundary layer and inhibit transition.

9.5.1 INTEGRAL EQUATIONS FOR THE BOUNDARY LAYER

As indicated in Chapter 5, the method of Drela and Giles [49] based on the formulation of two integral equations complemented with a set of closure equations, may be used to solve for the momentum and displacement thicknesses of the boundary layer at each streamwise location. To briefly recapitulate, in Chapter 5, the displacement thickness is defined as:

$$\delta^* = \int_0^\delta \left(1 - \frac{\rho u}{\rho_e u_e}\right) dy. \tag{9.57}$$

In Equation (9.57), y is the normal coordinate across the thickness of the boundary layer. The momentum thickness is defined as:

$$\theta = \int_0^\delta \frac{\rho u}{\rho_e u_e}\left(1 - \frac{u}{u_e}\right) dy. \tag{9.58}$$

The former is expressed as

$$\delta^* = H\theta. \tag{9.59}$$

The energy thickness of the boundary layer θ^*, is defined as

$$\theta^* = \int_0^\delta \frac{\rho u}{\rho_e u_e}\left(1 - \frac{u^2}{u_e^2}\right) dy. \tag{9.60}$$

The two integral equations, including the effects of suction are solved for the momentum thickness θ and the shape factor H. Initially, the second integral equation is formulated in terms of the energy thickness shape factor

$$H^* = \theta^*/\theta. \tag{9.61}$$

However, eventually H^* is expressed in terms of H by the use of a closure equation and the equation is solved for H.

The integral momentum equation with blowing, suction or porosity is:

$$\frac{d\theta}{d\xi} + \left(2 + H - M_e^2\right)\frac{\theta}{u_e}\frac{du_e}{d\xi} = \frac{C_f}{2} + m_s, \tag{9.62}$$

where the suction coefficient m_s, is defined in terms of the wall suction velocity v_w and the density of the suction flow ρ_s as,

$$m_s = \int_0^\delta \frac{\rho_s v_w}{\rho_e u_e}\, dy. \tag{9.63}$$

Additionally, in Equation (9.62), u_e is the velocity at the edge of the boundary layer, M_e is the corresponding Mach number, C_f the skin friction coefficient and ξ the non-dimensional curvilinear coordinate in the streamwise direction. The integral energy equation with blowing, suction or porosity can be expressed as,

$$\theta\frac{dH^*}{d\xi} + H^*\left(2\bar{H}^{**} + (1 - H)\right)\frac{\theta}{u_e}\frac{du_e}{d\xi} = 2C_D - H^*\left(\frac{C_f}{2} + \frac{1}{3}m_s\right). \tag{9.64}$$

In Equation (9.64), $\bar{H}^{**}$ is the ratio of the density shape parameter to the kinetic energy shape parameter and C_D the drag or dissipation coefficient. Ferreira [50] uses a slightly different approximation for the last term in Equation (9.64). However, the numerical values obtained by him are similar.

In both the equations, the integral momentum equation and the integral equation for the kinetic energy shape parameter of the skin-friction coefficient is modified independently, as mentioned in Chapter 5. Considering the integral momentum equation, the skin friction coefficient is modified to account for the effects of transpiration by using the relation by Kays and Crawford [51]. This is expressed implicitly as

$$\mathrm{Re}_\theta\frac{C_{fpm}}{2} = \mathrm{Re}_\theta\frac{C_f}{2}\left|\frac{\ln\left(1 + B_{fm}\right)}{B_{fm}}\right|^{1.25}\left(1 + B_{fm}\right)^{0.25} \tag{9.65}$$

where

$$B_{fm} = 2\,\mathrm{Re}_\theta\frac{m_s}{2}\bigg/\mathrm{Re}_\theta\frac{C_{fpm}}{2}. \tag{9.66}$$

It must be mentioned that the Equation (9.65) is an implicit relation and that it is only valid for small pressure gradients and small mass transfer values. The skin friction coefficient is determined by Newton iteration. Considering the integral equation for the kinetic energy shape parameter, the skin friction coefficient is again modified to account for the effects of transpiration by using the same relation given by Equation (9.65) with B_{fm} replaced by B_{fe} which is given by

$$B_{fe} = 2\,\text{Re}_\theta \frac{m_s}{6} \Big/ \text{Re}_\theta \frac{C_{fpe}}{2}. \tag{9.67}$$

Merchant [52] introduced the concept of the wall slip velocity and expressed the skin friction coefficient as

$$C_{fpm} = C_f - 2\,\rho v_w u_s \big/ \big(\rho_e u_e^2\big). \tag{9.68}$$

Introducing a parameter, the slip velocity ratio s_{vr}, the skin friction coefficient may be related to the suction coefficient m_s, by

$$C_{fpm} = C_f - 2 m_s s_{vr}. \tag{9.69}$$

Similarly, the modified skin friction coefficient in the energy equation is,

$$C_{fpe} = C_f - 2 m_s s_{vr}/3. \tag{9.70}$$

Thus, the momentum and energy integral equations are expressed respectively as

$$\frac{d\theta}{d\xi} + \big(2 + H - M_e^2\big) \frac{\theta}{u_e} \frac{du_e}{d\xi} = \frac{C_{fpm}}{2}, \tag{9.71}$$

$$\theta \frac{dH^*}{d\xi} + H^* \big(2\bar{H}^{**} + (1-H)\big) \frac{\theta}{u_e} \frac{du_e}{d\xi} = 2C_D - H^* \frac{C_{fpe}}{2}. \tag{9.72}$$

It may be observed that although in this section only laminar flow is considered, the integral analysis may also be applied to turbulent flow, in which case one has two more differential equations for the entrainment coefficient and the lag entrainment equation. The entrainment coefficient incorporates the edge flow from the outer inviscid region into the boundary layer, while the lag entrainment equation accounts for the history effects in a non-equilibrium turbulent boundary layer. Although the entrainment concept is generally valid in laminar flow as well, the entrainment coefficient and lag entrainment equations are not considered in this section.

9.5.2   THE INVERSE BOUNDARY LAYER METHOD: UNIFORM SOLUTIONS

In the case of constant θ and δ^* with uniform suction, from the momentum integral equation

$$\big((2 - M_e^2)\theta + \delta^*\big) \frac{1}{u_e} \frac{du_e}{d\xi} = \frac{C_{fpm}}{2}. \tag{9.73}$$

Hence,

$$C_{fpm} = \left(\left(2 - M_e^2 \right) \theta + \tilde{\delta}^* \right) \frac{2}{u_e} \frac{du_e}{d\xi}. \tag{9.74}$$

With uniform suction at the boundary in the recovery region where θ and δ^* are maintained as constants, the momentum equation reduces to

$$\rho v_{\text{suction}} \frac{du}{dy} = -\frac{d\hat{p}}{dx} + \mu \frac{d^2 u}{dy^2}, \tag{9.75}$$

where $\hat{p}$ is the total hydrostatic pressure. Moreover, in the recovery region the pressure gradient is small and positive and it follows that

$$v_{\text{suction}} = \left(\mu \frac{d^2 u}{dy^2} - \frac{d\hat{p}}{dx} \right) \bigg/ \rho \frac{du}{dy}. \tag{9.76}$$

Retaining the dominant term only

$$v_{\text{suction}} \approx \mu \frac{d^2 u}{dy^2} \bigg/ \rho \frac{du}{dy}. \tag{9.77}$$

Thus, the suction velocity may be expressed as

$$v_{\text{suction}} \approx \frac{\mu}{\rho} \left(\frac{d^2 u}{dy^2} \bigg/ \frac{du}{dy} \right), \quad \frac{dp}{dx} << -\mu \frac{d^2 u}{dy^2}. \tag{9.78}$$

Thus, the minimum suction velocity magnitude, $v_w = v_{\text{suction}}$ may be expressed as a Reynolds number as

$$\text{Re}_{\text{suction}} = \text{Re}_\delta \, v_w / u_e = \frac{\rho v_{\text{suction}} \delta}{\mu} = \left(\frac{d^2 u}{dy^2} \delta \bigg/ \frac{du}{dy} \right). \tag{9.79}$$

9.5.3 UNIFORM AND PRESCRIBED SHAPE FACTOR

Following Wauquiez [53] and introducing the variable $\omega = \rho \theta^2 / \mu$ and

$$\text{Re}_\theta = \rho u_e \theta / \mu = u_e \sqrt{\rho \omega / \mu}, \tag{9.80}$$

from the momentum integral equation,

$$\frac{u_e}{2} \frac{d\omega}{d\xi} = \frac{\rho u_e \theta}{\mu} \frac{d\theta}{d\xi} = \text{Re}_\theta \frac{C_{fpm}}{2} - \left(2 + H - M_e^2 \right) \frac{\rho \theta^2}{\mu} \frac{du_e}{d\xi}. \tag{9.81}$$

Hence,

$$\frac{u_e}{2} \frac{d\omega}{d\xi} + \left(2 + H - M_e^2 \right) \frac{du_e}{d\xi} \omega = \text{Re}_\theta \frac{C_{fpm}}{2}. \tag{9.82}$$

From the energy integral equation,

$$\frac{\theta}{H^*}\frac{dH^*}{d\xi}+\left(2\bar{H}^{**}+\left(1-H\right)\right)\frac{\theta}{u_e}\frac{du_e}{d\xi}=\frac{2C_D}{H^*}-\frac{C_{fpe}}{2}. \tag{9.83}$$

Hence,

$$\mathrm{Re}_\theta\,\theta\,\frac{d\log\left(H^*\right)}{dH}\frac{dH}{d\xi}+\left(2\bar{H}^{**}+\left(1-H\right)\right)\mathrm{Re}_\theta\,\frac{\theta}{u_e}\frac{du_e}{d\xi}=\mathrm{Re}_\theta\,\frac{2C_D}{H^*}-\mathrm{Re}_\theta\,\frac{C_{fpe}}{2}, \tag{9.84}$$

which may be expressed as

$$u_e\,\frac{d\log\left(H^*\right)}{dH}\,\omega\,\frac{dH}{d\xi}+\frac{du_e}{d\xi}\left(2\bar{H}^{**}+\left(1-H\right)\right)\omega=\mathrm{Re}_\theta\,\frac{2C_D}{H^*}-\mathrm{Re}_\theta\,\frac{C_{fpe}}{2}. \tag{9.85}$$

In the case of constant $H=\delta^*/\theta$,

$$\frac{du_e}{d\xi}\left(2\bar{H}^{**}+\left(1-H\right)\right)\omega=\mathrm{Re}_\theta\,\frac{2C_D}{H^*}-\mathrm{Re}_\theta\,\frac{C_{fpe}}{2}. \tag{9.86}$$

Hence,

$$\mathrm{Re}_\theta\,\frac{C_{fpe}}{2}=\mathrm{Re}_\theta\,\frac{2C_D}{H^*}-\frac{du_e}{d\xi}\left(2\bar{H}^{**}+\left(1-H\right)\right)\omega. \tag{9.87}$$

From the momentum integral equation,

$$\frac{u_e}{2}\frac{d\omega}{d\xi}+\left(2+H-M_e^2\right)\frac{du_e}{d\xi}\,\omega=\mathrm{Re}_\theta\,\frac{C_{fpm}}{2}. \tag{9.88}$$

For a given ω, since,

$$\mathrm{Re}_\theta\,\frac{C_{fpe}}{2}=\mathrm{Re}_\theta\,\frac{2C_D}{H^*}-\frac{du_e}{d\xi}\left(2\bar{H}^{**}+\left(1-H\right)\right)\omega, \tag{9.89}$$

one can then directly determine the required suction in terms of $\mathrm{Re}_\theta\,m_s/2$, as well as $\mathrm{Re}_\theta\,C_{fpm}/2$, which is then used to integrate equation for ω.

It may be noted that typically for laminar flow $H=2.5$, while for turbulent flow $H=1.3$. However, for large H, $H>3.5$ for laminar flow and $H>2.4$ for turbulent flow, one can expect separation. Thus, to maintain laminar flow, an optimum choice for a constant H boundary layer is $H=2.2$.

In the case when $dH/d\xi$ is prescribed, it is assumed that $dH/d\xi$ is known and defined as

$$dH/d\xi=\left(H_n-H\right)/L_H \tag{9.90}$$

where H_n and L_H are assumed to be known. In this case, H settles exponentially to a uniform value. Hence $dH/d\xi$ is replaced by $(H_n-H)/L_H$ in Equation (9.85). Thus, the equation for the shape factor may be expressed as

$$\mathrm{Re}_\theta \frac{C_{fpe}}{2} = \mathrm{Re}_\theta \frac{2C_D}{H^*} - u_e \frac{d\log(H^*)}{dH} \omega\left(\frac{H_n - H}{L_H}\right) - \frac{du_e}{d\xi}\left(2\bar{H}^{**} + (1 - H)\right)\omega. \quad (9.91\mathrm{a})$$

Rearranging the preceding equation for modified friction coefficient in the energy equation,

$$C_{fpe} = \frac{4C_D}{H^*} + \frac{2\omega H}{\mathrm{Re}_\theta}\left(\frac{u_e}{L_H}\frac{d\log(H^*)}{dH} + \frac{du_e}{d\xi}\right) - \omega H_n$$

$$\times \left(\frac{u_e}{L_H}\frac{d\log(H^*)}{dH} + \frac{du_e}{d\xi}\frac{(2\bar{H}^{**} + 1)}{H_n}\right).$$

$$(9.91\mathrm{b})$$

Again one can then directly determine the required suction in terms of $\mathrm{Re}_\theta\, m_s/2$, as well as $\mathrm{Re}_\theta\, C_{fpm}/2$, which is then used to integrate equation for ω and the controlled equation for H.

9.5.4 THE VORTICITY–VELOCITY FORMULATION WITH CONTROL FLOW INPUTS

The scalar vorticity equation in two-dimensional flow is,

$$(\partial/\partial t)\omega + \mathbf{u}\cdot\nabla\omega - \nu\nabla^2\omega = 0. \quad (9.92)$$

The velocity components are assumed to be obtained from a stream function, ψ. Hence if,

$$\mathbf{u} = (u_1, u_2) = (\partial/\partial y, -\partial/\partial x)\psi, \quad (9.93)$$

the stream function, ψ, must satisfy $\nabla^2\psi = \omega$. The base flow when $\mathbf{u} = \mathbf{u}_B = (U_B, 0)$ is assumed to satisfy the steady NS equations.

Consider a perturbation to the base flow. The velocity and vorticity may be expressed as the sum of the base flow quantities and the perturbation quantities. Hence dropping the subscripts "p", $\mathbf{u} \to \mathbf{u}_p + \mathbf{u}_B \equiv \mathbf{u} + \mathbf{u}_B$, $\omega \to \omega_p + \omega_B \equiv \omega + \omega_B$. Thus, from here on, $\mathbf{u} = \mathbf{u}_p$ is the perturbation velocity, $\omega = \omega_p$ is the perturbation vorticity. The perturbation vorticity satisfies

$$\frac{\partial}{\partial t}\omega + (\mathbf{u} + \mathbf{u}_B)\cdot\nabla\left(\omega + \frac{\partial}{\partial y}U_B\right) - \nu\nabla^2\left(\omega + \frac{\partial}{\partial y}U_B\right) = 0, \quad (9.94)$$

where

$$\omega = (\partial/\partial y, -\partial/\partial x)\cdot(u_1, u_2) \quad (9.95)$$

and the perturbation stream function ψ is defined as

$$\mathbf{u} = (u_1, u_2) = (\partial/\partial y, -\partial/\partial x)\psi \quad (9.96)$$

and satisfies

$$\nabla^2 \psi = \omega. \tag{9.97}$$

If the base flow satisfies NS equations, the linearized perturbation vorticity equation is

$$\frac{\partial}{\partial t}\omega + \left(u_1\frac{\partial}{\partial x} + u_2\frac{\partial}{\partial y}\right)\frac{\partial}{\partial y}U_B + U_B\frac{\partial}{\partial x}\omega - \nu\nabla^2\omega = 0, \tag{9.98}$$

where ω satisfies

$$\omega = \frac{\partial}{\partial y}u_1 - \frac{\partial}{\partial x}u_2 \text{ and } \nabla^2\psi = \omega, \tag{9.99}$$

ψ is the perturbation stream function.

Making the further assumption that the base flow and its normal derivative are parallel to first order, the term, $\left(u_1\dfrac{\partial}{\partial x}\right)\dfrac{\partial}{\partial y}U_B$ may be assumed to be a second order term and ignored. Hence to first order,

$$\frac{\partial}{\partial t}\omega + u_2\frac{\partial^2}{\partial y^2}U_B + U_B\frac{\partial}{\partial x}\omega - \nu\nabla^2\omega = 0, \tag{9.100}$$

where u_2 satisfies

$$\nabla^2 u_2 = v = -\nabla^2\frac{\partial}{\partial x}\psi = -\frac{\partial}{\partial x}\omega. \tag{9.101}$$

Now consider the perturbation control flow which is assumed to satisfy the NS equations. The velocity and vorticity associated with control flow are respectively given by $u_C = (U_C, 0)$, $\omega_C = \partial U_C/\partial y$. The control flow is assumed to be a perturbation in addition to the perturbation flow. Thus, it can be shown that the vorticity equation is modified to

$$\frac{\partial}{\partial t}\omega + u_2\frac{\partial^2}{\partial y^2}U_{BC} + U_{BC}\frac{\partial}{\partial x}\omega - \nu\nabla^2\omega = 0, \tag{9.102}$$

where $U_{BC} = U_B + U_C$, u_2 satisfies

$$\nabla^2 u_2 = v = -\nabla^2\frac{\partial}{\partial x}\psi = -\frac{\partial}{\partial x}\omega. \tag{9.103}$$

Because it was assumed that the base flow satisfies the NS equations, there are no source terms in the right-hand side of Equation (9.102), as in the case of Davies and Carpenter [33]. The Equations (9.102) and (9.103) are solved by the method proposed by Davies and Carpenter [33]. These equations are integrated twice in the wall normal direction and the vorticity ω and the wall normal velocity component, $u_2 \equiv v$, are expressed as weighted linear combinations of Chebyshev polynomials in "y". The integral operator along the wall normal direction is approximated using a pseudo-spectral method (Davies and Carpenter [33]). Furthermore, the streamwise derivatives $\partial/\partial x$ and $\partial^2/\partial x^2$ are approximated by central difference formulae.

9.5.5 ACTIVE CONTROL OF VELOCITY PROFILES

In the methods discussed in Sections 9.3–9.4, it is often the case that the required suction is not physically realizable. In such situations, one must necessarily supplement the active suction controller with an additional actuator that can control the base flow velocity distribution. The active control of the base flow velocity profiles may be achieved by employing plasma actuators.

Dielectric barrier discharge plasma actuators have been used by several researchers to shape the flow velocity profiles within a boundary layer so as to achieve a desired velocity profile in a particular direction. The flow induced is similar to the flow induced by wall jets and can be shaped by suitable choice of the geometry of the electrodes that are used to create an electrostatic field of ions in a region close to the electrodes. By applying a suitable voltage to the electrodes, the ions, and hence the flow, are propelled in a desired direction. The flow velocity in the normal direction is then determined from considerations of continuity. It is often required to reduce the slope of the streamwise velocity in the vicinity of the wall or negate the existence of a point of inflection in the velocity profile within the boundary layer. Uniform suction of the boundary layer must however be applied at carefully selected locations so as not to alter substantially the receptivity mechanisms and induce secondary instabilities. The application of plasma actuation of the flow field within a boundary layer has been thoroughly investigated for over a decade (Roth, Sherman and Wilkinson [54]) and a recent review by Corke, Enloe and Wilkinson [55] summarizes the progress made in the field. More recently, Gibson, Arjomandi and Kelso [56] found that significant thinning of the boundary layer can be realized with an orthogonally arranged actuator over a short distance downstream of the device. They also found that when used in conjunction with a subtle suction control, the thinning can be exacerbated. However, further downstream, they found rapid thickening of the layer, supported by a decrease in the shape factor of the flow suggesting that the layer becomes unstable, in an accelerated fashion, in the presence of the actuation. This type of destabilization must be avoided when using active controls. In another study, Duchmann et al. [57] reported a significant reduction in spatial amplitude growth rates and thus achieved considerable delay of transition. In a recent dissertation, Belson [58] applied feedback control to successfully delay transition. The delay in transition was achieved via active feedback control, by measuring the localized flow velocity near the wall and used these measurements to synthesize a flow controller to manipulate the base flow velocity profiles. Riherd and Roy [59] and Riherd, Roy and Balachandar [60] study how momentum injection and wall-jet-like effects stabilize or destabilize a boundary layer. Thus, it is concluded that when actively modifying the base flow velocity profiles, the analysis of the stability and the assessment of the transition are essential.

Thomas et al. [61] established the typical velocity profiles generated by dielectric barrier discharge plasma actuators. Zito et al. [62] and Santhanakrishnan, Reasor and LeBeau [63] have also measured these velocity profiles using PIV imaging techniques. Based on these models and by the process of curve fitting, generic models of the base flow velocity distributions induced by plasma actuators, have been established. Two examples of such models are shown in Figure 9.20, where the base flow velocity profiles in the streamwise direction (u) are also compared with a simple polynomial approximation. Also plotted on the same figure are corresponding velocities induced in the normal direction (v).

In this section, models of the type shown in Figure 9.20 are used to manipulate the base flow velocities to control the stability and transition within the boundary layer.

9.5.6 HYBRID ACTIVE LAMINAR FLOW CONTROL WITH PLASMA ACTUATION

The research of a number of authors, notably Roth, Sherman and Wilkinson [54], Chan, Zhang and Gabriel [64], Peers, Huang and Luo [65], Moreau et al. [66], Jolibois, Forte and Moreau [67], Magnier et al. [68], Huang and Zhang [69], Opaits [70], Roth [71] and Kurz et al. [72], has contributed significantly to the understanding of the flow velocity distributions induced by plasma actuators. A distribution of plasma actuators is selected to implement the control law obtained in the preceding

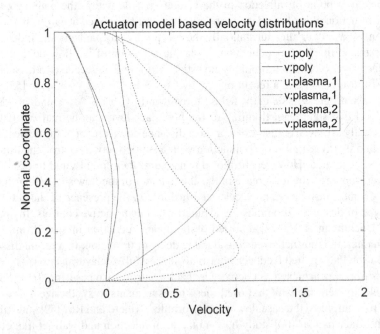

FIGURE 9.20 Base velocity profiles of flow velocities induced by typical plasma actuators compared with polynomial models.

section. Each plasma actuator's contribution to the flow velocity is modeled by a typical velocity distribution function given by

$$f(x,y) = P(x,\bar{y})\left(\gamma \bar{y}^{(\gamma-1)}/\alpha(x)\right)\exp\left(-\left(x^\gamma + \bar{y}^\gamma\right)\right), \qquad (9.104)$$

where

$$\bar{y} = \left(y - \mu(x)\right)/\alpha(x), \quad P(x,\bar{y}) = \sum_{i=1}^{M} A_i(x)\bar{y}^{i-1}, \quad \gamma = 2, \qquad (9.105)$$

$$\alpha(x) = abs\left(\alpha_0 + \alpha_1 x + \alpha_2 x^2 + \cdots + \alpha_{nq} x^{nq}\right), \quad \alpha_0 = 2, \qquad (9.106)$$

$$\mu(x) = \mu_0 + \mu_1 x + \cdots + \mu_{nq-1} x^{nq-1}, \qquad (9.107)$$

$$A_i(x) = a_{i0} + a_{i1} x + a_{i2} x^2 + \cdots + a_{i,nq} x^{nq}. \qquad (9.108)$$

The number of independent coefficients in the above model are $N = (nq+1)M + 2nq$. These coefficients of the model are found by matching the distribution with results of the velocity distributions obtained by Ramakumar and Jacob [73]. The optimum distribution of plasma actuators over different locations is then selected by minimizing the weighted distribution of velocity error between the desired distribution and the actual distribution obtained by the use of plasma actuators at different locations. The results are promising and indicate the feasibility of using a distribution of plasma actuators at different locations to modify the flow velocity profiles within the boundary layers. These results are expected to be published elsewhere.

The approaches adopted in this section to control the stability and transition within the boundary layer are either based exclusively on suction or are a hybrid combination of suction and plasma actuation of the base flow velocity profiles. Such an approach allows one to optimize the use of active controls to minimize the skin friction drag.

The methodology adopted to validate the active control approach is based on calculating the N-factor distribution from the time domain vorticity–velocity response. These responses are in turn used to estimate the receptivity and the N-factor distribution is obtained from the former.

9.5.7 APPLICATION OF THE CONTROL LAWS TO A TYPICAL AIRFOIL

Some typical results are shown in the first instance. Figure 9.21 shows the NACA 64(2)-415 6-digit airfoil section considered in a flow field with the freestream Mach number equal to 0.54 at sea level, corresponding to a Reynolds number of about 25 million, at zero angle of attack. Figure 9.22 shows the boundary layer shape factor

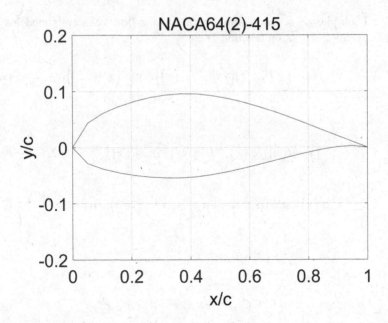

FIGURE 9.21   The unsymmetrical airfoil section considered in this study.

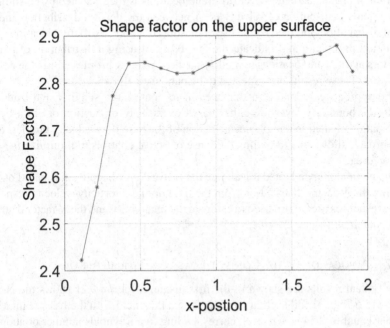

FIGURE 9.22   The boundary layer shape factor.

on the upper surface of the section. Figure 9.23 shows the spatial distribution of the N-factor, obtained from time domain receptivity response of the vorticity–velocity analysis, across the chord of the airfoil. The N-factor distribution results in a transition point near the leading edge at 0.0837 of the chord.

In order to stabilize the boundary layer, and push the point of the onset of transition towards the trailing edge (TE), it was decided to inhibit the unsteadiness in the shape factor in the vicinity of the TE. Thus, the shape factor was forced to be steady over the last four stations by introducing suction. This approach proved to be the most successful of all the methods of control that were attempted in reducing the spatial growth rates of the boundary layer.

The shape factor obtained by integrating the boundary layer equations is shown in Figure 9.24, and the corresponding N-factor distribution obtained by the vorticity–velocity analysis in Figure 9.25. The distribution in Figure 9.25 is similar to the one in Figure 9.23, although there is no change in the position of the transition point. Moreover, there is substantial fall in the N-factor distribution in the region near the trailing edge.

Suction was also introduced so as to force the shape factor to be steady, from mid-chord downstream to the trailing edge. The shape factor corresponding to this case is shown in Figure 9.26 and the corresponding N-factor distribution obtained by the vorticity–velocity analysis in Figure 9.27. While the transition point does not move back, there is a substantial increase in the growth rates in the region behind mid-chord. For this reason, this form of suction was no longer considered.

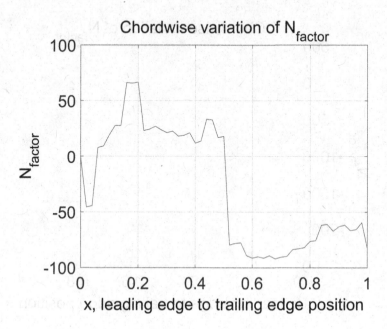

FIGURE 9.23 N-factor distribution obtained from time domain vorticity–velocity analysis and the pressure receptivity response.

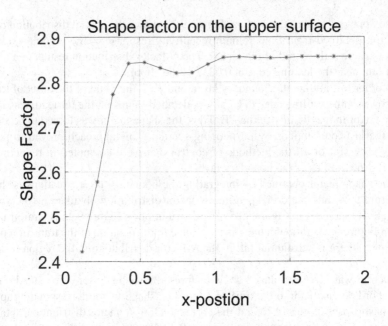

FIGURE 9.24 Boundary layer shape factor with suction near the TE.

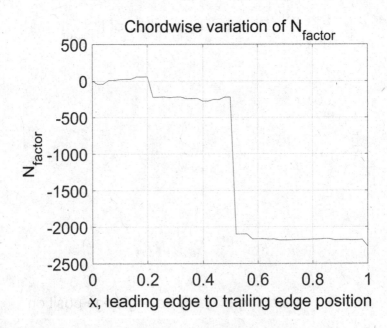

FIGURE 9.25 *N*-factor distribution obtained from time domain vorticity–velocity analysis of the case with suction near the TE.

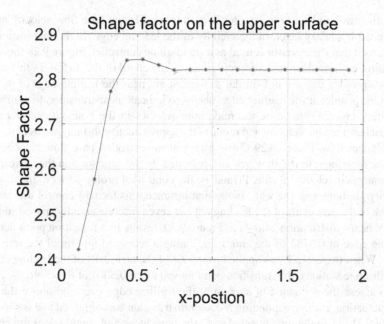

FIGURE 9.26   Boundary layer shape factor with suction aft of mid-chord.

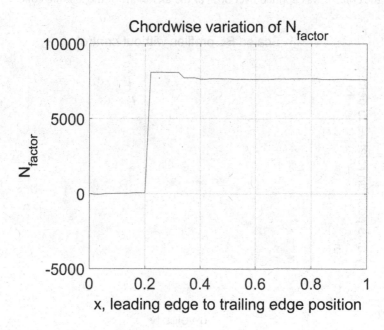

FIGURE 9.27   *N*-factor distribution obtained from time domain vorticity–velocity analysis of the case with suction aft of mid-chord.

In the next study, a flow actuator is used to modify the base flow velocity profile within the boundary layer, in the vicinity of the leading edge. In the first instance, it is assumed that the actuator is used as a stand-alone controller. Figure 9.28 shows the uncontrolled base flow velocity profiles across the chord of the airfoil. In this figure, the two profiles that are not similar to the rest are near the leading edge of the section. One profile, at the leading edge, seems to indicate near-transition to a turbulent boundary layer. In this case the maximum velocity in the boundary layer rapidly approaches the edge velocity and remains the approximate constant, just as in the case of turbulent flow. Figure 9.29 shows a typical uncontrolled base flow velocity profile, the increment to the flow velocity provided by the actuator and the corresponding controlled velocity profile. Primarily, the controlled profile seeks to increase the velocity gradient near the wall. In the first instance, this form of control was applied to 40% of the airfoil aft of the leading edge at seven discrete equally spaced stations. The N-factor distribution, shown in Figure 9.30, results in a transition point near the leading edge at 0.3733 of the chord, indicating a rearward motion of the transition point. When the control was applied over 50% of the airfoil aft of the leading edge at nine discrete stations, the transition point moved aft to 0.5001 of the chord.

To assess the influence of suction at the trailing edge over and above the base flow actuation, the corresponding N-factor distribution was obtained and is shown in Figure 9.31. As in the uncontrolled case, the introduction of suction does not change the position of the transition point, although it does reduce the spatial growth rates substantially aft of the transition point. Similar features were observed in the case when the control was applied over 50% of the airfoil aft of the leading edge.

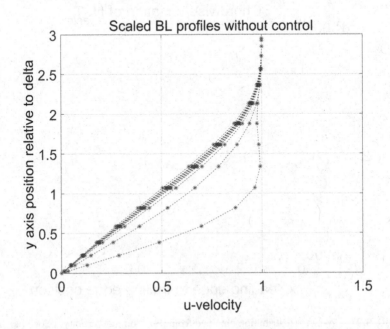

FIGURE 9.28 Base velocity profiles at different chordwise locations in the boundary layer, without control.

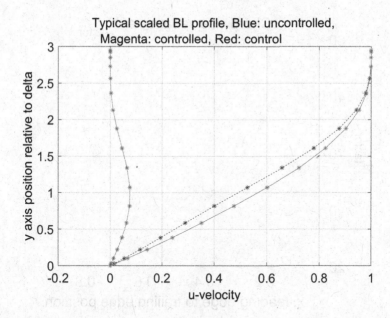

FIGURE 9.29 Typical base velocity profile at a chordwise location, with and without control, and the velocity increment due to the control actuator.

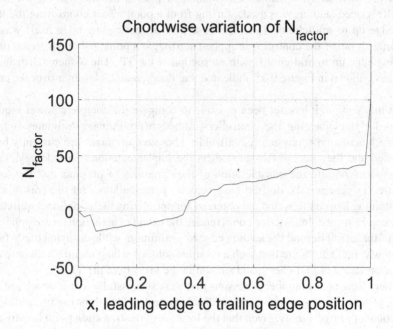

FIGURE 9.30 *N*-factor distribution obtained in the case with base flow actuation over 40% of the airfoil aft of the leading edge.

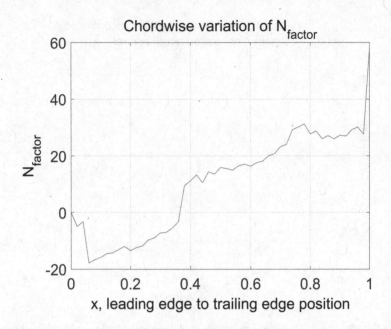

FIGURE 9.31   *N*-factor distribution obtained with base flow actuation over 40% of the airfoil aft of the leading edge and suction in the vicinity of the TE.

In the final study, a limited distribution of plasma actuators at eight discrete equally spaced stations was used, starting from a point 5% of chord from the leading edge up to mid-chord. Thus, the base flow at the leading edge itself was not influenced; rather the control was applied starting at a point just downstream of the leading edge up to mid-chord, with no suction at the TE. The *N*-factor distribution obtained, shown in Figure 9.32, indicated that flow remained laminar over the entire chord.

In this section, it has not been possible to compare the complete power requirements for implementing the controllers, although preliminary estimates indicate that such actuation is physically realizable. These requirements are currently being assessed and they necessarily depend on the implementation methodology. These assessments require the consideration of the dynamics of plasmas and the forces required to generate the desired base velocity perturbations. Yet the feasibility of maintaining laminar flow and the potential for optimizing the actuation required has been demonstrated. In fact, one could reduce the extent of actuation on the upper surface of the airfoil beyond the leading edge to a minimum while ensuring the *N*-factor is below 9. Indications are that such a minimal solution would require both base flow actuation ahead of mid-chord, and suction in the vicinity of the TE.

There were also a number of lessons that were successfully learnt which are now being used to implement an active feedback controller for inhibiting instabilities in the boundary layer. Firstly, given that the location of the transition point is extremely sensitive to the *N*-factor distribution, the importance of predicting accurately the

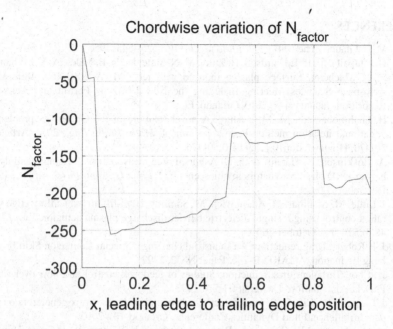

FIGURE 9.32   *N*-factor distribution obtained with base flow actuation over 50% of the airfoil aft of, but not at, the leading edge.

N-factor distribution is vital. Secondly the importance of either measuring or estimating the shape factor of the boundary layer is recognized. Next, by understanding the features of the flow in the boundary layer and the base flows required to inhibit instabilities, one is able to decide on what features must be measured or estimated within the boundary layer. Finally, by knowing the features of the flow within the boundary layer that must be measured or estimated, one is able to design state estimators, filters and custom flow controls that would be capable of generating the required control flows. An integral equation approach is adopted even in the case of plasma actuation which can also be modeled as an equivalent suction flow without being limited by the Kays and Crawford [51] relation. These lessons are currently being used to design practical flow controllers for wings in three-dimensional flow, which is being simulated by the vorticity–velocity approach. A complete understanding and knowledge of the flow in three dimensions is essential, if one wishes to establish a minimal distribution of actuators for controlling the base flow.

CHAPTER SUMMARY

Plasma actuators are one of the most important means of reducing the drag of electric aircraft, which is extremely important in estimating the total propulsive power required to propel an aircraft forward. For this reason, the underlying theories of passive and active devices that could be used for drag reduction of an electric aircraft and the underlying principles of the application of active control to drag reduction are discussed in this chapter.

REFERENCES

1. M. B. Glauert, The wall jet, *J. Fluid Mech.*, 1(6):625–643, 1956.
2. D. F. Opaits, M. R. Edwards, S. H. Zaidi, M. N. Shneider, R. B. Miles, A. V. Likhanskii, S. O. Macheret, Surface plasma induced wall jets, AIAA 2010-469, 48th AIAA Aerospace Sciences Meeting Including the New Horizons Forum and Aerospace Exposition, January 4–7, 2010, Orlando, FL.
3. H. Shahmohamadi, M. M. Rashidi, A novel solution for the Glauert-jet problem by variational iteration method-Padé approximant, *Math. Probl. Eng.*, 2010, Article ID 501476, Hindawi, doi:10.1155/2010/501476.
4. A. Hoskinson, N. Hershkowitz, D. Ashpis, Force measurements of single and double barrier DBD plasma actuators in quiescent air, *J. Phys. D: Appl. Phys.*, 41(24):245209, December 2008.
5. J. Little, M. Nishihara, I. Adamovich, M. Samimy, High-lift airfoil trailing edge separation control using a single dielectric barrier discharge plasma actuator, *Exp. Fluids*, 48(3):521–537, October 2009.
6. J. P. Robert, Drag reduction: An industrial challenge. Special Course on Skin Friction Drag Reduction, AGARD-R-786, Paper No. 2, 1992.
7. H. Choi, Turbulent drag reduction: Studies of feedback control and flow over riblets. PhD Thesis, Stanford University, USA, 1992.
8. J. Casper, J. C. Lin, C. S. Yao, Effect of sub-boundary layer vortex generators on incident turbulence, Fluid Dynamics Conference, Orlando, FL, 2003.
9. H. Mai, G. Dietz, W. Geissler, K. Richter, J. Bosbach, H. Richard, K. de Groot, Dynamic stall control by leading-edge vortex generators, *J. Am. Helicopter Soc.*, 62:26–36, 2006.
10. H. Babinsky, E. Loth, S. Lee, Vortex generators to control boundary layer interactions. US Patent US 8,656,957 B2, February 25, 2014.
11. L. Gao, H. Zhang, Y. Liu, S. Han, Effects of vortex generator on a blunt trailing-edge airfoil for wind turbine, pp 304–305. In *Effects of Vortex Generator on a Blunt Trailing-Edge Airfoil for Wind Turbine*, Elsevier Ltd, Beijing, 2014.
12. K. Yang, L. Zhang, J. Z. Xu, Simulation of aerodynamic performance affected by vortex generators on blunt trailing-edge airfoils, *Sci. China Technol. Sci.*, 53(1):1–7, 2010.
13. G. Godard, M. Stanislas, Control of a decelerating boundary layer. Part 1: Optimization of passive vortex generators. *Aero. Sci. Tech.*, 10(3):181–191, 2006.
14. J. C. Lin, Control of turbulent boundary-layer separation using micro-vortex generators, *AIAA J.*, 3404:99, 1999.
15. F. Lu, Q. Shih, A. Pierce, C. Liu, Review of micro vortex generators in high-speed flow, AIAA-2011-31, 49th AIAA Aerospace Sciences Meeting including the New Horizons Forum and Aerospace Exposition, Orlando, FL, January 4–7, 2011.
16. H. Taylor. *The Elimination of Diffuser Separation by Vortex Generator*, United Aircraft Corporation, 1947.
17. R. S. Shevell, R. D. Schaufele, R. L. Roensch. *Stall Control Device for Swept Wings*, US Patent 524,024, February 1, 1966.
18. L. H. Feng, T. N. Jukes, K.-S. Choi, J. J. Wang, Flow control over a NACA 0012 airfoil using dielectric-barrier-discharge plasma actuator with a gurney flap, *Exp. Fluids*, 52(6):1533–1546, 2012.
19. T. N. Jukes, K.-S. Choi, On the formation of streamwise vortices by plasma vortex generators, *J. Fluid Mech.*, 733:370–393, 2013.
20. K.-S. Choi, T. N. Jukes, R. D. Whalley, L. Feng, J. J. Wang, T. Matsunuma, T. Segawa, Plasma virtual actuators for flow control, *J. Flow Control Meas. Vis.*, 3:22–34, 2015.
21. P. F. Zhang, A. B. Liu, J. J. Wang, Aerodynamic modification of a NACA 0012 airfoil by trailing-edge plasma gurney flap, *AIAA J.*, 47(10):2467–2474, 2009.

22. T. N. Jukes, T. Segawa, H. Furutani, Flow control on a NACA 4418 using dielectric-barrier-discharge vortex generators, *AIAA J.*, 51(2):452–464, 2013.

23. N. Tetervin, Laminar flow of a slightly viscous incompressible fluid that issues from a slit and passes over a flat plate, TN 1644, NACA, 1948.

24. T. R. Bewley, S. Liu, Optimal and robust control and estimation of linear paths to transition, *J. Fluid Mech.*, 365:305–349, 1998.

25. T. R. Bewley, Flow control: New challenges for a new renaissance, *Prog. Aerosp. Sci.*, 37(1):21–58, 2001.

26. T. R. Bewley, P. Moin, R. Temam, DNS-based predictive control of turbulence: An optimal benchmark for feedback algorithms, *J. Fluid Mech.*, 447:179–225, 2001.

27. M. Chevalier, J. Hœpffner, E. Åkervik, D. S. Henningson, Linear feedback control and estimation applied to instabilities in spatially developing boundary layers, *J. Fluid Mech.*, 588:163–187, 2007.

28. S. Ahuja, Reduction methods for feedback stabilization of fluid flows. PhD Thesis, Princeton University, New Jersey, 2009.

29. S. Ahuja, C. W. Rowley, Feedback control of unstable steady states of flow past a flat plate using reduced-order estimators, *J. Fluid Mech.*, 645:447–478, 2010.

30. R. Dadfar, O. Semeraro, A. Hanifi, D. S. Henningson, Output feedback control of Blasius flow with leading edge using plasma actuator, *AIAA J.*, 51(9):2192–2207, 2013, doi:10.2514/1.J052141.

31. A. Monokrousos, F. Lundell, L. Brandt, Feedback control of boundary-layer bypass transition: Comparison of simulations with experiments, *AIAA J.*, 48(8):1848–1851, 2010.

32. O. Semeraro, S. Bagheri, L. Brandt, D. S. Henningson, Feedback control of three-dimensional optimal disturbances using reduced-order models, *J. Fluid Mech.*, 677:63–102, 2011, doi:10.1017/S0022112011000620.

33. C. Davies, P. W. Carpenter, A novel velocity-vorticity formulation of the Navier-Stokes equations with applications to boundary layer disturbance evolution, *J. Comp. Phys.*, 172(9):119–165, 2011.

34. D. F. Opaits, Dielectric barrier discharge plasma actuator for flow control, NASA/CR—2012-217655, National Aeronautics and Space Administration Glenn Research Center, Cleveland, OH, September 2012.

35. M. V. Wickerhauser. *Adaptive Wavelet Analysis: From Theory to Software*, IEEE Press, New York, 1994.

36. B. Vidakovic. *Statistical Modeling by Wavelets*, Wiley, New York, 1999.

37. S. Mallat, A. Wavelet. *Tour of Signal Processing*, Academic Press, San Diego, 1998.

38. G. Strang, T. Nguyen. *Wavelets and Filter Banks*, Cambridge Press, Wellesley, MA, 1996.

39. A. Graps, An Introduction to wavelets, *IEEE Comp. Sci. Eng.*, 2(2), Summer, 50–61, 1995.

40. M. Misiti, Y. Misiti, G. Oppenheim, J.-M. Poggi, *MATLAB Wavelet Toolbox User's Guide*, The MathWorks, Inc., Natick, MA, 1997.

41. R. Vepa, An inverse boundary layer approach to laminar flow control, Paper Y1, Advanced Aero Concepts, Design and Operations, Applied Aerodynamics Conference 2014, Bristol, July 22–24, 2014.

42. M. Gaster, Active control of boundary layer instabilities using MEMS, *Curr. Sci.*, 79(6):774–780, 2000.

43. M. Högberg, D. S. Henningson, Linear optimal control applied to instabilities in spatially developing boundary layers, *J. Fluid Mech.*, 470:151–179, 2002.

44. J. Kim, T. R. Bewley, A linear systems approach to flow control, *Annu. Rev. Fluid Mech.*, 39(1):383–417, 2007.

45. M. Gaster, Active control of laminar boundary layer disturbances, IUTAM Symposium on Flow Control and MEMS, IUTAM Book Series 7, pp 281–292, 2008.

46. B. A. Belson, S. Onofrio, C. W. Rowley, D. S. Henningson, Feedback control of instabilities in the two-dimensional Blasius boundary layer: The role of sensors and actuators, *Phys. Fluids*, 25(5):054106-0–054106-17, 2013.

47. R. E. Hanson, P. Lavoie, K. M. Bade, A. M. Naguib, Steady-state closed-loop control of bypass boundary layer transition using plasma actuators, AIAA 2012-1140, 50th AIAA Aerospace Sciences Meeting including the New Horizons Forum and Aerospace Exposition, January 9–12, 2012, Nashville, TN.

48. P. Schmid, D. S. Henningson, *Stability and Transition in Shear Flows*, Applied Mathematical Sciences Series, Vol. 142, Springer-Verlag, New York, 2000.

49. M. Drela, M. B. Giles, Viscous-inviscid analysis of transonic and low Reynolds number airfoils, *AIAA J.*, 25(10):1347–1355, 1987.

50. C. Ferreira, Implementation of boundary layer suction in XFOIL and application of suction powered by solar cells at high performance sailplanes, Master's Thesis, Delft University of Technology, 2002.

51. W. M. Kays, M. E. Crawford, *Convective Heat and Mass Transfer*, 3rd Edition, McGraw-Hill, New York, 1993.

52. A. A. Merchant, Design and analysis of supercritical airfoils with boundary layer suction. Master's Thesis, Massachusetts Institute of Technology, 1996.

53. C. Wauquiez, Shape optimization of low speed airfoils using MATLAB and automatic differentiation, Licentiate's Thesis, Department of Numerical Analysis and Computing Science, Royal Institute of Technology, Stockholm, Sweden, 2000.

54. J. R. Roth, D. M. Sherman, S. P. Wilkinson, Boundary layer flow control with a one atmosphere uniform glow discharge surface plasma, AIAA Paper 1998-0328, 1998.

55. T. C. Corke, C. L. Enloe, S. P. Wilkinson, Dielectric barrier discharge plasma actuators for flow control, *Annu. Rev. Fluid Mech.*, 66(1):505–529, 2010.

56. B. A. Gibson, M. Arjomandi, R. M. Kelso, The response of a flat plate boundary layer to an orthogonally arranged dielectric barrier discharge actuator, *J. Phys. D: Appl. Phys.*, 45(2):025202, 2012.

57. A. Duchmann, A. Reeh, R. Quadros, J. Kriegseis, C. Tropea, Linear stability analysis for manipulated boundary-layer flows using plasma actuators, Seventh IUTAM Symposium on Laminar-Turbulent Transition, *IUTAM Bookseries*, 18:153–158, 2010.

58. B. A. Belson, Control of the transitional boundary layer, PhD Dissertation, Department of Mechanical and Aerospace Engineering, Princeton University, Princeton, NJ, 2014.

59. M. Riherd, S. Roy, Linear stability analysis of a boundary layer with plasma actuators, AIAA-2012-0290, 50th AIAA Aerospace Sciences Meeting, Nashville, TN, 2012.

60. M. Riherd, S. Roy, S. Balachandar, Local stability effects of plasma actuation on a zero pressure gradient boundary layer, *Theor. Comp. Fluid Dyn.*, 28(1):65–87, 2014.

61. F. O. Thomas, T. C. Corke, M. Iqbal, A. Kozlov, D. Schatzman, Optimization of dielectric barrier discharge plasma actuators for active aerodynamic flow control, *AIAA J.*, 47(9):2169–2178, 2009.

62. J. C. Zito, D. P. Arnold, T. Houba, J. Soni, R. J. Durscher, S. Roy, Microscale dielectric barrier discharge plasma actuators: Performance characterization and numerical comparison, AIAA Paper 2012-3091, 43rd AIAA Plasma dynamics and Lasers Conference, New Orleans, LA, June 25–28, 2012.

63. A. Santhanakrishnan, D. A. Reasor, Jr., R. P. LeBeau, Jr., Characterization of linear plasma synthetic jet actuators in an initially quiescent medium, *Phys. Fluids*, 21(043602):1–18, 2009.

64. S. Chan, X. Zhang, S. Gabriel, Attenuation of low-speed flow-induced cavity tones using plasma actuators, *AIAA J.*, 45(7):1525–1535, 2007.

65. E. Peers, X. Huang, X. Luo, A numerical model of plasma-actuator effects in flow-induced noise control, *IEEE Trans. Plasma Sci.*, 37(11):2250–2255, 2009.

66. E. Moreau, A. Debien, N. Bénard, T. N. Jukes, R. Whalley, K.-S. Choi, A. Berendt, J. Podlinski, J. Mizeraczyk, Surface dielectric barrier discharge plasma actuators. ERCOFTAC Bulletin 94, pp 5–10, 2012.

67. J. Jolibois, M. Forte, E. Moreau, Separation control along a NACA 0015 Airfoil using a dielectric barrier discharge actuator, pp 175–182. In Morrison J. F., Birch D. M., Lavoie P. (Eds.). Proceedings of the IUTAM Symposium on Flow Control and MEMS, IUTAM Book Series, Vol. 7, 2008, ISBN 978-1-4020-6857-7.

68. P. Magnier, B. Dong, D. Hong, A. Leroy-Chesneau, J. Hureau, Control of subsonic flows with high voltage discharges, pp 199–202. In Morrison J. F., Birch D. M., Lavoie P. (Eds.). Proceedings of the of the IUTAM Symposium on Flow Control and MEMS, IUTAM Book Series, Vol. 7, ISBN 978-1-4020-6857-7.

69. X. Huang, X. Zhang, Streamwise and spanwise plasma actuators for flow-induced cavity noise control, *Phys. Fluids*, 20(3):1–7, 2008.

70. D. F. Opaits, Dielectric barrier discharge plasma actuator for flow control, NASA/CR—2012-217655, pp 3–44, 2012.

71. J. R. Roth, Aerodynamic flow acceleration using para-electric and peristaltic electrohydrodynamic effects of a one atmosphere uniform glow discharge plasma, *Phys. Plasmas*, 10(5):2117, 2003.

72. A. Kurz, S. Grundmann, C. Tropea, M. Forte, A. Seraudie, O. Vermeersch, D. Arnal, N. Goldin and R. King, Boundary layer transition control using DBD plasma actuators, *J. Aerosp. Lab*, AL06-02:1-8 (HAL-id: hal-01184462), 2013.

73. K. Ramakumar, J. D. Jacob, Flow control and lift enhancement using plasma actuators, AIAA Paper 2005-4635, 35th Fluid Dynamics Conference, Toronto, ON, Canada, June 6–9, 2005.

10 Photovoltaic Cells

10.1 HISTORY OF THE PHOTOELECTRIC EFFECT

Although the photoelectric effect was discovered by Henri Becquerel in 1839, photo conductivity of selenium was established in 1873, which was followed by the first semiconductor point-contact rectifier being manufactured a year later, the first silicon solar cell appeared only in 1954, as a result of the upcoming semiconductor technology, and it had an efficiency as low as ETA = 5%. The photoelectric effect is the generation of charge following the absorption of photons by a material. Photons are first captured by a material and the charges that are generated are conducted away by the external circuit, thus resulting in a current flow. Silicon photo diodes are light sensitive "*p-n*" junction diodes encased in a transparent material. The junction between a "*p*"-type or positive doped semiconductor with an excess of holes and an "*n*"-type or negatively doped semiconductor with an excess of electrons, is usually reverse-biased when operating as a photoconductive cell. The reverse leakage current present when the diode is reverse-biased is sensitive to light. However, when operating as a photovoltaic cell, a similar junction is unbiased.

10.2 SEMICONDUCTORS: SILICON PHOTO DIODES

When enough energy is supplied by the effect of the light radiation to a photo-diode to separate the valence electrons from the atom and reduce the thickness of the depletion layer at the junction of the p-type and n-type semiconductor layers, the effect of the reduction in the thickness is to increase the reverse current by as much as 100%. Figure 10.1 shows the construction of such a silicon photo diode of the p-n junction type. A thin plastic lens is usually incorporated into the glass housing to focus the light onto the active region of the junction. A p-n junction bar with light impinging on it generates holes only on the n side. The generation of a hole or the absence of an electron is accompanied by the generation of an electron. Only those holes within the hole diffusion length, L_p, of the space charge region contribute to the photocurrent.

The hole diffusion current can be expressed by the product of the charge

$$I_p = qAL_pG, \qquad (10.1)$$

where A is the junction cross-sectional area, G is the hole–electron pair generation rate, which is proportional to the incident light flux. The product AL_p represents the effective volume in which the relevant holes are generated. Only those photons which possess an energy exceeding the band gap can produce hole–electron pairs. The photocurrent is proportional to the number of these incident photons per unit time. It can be written as:

$$I = I_0\left(e^{qV/kT} - 1\right) - qAG\left(L_p + L_n\right). \qquad (10.2)$$

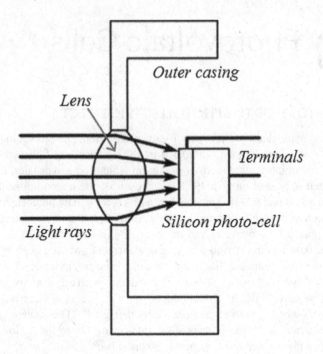

FIGURE 10.1   Construction of such a silicon photo diode of the *p-n* junction type.

The first term is the *p-n* junction dark current under the influence of a voltage, the second is the combined photocurrent due to hole generation in the *n* side and electron generation in the *p* side of the junction.

The equivalent circuit of a silicon photo-diode is shown in Figure 10.2. The effect of the current generated by the illumination is shown as a current source in parallel with the non-linear light-free or dark current characteristic.

The typical current voltage characteristics of photodiode both in the dark and in the presence of light are shown in figure on the next slide for different values of the input power. When the diode is short-circuited, $V=0$ and $I=I_{sc}$. The optically generated current I_p, is the current when the reverse bias voltage is very large. The performance of a photo-diode and its operating characteristics are determined by

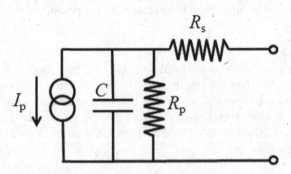

FIGURE 10.2   Equivalent circuit of a silicon photo-diode.

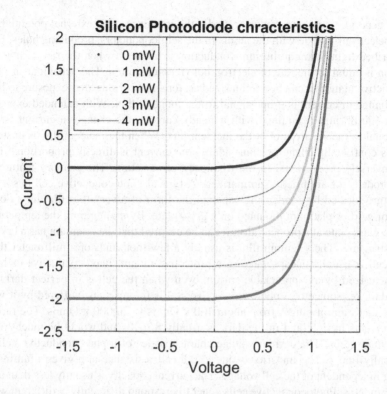

FIGURE 10.3 The current versus voltage characteristics of a silicon photo-diode.

the current versus voltage behavior. In Figure 10.3 are plotted current versus voltage characteristics of a typical photo-diode.

10.3 PHOTOCONDUCTIVE CELLS

Photoconductive cells or light-dependent resistors are based on the photo conductive effect. This is the change in resistance of certain semiconductor materials when exposed to light. The basic process of photoconductivity is the absorption of photons by the semiconductor, resulting in the production of free carriers. Photoconductors exploit the resulting change of conductance. Both intrinsic and extrinsic photoconduction is possible. An intrinsic or pure semiconductor is a semiconductor without any significant doping present. In such photoconducting materials, the number of holes and electrons are the same and the number of photoconductors is intrinsic to the material and does not depend on the doping. There are a number of intrinsic photo conductive materials such as: selenium, cadmium sulfide, lead sulfide, selenide and indium antimonide. In an intrinsic photoconductive material, incident light results in electron emission within the material. In an extrinsic material, impurity dopants of both types ae added, and additional energy levels are introduced by the dopants which provide additional excitation pathways. Examples of extrinsic materials are silicon and germanium, doped with arsenic, copper, gold or indium. They are mostly used for detecting long wavelength infrared light.

In general, in a photoconductor, under the influence of an external potential, the photoelectrons collide with the atoms in the semiconductors, producing holes. For a prescribed light flux, equilibrium conduction is obtained when the rate of hole formation is equal to the rate of electron hole recombination. The behaviour of photo conductive materials can be summarized as: for short periods after exposure to light, the change in current is proportional to the incident flux. With extended exposure, when equilibrium is attained with a steady current, the change in current is proportional to the square root of the incident flux. When the incident flux is unsteady and is continually changing sinusoidally, the current is directly proportional to it, but inversely proportional to the frequency of variation. The primary features of a photodetector are briefly summarized. A typical photoconductive cell has a grid structure formed by condensing vapor in the form of a very thin film on a double grid of gold or platinum mounted on a glass plate. By maintaining the temperature of the glass plate at a proper value, it can be ensured that the selenium has a crystalline structure. The selenium film is usually a few thousands of a millimeter thick. The cell is usually encapsulated in a protective mount. Photoconductive cells are characterized by an appreciable current even when the cell is in perfect darkness. The dark resistance of a photo conductive cell is the resistance offered by it when in the dark, and it ranges from about 100 k Ohms to 20,000 k Ohms. The ratio of the currents in light and in the dark is usually 8 to 10 and would be much higher in certain cases. The current voltage characteristics of a photoconductive cell are generally linear. The linearity is due to the cell-resistance at a given illumination being independent of the cell voltage. The current capacity is usually less than a few milliamperes. Photoconductive cells generally respond differently at different wavelengths. Selenium cells are particularly sensitive in the red and infrared region and are therefore used as infrared detectors.

10.4   THE PHOTOVOLTAIC EFFECT

The photovoltaic (PV) effect is the generation of a potential difference across the junction of two semiconductor substrates. The main difference between the photovoltaic cell and the photo-diode is the fact that the external voltage is used to bias the junction in a photo-diode. Photovoltaic cells can be constructed from semiconductor *p-n* junctions as well as semiconductor metal junctions. In the former type, the effect of the reverse bias is generated by the incident light or thermal radiation. To be effective the junction areas must be quite large compared to those of an ordinary rectifier. A typical silicon photovoltaic cell constructed from semiconductor *p-n* junctions is illustrated in Figure 10.4.

Silicon, with an atomic number of 14, has 14 protons and has four electrons in the outermost shell, eight in the next shell and two in the innermost shell. The electrons in the outer shell, known as the valence electrons, connect with four other neighboring atoms, giving it a diamond crystalline structure. In a crystalline solid, each silicon atom normally shares one of its four valence electrons in a "covalent" bond with each of four neighboring silicon atoms. When a dopant such as phosphorus is added to silicon, there is an additional free electron and with a dopant such as boron there

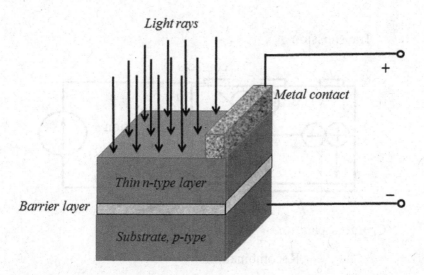

FIGURE 10.4 Typical silicon photovoltaic cell constructed from semiconductor *p-n* junctions.

is an additional hole. The former is an *n*-type semiconductor and the latter a *p*-type semiconductor, while silicon itself is a non-conductor of electricity.

A typical silicon PV cell is composed of a thin wafer consisting of an ultra-thin layer of phosphorus-doped (*n*-type) silicon on top of a thicker layer of boron-doped (*p*-type) silicon. An electrical field is created in the *n*-type layer when photons enter the *p*-type layer with an energy exceeding the band gap of the layer. When photons from a light source enter the *p*-type layer, a number of free electron–hole pairs are established, resulting in a flow of current when the solar cell is connected to an electrical load. At the *p-n* junction, a barrier layer is formed with very few charge carriers. This region is also known as the depletion zone. The important feature of these barrier layers is that there is no need to reverse-bias the junction. The potential barrier is almost as wide and there is a potential hill. The incidence of energy from sunlight creates an electron–hole pair by breaking the covalent bonds between atoms in the barrier layer. Holes and electrons are swept to the *p*- and *n*-sides. Thus, the generation of electric current following the incidence of photons happens inside the barrier layer or depletion zone of the *p-n* junction. Only incident photons with an energy greater than the band gap energy E_G, generate electron–hole pairs and using Planck's law, it follows that the photon frequency must satisfy the relation $hf > E_G$ in order to generate any photocurrent. The energy gap is itself a function of the voltage generated by the incident photons.

10.4.1 The Photovoltaic Cell: The Solar Cell

The process of electron–hole pair generation in a silicon photocell (solar cell) is illustrated in Figure 10.5. The ideal efficiency of solar cells at 300 K exposed to maximum radiation and with different band gap energy levels is shown in Figure 10.6.

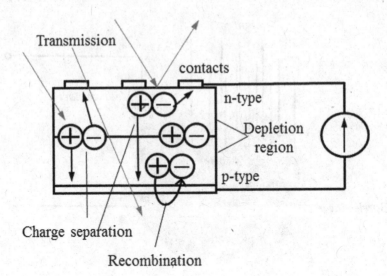

FIGURE 10.5 Electron–hole pair generation in a silicon photocell.

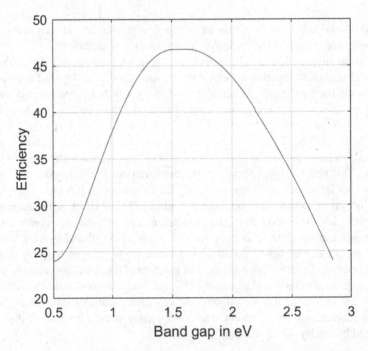

FIGURE 10.6 The ideal efficiency of solar cells at 300 K with different band gap energy levels.

Different semiconductor materials which are characterized by different band gaps have differing efficiencies in generating electron–hole pairs when exposed to sunlight. Figure 10.6 illustrates the estimated ideal solar cell efficiency at 300 K. Atmospheric absorption by the Earth's atmosphere corresponding to an air mass of 1.5 is assumed in Figure 10.6.

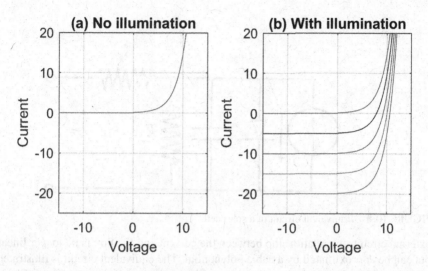

FIGURE 10.7 (a) without illumination (b) compared with different levels of illumination.

10.4.2 SOLAR CELL CHARACTERISTICS

The current-voltage characteristic of a solar cell can be obtained from the bottom right-hand part of the photo-diode characteristic. The characteristics of a typical solar cell are illustrated in the Figure 10.7.

The objective is to obtain the maximum power from such a cell. Hence, the operating current and voltage must be chosen so that the area enclosed by it is a maximum. The ratio of the product of the operating current and voltage $I_m V_m$ to the product of the short circuit current and the open circuit voltage $I_{sc} V_{oc}$ is always less than one and is referred to as the "fill factor" (FF). Hence, the maximum power output of a silicon photocell may be written as $P_{max} = FF I_{sc} V_{oc}$. Such solar cells are extensively used for low power applications in unmanned satellites. They are extremely low in weight and yet can be connected in series-parallel combinations to deliver the necessary power in space. While silicon photocells are more common, gallium arsenide (GaAs) cells, which can operate at higher temperatures than silicon, can be used more efficiently by using solar concentrators. Solar sails use a large amount of semiconductor material and the use of GaAs may not always be cost effective.

The semiconductor metal type of barrier layer cells consists of a layer of semiconductors on a metal base. Thus, this may consist of a copper oxide layer deposited on a metal base plate of copper, gold or platinum. Alternatively, the cell may consist of a thin layer of ferrous selenide deposited on an iron plate. Illumination of the boundary of the barrier plane, between the semiconductor and the metal, results in a potential across the barrier. A current will flow if an external circuit connects the semiconductor to the metal. The incident light energizes the electrons in the semiconductor beyond the band gap. They are then able to leave the boundary and flow into the metal and return to the external circuit. If the external circuit is open, a potential difference is set up until equilibrium is established and no electron flow is possible. The potential difference so generated is usually of the order of 4 mv per foot candle of illumination. The short circuit current of a barrier layer type cell is directly proportional to the incident light flux. When a resistance is introduced in the

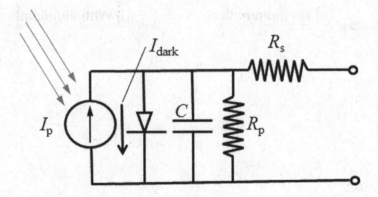

FIGURE 10.8 Equivalent circuit of a solar cell.

external circuit, the relationship between the current and the flux is no longer linear but can be approximated by a cubic polynomial. The equivalent circuit is illustrated in Figure 10.8. Photovoltaic cells are sensitive to high temperature. The selenide cells get damaged if operated above 60° C. The main advantage of these cells is that no external bias voltage is required to operate them. They are extensively used in photographic exposure meters.

10.4.3 Modeling the Power Output of a Solar Cell

The two important characteristics of a solar cell are the open circuit voltage output with an infinite load across its terminals and the short circuit current with no load across its terminals. The short circuit current is directly proportional to the illumination ϕ. Since the short circuit current I_{sc} is approximately proportional to the area of the solar cell exposed to light, the short circuit current density $J_{sc} = I_{sc}/A$ is the characteristic that is often measured. When the solar cell is connected to a load, a current will flow across its terminals and a voltage develops as there is a build-up of charge at the terminals. The current is a superposition of the short circuit current, caused by the absorption of photons, and a dark current, which is caused by the charge build up at the terminals and flows to oppose the short circuit current. The dark current may be expressed in terms of the electronic charge q, the voltage across the terminals V, the absolute temperature T and a constant J_0 as

$$J_{dark}(V) = J_0\left(e^{qV/kT} - 1\right).\qquad(10.3)$$

Consequently, the total current is

$$J = J_{sc} - J_0\left(e^{qV/kT} - 1\right).\qquad(10.4)$$

To find the open circuit voltage, $J = 0$. Solving for the corresponding voltage the

$$V_{oc} = \frac{kT}{q}\ln\left(\frac{J_{sc}}{J_0} + 1\right).\qquad(10.5)$$

The power density is simply the product of the load current density J_L, and the load voltage V_L. Hence,

$$P_{den} = J_L V_L. \tag{10.6}$$

The efficiency of a solar cell is defined as the power (density) output divided by the power (density) output. Thus,

$$\eta = J_{max} V_{max} / P_{in}. \tag{10.7}$$

In terms of the fill factor, the short circuit current density and the open circuit terminal voltage, the efficiency of the solar cell is

$$\eta = \left(J_{sc} V_{oc} / P_{in} \right) FF. \tag{10.8}$$

When considering a practical equivalent circuit of a solar cell, it is customary to include a shunt resistance in parallel with the cell across its terminals and a series resistance. When this is done the current density in the load may be expressed in terms of the load voltage V_L as

$$J_L = J_{sc} - J_0 \left(e^{q(V_L + J_L A R_s)/kT} - 1 \right) - \left(V_L + J_L A R_s \right) / A R_p. \tag{10.9}$$

Linearizing this relation, combining all the terms that are functions of current density J_L on the right-hand side of the equation, and solving for J_L:

$$J_L \approx \left(J_{sc} - J_0 \left(e^{q V_L / kT} - 1 \right) - \frac{V_L}{A R_p} \right) \Big/ \left\{ 1 + \frac{R_s}{R_p} \left(1 + \frac{R_p}{V_L} J_0 A e^{q V_L / kT} \right) \right\}. \tag{10.10}$$

Assuming that the dark current is constant, one expresses the load current as

$$I_L = I_{sc} - I_{dark} - \left(V_L + I_L R_s \right) / R_p. \tag{10.11}$$

Hence,

$$I_L = \left(I_{sc} - I_{dark} - \frac{V_L}{R_p} \right) \Big/ \left(1 + \frac{R_s}{R_p} \right). \tag{10.12}$$

Hence the expression for the load current may be modified as

$$I_L = \left(I_{sc} - I_{dark} - \frac{V_L}{R_p} \right) \Big/ \left(1 + \frac{R_s}{R_p} \left(1 + \frac{R_p}{V_L} J_0 A e^{q V_L / kT} \right) \right). \tag{10.13}$$

Since the short circuit current is proportional to the illumination, if I_{scn} is the short circuit current when the illumination $\phi = \phi_n$, the equation for the load current is

$$I_L = \left(I_{scn} \left(\frac{\phi}{\phi_n} \right) - I_{dark} - \frac{V_L}{R_p} \right) \Big/ \left(1 + \frac{R_s}{R_p} \left(1 + \frac{R_p}{V_L} J_0 A e^{q V_L / kT} \right) \right). \tag{10.14}$$

The power delivered to the output is

$$P_{\text{deli}} = V_L I_L$$

$$= V_L \left(I_{\text{scn}} \left(\frac{\phi}{\phi_n} \right) - I_{\text{dark}} - \frac{V_L}{R_p} \right) \bigg/ \left(1 + \frac{R_s}{R_p} \left(1 + \frac{R_p}{V_L} J_0 A e^{qV_L/kT} \right) \right), \tag{10.15}$$

which is a maximum when

$$V_L = \frac{R_p}{2} \left(I_{\text{scn}} \left(\frac{\phi}{\phi_n} \right) - I_{\text{dark}} \right)$$

$$\times \left(\frac{1 + \dfrac{R_s}{R_p} \left(1 + \dfrac{R_p}{V_L} J_0 A e^{qV_L/kT} \right) - \left(R_s J_0 A e^{qV_L/kT} \left(\dfrac{q}{kT} - \dfrac{1}{V_L} \right) \right)}{\left(1 + \dfrac{R_s}{R_p} \left(1 + \dfrac{R_p}{V_L} J_0 A e^{qV_L/kT} \right) - 0.5 \left(R_s J_0 A e^{qV_L/kT} \left(\dfrac{q}{kT} \cdot \dfrac{1}{V_L} \right) \right) \right)} \right). \tag{10.16}$$

An approximate maximum is given at

$$V_{L\max} \approx \frac{R_p k}{2} \left(I_{\text{scn}} \left(\frac{\phi}{\phi_n} \right) - I_{\text{dark}} \right),$$

$$I_{L\max} = \frac{\dfrac{2-k}{2} \left(I_{\text{scn}} \left(\dfrac{\phi}{\phi_n} \right) - I_{\text{dark}} \right)}{\left(1 + \dfrac{R_s}{R_p} \left(1 + \dfrac{R_p}{V_{L\max}} J_0 A e^{qV_{L\max}/kT} \right) \right)}. \tag{10.17}$$

where k may be considered to be a constant and is normally $1.2 \le k \le 1.4$. Hence, the maximum power delivered by the solar cell is given by

$$P_m = \frac{\dfrac{R_p k (2-k)}{4} \left(I_{\text{scn}} \left(\dfrac{\phi}{\phi_n} \right) - I_{\text{dark}} \right)^2}{\left(1 + \dfrac{R_s}{R_p} \left(1 + \dfrac{R_p}{V_{L\max}} J_0 A e^{qV_{L\max}/kT} \right) \right)}. \tag{10.18}$$

Further details of the derivation may be found in Vepa [1].

10.4.4 Maximum Power Point Tracking

The ability of a solar cell to produce power is limited, so it is necessary to force the cell to operate in conditions which match up with the cell's maximum power point (MPP). An MPP tracking (MPPT) controller which utilizes an MPPT algorithm can undertake this duty. When a solar cell operates at the MPP, the MPP is unique and is a fixed operating point. The load power requirements are not taken into account and output voltage will vary with variations of load resistance. Thus, the MPPT controller works in the way that the solar cell operates at its MPP while the output voltage is kept constant. The solar cell is usually simulated in two steps, one for verifying performance of the MPPT controller and another for establishing the stability of the voltage-regulating controller. A solar cell's output power depends on the non-linearly applied current or voltage, and there is a unique MPP. Thus, it is possible to track the MPPs by a maximum seeking controller. The locus of MPPs varies non-linearly with the unpredictable variations in the solar cell's operation conditions. An MPP tracking (MPPT) controller is essential to continuously deliver the highest possible power to the load when variations in operation conditions occur. Usually the MPPT controller is designed as a multi-loop controller with at least two loops. The inner loop is designed to operate the solar cell at its MPP. A fast and efficient MPPT inner loop control scheme for solar cells is based on the methodology of the sliding-mode control approach. Typically, the closed loop system includes the solar cell, a boost chopper, a battery and the controller. Sliding mode control is used to control the duty cycle of the chopper in order to achieve MPPT. Detailed algorithms for sliding mode control may be found in Vepa [2]. The outer loop uses a maximum seeking algorithm to estimate the MPP in real-time and then gives the estimated value to the inner loop as the set-point, at which the inner loop forces the solar cell to operate. The MPPT controller provides a control law that can keep the cell at the MPPs in real time. Simulation is used as a tool to show the stable operation of the controller in the presence of variations in the solar cell operation conditions.

10.4.5 THE SHOCKLEY–QUEISSER LIMIT

Shockley and Queisser [3] established a theoretical maximum efficiency limit of 33.7% for the performance of solar cells (assuming a single *p-n* junction with a band gap of 1.4 eV). The Shockley–Queisser limit was calculated by examining the amount of electrical energy that is extracted per incident photon. The Shockley–Queisser limit is a function of band gap energy and is given by a curve similar to Figure 10.4. The Shockley–Queisser limit is based on four assumptions: i) there is a single *p-n* junction, ii) the illumination is with sunlight that is not concentrated, iii) a single electron–hole pair is excited by each incoming photon and iv) thermal relaxation of the electron–hole pair energy is in excess of the band gap energy. The calculations are based on three considerations: i) Blackbody radiation from the solar cell is lost as heat (7%). ii) Recombination places an upper limit on the rate of electron–hole production which results in losses up to 10%. Recombination limits the maximum open-circuit current of a solar cell. iii) Spectrum losses resulting from the assumption that only each incident photon gets converted into an electron which limits the short-circuit current density (i.e. current density at zero voltage). Open circuit voltage limit due to electron–hole recombination is illustrated in Figure 10.9. The dotted line in the figure is $V_{oc} = E_g$.

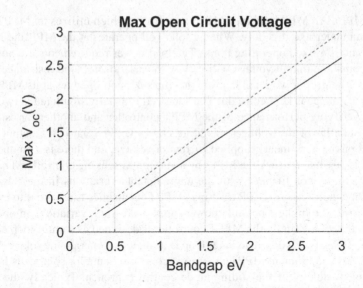

FIGURE 10.9 The maximum open circuit voltage due to electron–hole recombination.

At higher band gaps, there are fewer photons above the band gap, and the current density decreases with increasing band gap energy. There have been several generations of solar cells which have on one hand become less expensive and on the other hand, more efficient. The first-generation of cells were uni-junction cells based on expensive silicon wafers; they make up about 85% of the current commercial market. The second-generation cells are thin films of materials such as amorphous silicon, nanocrystalline silicon, cadmium telluride, or copper indium selenide. These cells are grouped into three families namely: amorphous silicon (a-Si) and micro-amorphous silicon (μa-Si); cadmium telluride (CdTe); and finally, copper indium selenide (CIS) and copper-indium gallium di-selenide (CIGS). The materials are less expensive, but research is being conducted to raise the cells' efficiency. Third-generation cells include organic PV cells which are currently still in development. However, new generations of solar cells have dramatic increases in efficiency and maintain the cost advantage of thin films. These modern designs of third- and fourth-generation cells make use of sunlight concentration, carrier multiplication, hot electron extraction, multiple junctions or new materials. Concentrating sunlight allows for a greater contribution from multi-photon processes which increase the theoretical efficiency limit to 41% for a single-junction cell with thermal relaxation. Carrier multiplication is a quantum-dot phenomenon that results in multiple electron–hole pairs for a single incident photon. Thermal relaxation of the electron–hole pair energy in excess of the band gap is by hot-electron extraction. Hot-electron extraction provides a way to increase the efficiency of nanocrystal-based solar cells by tapping off energetic electrons and holes before they have time to thermally relax.

10.5   MULTI-JUNCTION SILICON PV CELLS

A cell with a single *p-n* junction captures only a fraction of the solar spectrum: photons with energies less than the band gap are not captured, and photons with energies greater than the band gap have their excess energy lost to thermal relaxation. Stacked cells with different band gaps capture a greater fraction of the solar spectrum. A multi-junction solar PV cell is a stack of individual single-junction cells in descending order of band gap energy (E_g). The top cell captures the highest-energy photons. It passes the remaining photons on to be absorbed by next band gap cell. The next cell captures the higher-energy photons and passes the rest to the next cell.

The efficiency limit for two junction cells is 43%, 49% for three junctions and 66% for infinite junctions, when illuminated with sunlight that is not concentrated [4]. Concentrators use either Fresnel lenses or Cassegrain mirrors, to obtain an incident beam that is hundreds of times higher than ambient solar flux and increases the overall efficiency, although it is limited by the high temperature generated and therefore requires substantial and continuous cooling. Automatic tracking solar concentrator photovoltaic systems have also been developed.

10.5.1 MODELING THE POWER OUTPUT OF MULTI-JUNCTION CELLS

Figure 10.8 shows an equivalent circuit for a *PV* or solar cell with a single *p-n* junction where a single diode is used to model the dark current branch. The circuit could be extended to model a multi-junction cell *PV* or a solar cell with multiple diodes in parallel to model the dark current branches. Figure 10.10 shows a typical equivalent circuit of a multi-junction solar cell where multiple diodes are used to model the dark current branch.

One may express the current i in the series resistor R_S as

$$i = i_{pv} - i_{01}\left\{\exp\left(\frac{v_o + R_S i}{n_1 v_T}\right) - 1\right\} - i_{02}\left\{\exp\left(\frac{v_o + R_S i}{n_2 v_T}\right) - 1\right\} - \left(\frac{v_o + R_S i}{Z_P}\right), \quad (10.19)$$

where the photocurrent due to the incident light i_{pv} depends primarily on the solar irradiance λ_{si} and the operating temperature T_{op} and can be expressed as

$$i_{pv} = \lambda_{si}\left(i_{sc} + K_i\left(T_{op} - T_{ref}\right)\right). \quad (10.20)$$

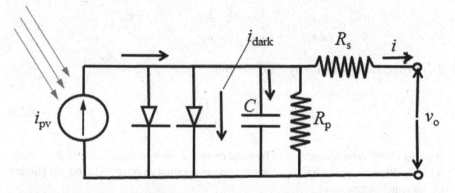

FIGURE 10.10   Equivalent circuit of a multi-junction solar cell.

The quantity v_T is the thermal voltage which is related to the Boltzmann constant k and the electronic charge q by the relation $v_T = n_s k T_{op}/q$, where n_s is the number of diodes connected in series and n_i is the ideality factor of the diodes in each parallel line. The quantity i_{0i} in Equation (10.19) is the reverse saturation current of the diodes in each parallel line which is a function of the operating temperature T_{op}. The voltage v_o is the output voltage of the equivalent circuit in Figure 10.10.

The reverse saturation current of the diodes can be expressed in terms of the reverse saturation current at a reference temperature T_{ref} as

$$i_{0i} = i_{0i,\text{ref}} \left(\frac{T_{op}}{T_{ref}} \right)^{\chi} \exp\left(\beta \left(\frac{1}{T_{ref}} - \frac{1}{T_{op}} \right) \right), \tag{10.21}$$

where χ and β are constants. Furthermore, in the expression for the current i, from the equivalent circuit in Figure 10.10, and in the transform domain, in Equation (10.19), the equivalent parallel output impedance is

$$Z_P = R_P / (1 + s C R_P). \tag{10.22}$$

In Equation (10.22), assuming C is small, and hence that $C=0$, $Z_P=R_P$ is a pure resistance.

Observe that the expression for the current is an implicit equation, with the current appearing on both sides of it. To solve for the current, we may adopt an iterative scheme such as the Newton-Raphson method. The solution for the current is obtained by iteration. Thus, we define

$$f(i) = i - i_{pv} + i_{01} \left\{ \exp\left(\frac{v_o + R_S i}{n_1 v_T} \right) - 1 \right\}$$

$$+ i_{02} \left\{ \exp\left(\frac{v_o + R_S i}{n_2 v_T} \right) - 1 \right\} + \left(\frac{v_o + R_S i}{Z_P} \right), \tag{10.23}$$

then it follows that

$$\frac{df}{di} = 1 + \frac{R_S i_{01}}{n_1 v_T} \exp\left(\frac{v_o + R_S i}{n_1 v_T} \right) + \frac{R_S i_{02}}{n_2 v_T} \exp\left(\frac{v_o + R_S i}{n_2 v_T} \right) + \frac{R_S}{Z_P}. \tag{10.24}$$

Thus, the current is iteratively obtained from

$$i_{k+1} = i_k - \frac{f(i_k)}{df(i_k)/di_k}. \tag{10.25}$$

A panel of PV cells is assumed to be constructed from N_p strings connected in parallel and with each string having N_s solar cells connected in series. The output current of the panel is given by

$$I = N_p I_{ph} - N_p I_0 \left\{ \exp\left(\frac{(V/N_s) + (I/N_p) R_s}{A V_t} \right) - 1 \right\}$$

$$- N_p \frac{(V/N_s) + (I/N_p) R_s}{R_{sh}}.$$

(10.26)

Other important parameters are the dependence of the solar cell temperature on the intensity of the solar irradiation, the dependence of the solar cell temperature on the ambient temperature, the 24-hour variation of the ambient temperature and the wind speed. The wind provides a natural convective cooling for the panel.

Multi-junction solar cells are one of the most promising technologies achieving high sunlight to electricity conversion efficiency. Resistive losses constitute one of the main underlying mechanisms limiting their efficiency under high illumination. Given that the smaller the band gap, the easier it is to promote movement of electrons and there is a corresponding reduction in the resistivity of the material, fine-tuning of the different electronic band gaps (engineering the band gaps) involved in multi-junction stacks will mitigate the detrimental effects of losses for both concentration-dependent and independent series resistances.

CHAPTER SUMMARY

Photovoltaic cells are one of the most important means of reducing the total externally sourced power requirements of an electric aircraft. For this reason, the underlying theories of photovoltaic cells, as well as several recent developments associated with their design that could be used for reducing the overall power drawn by an electric aircraft from a power storage source, such as a battery, are discussed in this chapter.

REFERENCES

1. R. Vepa, Non-conventional energy generation: Solar, wave and tidal energy generation, Chapter 8, pp 349–364. In *Dynamic Modelling, Simulation and Control of Energy Generation*, Lecture Notes on Energy Series, Vol. 20, Springer, ISBN 978-1-4471-5399-3, 2013.
2. R. Vepa, *Nonlinear Control of Robots and Unmanned Aerial Vehicles: An Integrated Approach*, CRC Press, 2016.
3. W. Shockley, H. J. Queisser, Detailed balance limit of efficiency of p-n junction solar cells, *J. Appl. Phys.*, 32(3):510–519, 1961.
4. G. W. Crabtree, N. S. Lewis, Solar energy conversion, *Phys. Today*, 60(3):37–42, 2007.

11 Semiconductors and Power Electronics

11.1 SEMICONDUCTORS AND TRANSISTORS

11.1.1 SEMICONDUCTORS AND SEMICONDUCTOR DIODES

Of the different types of electronic bonding, namely, ionic, covalent, metallic and van der Waal bonding, semiconductors are characterized by covalent bonding which involves the bonding by sharing of the four electrons in the outer shell. There are seven major crystalline structures in materials, namely, cubic, tetragonal, rhombic, monoclinic, triclinic, trigonal and hexagonal, with 32 sub-categories. Semiconductors generally belong to the face-centered cubic category where the valency electrons are joined by covalent bonds. The distinguishing features of semiconductors are their crystalline structure which leads to the temperature sensitivity of resistance, sensitivity of conductivity to impurities or dopants, the availability of free electrons, leading to electron and hole conductivity and a narrow band gap between the outer conduction band of electrons and the inner valency band. Intrinsic semiconductors are naturally occurring semiconductors, while extrinsic semiconductors conduct electricity due to the presence of impurities. Thus semiconductors could be "doped" with an impurity, of which there are two types. The two types of impurities are the pentavalent or "n"-type impurities, such as antimony, arsenic and phosphorus that have a free electron to share, and trivalent or "p"-type impurities, such as indium, gallium and boron that have a free hole or one less electron. Thus, in a "p"-type semiconductor there is an excess of holes which can be considered to be the majority carriers of current, while in an "n"-type there is an excess of electrons. There are two types of carrier transport mechanisms in extrinsic semiconductors, namely, transport by diffusion, when electrons move from a region of high concentration to a region of low concentration by the process of diffusion, and transport by drift, when electrons travel under an applied voltage and exchange momentum by almost elastic collisions with other electrons.

A p-n junction or a p-n junction diode is formed by bring two layers of p-type and of n-type semiconductors in close contact with each other. Thus, there two different majority carriers in a p-n junction diode. Holes from the p-type region diffuse across the junction to combine with electrons in the n-type region, nearer to the junction. Thus, there is a small region across the junction known as the depletion region, where a barrier potential develops as a consequence of the diffusion of holes and electrons across the junction. In the depletion region, due to the negative acceptor ions, electrons are repulsed, while holes face repulsion due to the positive donor holes. The polarity of the potential barrier region encourages minority carriers (holes in the n-side and electrons in the p-side) to cross the junction. Thus, a sustained barrier potential difference exists at a *p-n* junction in typical diode, which is not measurable. When the diode is forward-biased, that is when the *p*-type semiconductor is connected to the positive terminal and the *n*-type semiconductor is connected to the negative terminal of a battery, both the depletion region and the potential barrier are reduced. The direction of the field due to the potential barrier is in the opposite direction to the electric field generated by the voltage applied externally to the battery. When a *p-n* junction is forward-biased, the current flow from the positive to the negative terminals is the hole flow and holes are flowing into the *p*-type semiconductor. Thus, there is a reduction in the net barrier potential. In a reverse-biased *p-n* junction, the external applied field enhances the barrier potential. The applied potential being in the same direction of the barrier potential increases the depletion region. Thus, in a forward-biased state, beyond about 0.7 V a very slight increase in the forward bias results in a large increase in the diode current, and the diode conducts while it acts as an insulator in the reverse-biased situation. However, there is a limit to the maximum current that can flow through the diode. Beyond the limit the diode will overheat and is destroyed. The leakage current reaches its maximum very quickly and does not change until the breakdown point is reached. There is then a sharp turning point in the characteristic within the reverse-biased region, as the diode breaks down. A Zener diode, like a normal diode, also permits current to flow in the reverse direction when its "Zener voltage" is reached.

11.1.2 Transistors

Typically, a transistor is a three-terminal device with three layers of extrinsic semiconductors, such as a *p*-type, *n*-type and *p*-type semiconductor layers, or a *n*-type, *p*-type and *n*-type semiconductor layers. There are two junctions between the central and the two adjoining layers. The two junctions of the transistor have depletion regions which are produced during manufacture, just as in a diode. The base or the central layer is made much narrower or thinner than in a diode and this applies to both *p-n-p* and *n-p-n* transistors. Both the depletion regions are of the same size when the junctions are unbiased. The central layer is known as the base region or simply the base, while the other two layers are the emitter and the collector. Generally, in an *n-p-n* transistor, the base-emitter junction is forward biased while the base-collector junction is reversed biased. In an *n-p-n* transistor, the depletion region of the base-emitter junction is reduced when it is forward biased, as in a forward-biased *n-p* junction of a diode while other is not affected. In an *n-p-n* transistor, the depletion region of the base-collector junction is increased when it is reverse biased, while the other is not affected. Only a small current flows through the base-collector junction in this state. When the base-emitter junction is forward biased and the base-collector junction is reverse biased, current enters the emitter layer from the negative terminal of the externally connected battery. It then splits into two and one flows into the base which is very small, while the other flows into the collector. When the base-emitter junction is forward biased and the base-collector junction is reverse biased, a net electron current flows into the collector region. In a *p-n-p* transistor the base-emitter junction is forward biased and a current flows into the emitter. The base-collector junction is reverse biased and no current flows through it. The overall operation is otherwise similar to an *n-p-n* transistor. The common emitter output characteristics of the transistor, that is, the emitter held at the ground potential relative to both the base and collector, indicate that a small change in the base current (10 s of μ amps) generates a big change in the collector current (in milliamps). Likewise, in a common collector configuration, the transistor characteristics indicate that a small change in the base current (10 s of μ amps) generates a big change in the emitter current (in milliamps). Thus, the transistor performs like an amplifier when suitably biased.

11.2   POWER ELECTRONIC DEVICES

The bipolar junction transistor (BJT) was a low current gain transistor that led to power electronics applications of transistors. It needs a high base current to obtain a reasonable current in the collector. However, it was the invention of the field effect transistors (FET) that heralded the development of power electronics applications. FETs have two big advantages over bipolar transistors: one is that they have a near-infinite input resistance and thus have near-infinite current and power gain; the other is that their switching action is fast and suitable for digital switching applications. An FET is a three-terminal amplifying device with its terminals known as the source, the gate and the drain, and they correspond respectively to the emitter, base and collector of a normal transistor. Two different families of FETs are currently in use and have several power control applications. The first of these families is known as the "junction-gate" type of FETs which are abbreviated to JFET. The second family is known as metal oxide semiconductor FETs, and are generally abbreviated to MOSFET. Both "*n*-channel" and "*p*-channel" versions of MOSFETS are available. From the MOSFET evolved the power MOSFET, and from the JFET, the static induction transistor (SIT) was developed. A balanced array of parallel-connected low-power MOSFETs are equivalent to a single high-power MOSFET. These high-power devices are known to result in a performance that is essential in power amplifier applications. While the traditional transistor was purely an amplifying device, the power electronic devices were capable of being used as switches and for sequential switching operations of high-power devices.

Power electronic devices may be classified as i) controlled semiconductor devices, ii) uncontrolled devices such as the power diode, iii) semi-controlled (which turn on) such as the thyristor and the silicon-controlled rectifier (SCR) and iv) fully-controlled devices, such as the MOSFET, the insulated gate bipolar transistor (IGBT) and the gate-turnoff thyristor (GTO). There are indeed other power electronic devices as well as other methods of classification such as current-controlled (e.g. the GTO), or voltage (field)-controlled (e.g. the MOSFET), pulse-triggered devices (e.g. the GTO) or level-sensitive devices (e.g. the MOSFET). The most common power electronic devices are summarized in Table 11.1. Some of these are now obsolete and are not manufactured any longer. Power electronic devices are the electronic devices that are used in the power-processing circuits to convert or control electric power. There are a number of features that distinguish power electronics from conventional information processing electronic devices. It is not primarily the power handling capacities. Power electronics all devices are operated in the switching mode—either fully "ON" or "OFF" states. Linear electronic devices generally concentrate on fidelity in signal

TABLE 11.1
Power Electronic Devices: Summary

Device	Year invented	Symbol
Diode	1955	
Thyristor	1958	
Triac	1958	
Gate turn-off thyristor	1980	
Bipolar power transistor	1975	
Power MOSFET	1975	
High electron mobility Transistor (HEMT)	1980	
Insulated gate bipolar transistor	1986	
Static induction transistor (SIT)	1985	
Integrated gate commutated thyristor	1996	

amplification, requiring transistors to operate strictly in the linear zone. Power electronic devices are characterized by significant levels of power inputs and outputs. The typical states of operation correspond to "ON" states or "OFF" states, and generally, linear operating states are avoided. They do not operate about an equilibrium point and only the base emitter configurations are preferred. Thus, control signals are fed to the base. There are indeed high losses due to the switching transients.

The practically important power semiconductor devices in relation to motor drives are diodes, thyristors (also called the silicon-controlled rectifier, SCR), the triode thyristors or Triacs, GTOs, IGCTs, MOSFETs, IGBTs and the BJT (although this device has largely been replaced by IGBTs). The IGBTs are being used for a large variety of applications and replacing most other power electronic devices [1, 2]. They are the most common devices used for the control of high-performance motors. The theory and design of the IGBT is explained in some detail in [3]. The IGBT has become an integral part of the power electronic subsystems which are used as building blocks in the design of motor controllers. An example of such a controller for a permanent magnet synchronous motor is shown in Figure 11.1, where a lithium ion battery, represented by its equivalent circuit, drives a DC-DC boost converter followed by a DC-AC voltage source converter which is then used to drive the motor. The purpose of the DC-AC converter is to provide an AC voltage at the right frequency and also meet the matching requirements in that the impedance of the converter must be matched to the motor. While a wide range of devices can be used in place of the three terminal power transistors in the circuit, IGBTs are currently the most popular choice. Future converters may be required to cater to the need for compact, low weight, variable frequency drives.

Conventional semiconductors are not suitable for power electronic device applications. Gallium nitride devices offer five key characteristics that are important for power electronic applications. These include high dielectric strength, high operating temperature, high current density, high speed switching and low resistance in switching "ON". Silicon carbide (SiC), also known as carborundum, is a compound of silicon and carbon, used in semiconductor electronic devices that operate at high temperatures or high voltages, or both. Other materials that are suitable for power electronic devices are gallium arsenide and silicon. With the new developments of wide band gap devices which are able to withstand high voltages in excess of 1,200 V, and currents using materials such as silicon carbide (SiC) and possibly gallium-arsenide (GaN), a new generation of power electronic devices are emerging. A particular area of interest for power electronics is that of electric aircraft, where the demand for

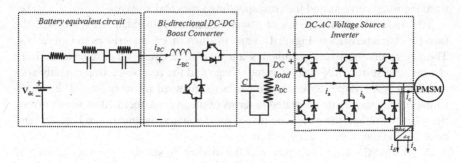

FIGURE 11.1 Building block approach for controlling a permanent magnet synchronous motor.

more compact, long endurance, high power density, high temperature operation, in harsh operating conditions, is of critical importance [4]. The extraordinary features of SiC [5] are definitely going to have an impact on future devices for all-electric aircraft. Wilamowski and Irwin [6] have provided a survey of the current state-of-the-art power electronic devices and their applications to electric motors such as those discussed by Emadi [7]. Power semiconductor devices and smart-control integrated circuits have been key technology drivers in the recent past. In the future, the focus will be on packaging and interconnection technologies, high-power density system integration, together with advances in silicon devices and system reliability, will drive the technical development of power electronics applications in general and power electronic controllers, converters, inverters and active power filters, in particular, for all-electric aircraft. There is also a need for optimizing the management and control of the onboard electrical power to minimize the non-propulsive use of electric power. Power electronic devices used for the control of high performance motors often require new approaches to their modeling and control, as well as new approaches for modeling power converters and filters, and there have been a number of recent studies directed towards these new applications [8–11]. In the following sections, the basic features of power electronic devices which make them useful for applications such the control of high-performance electric motors will be briefly reviewed.

11.2.1 Power Diodes: A Three-Layered Semiconductor Device

A power diode is a three-layer device where the top layer is a heavily doped $p+$ semiconductor layer, the middle layer is lightly doped $n-$ semiconductor layer and the lowest layer is a heavily doped $n+$ semiconductor layer. The heavily doped $p+$ layer is connected to the anode (positive terminal) while the lowest layer is connected to the cathode (negative terminal). The thickness of this layer is around 10 μm while the lowest layer is 25 times that. The middle layer is relatively thin and its actual thickness is engineered according to the desired breakdown voltage. The junction capacitance, including the diffusion and potential barrier capacitances, influences the switching characteristics of power diode. The key characteristics that determine the performance of the diode are the static current-voltage characteristics, the turn-off and turn-on current and voltage characteristics associated with reverse and forward recovery currents and the corresponding times and voltages.

The static I-V characteristics of power diodes are illustrated in Figure 11.2, the turn-off characteristics in Figure 11.3 and the turn-on characteristics in Figure 11.4. The key reverse-recovery parameters are reverse-recovery time, reverse-recovery charge, reverse-recovery peak current/voltage and the reverse-recovery steady voltage. The key forward recovery parameter is the forward recovery time. The specifications of a power diode are stated in terms of the average rectified forward current, the forward voltage, the peak repetitive reverse voltage, the maximum junction temperature and the reverse-recovery time.

Apart from the general purpose rectifier diode with standard recovery times (~50 to 200 μs), relatively fast recovery diodes with both fast and very fast recovery times (1 μs to 0.1 μs) are also available. Schottky diodes and Schottky barrier diodes with a low reverse voltage are also manufactured.

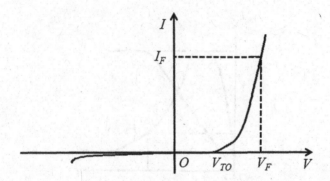

FIGURE 11.2 Static current-voltage characteristics of a power diode.

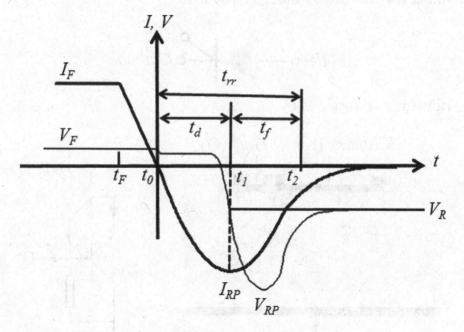

FIGURE 11.3 Turn-off characteristics of a power diode.

11.2.2 THYRISTORS AND SILICON CONTROLLED RECTIFIER (SCR)

The thyristor is an electronic switch, with two main terminals (an anode and a cathode) and a "turn-on" terminal gate). The non-conducting state can only be restored after the anode–cathode current has reduced to zero. It is also known as a silicon-controlled rectifier (SCR). The symbol of an SCR is as shown in Figure 11.5.

The thyristor has a four-layer structure, making it equivalent to two bipolar transistors interconnected together to act as regenerative pair, as shown in Figure 11.6. The gate, as for standard SCRs, is connected to the base of the NPN transistor. A thyristor can only be triggered by a positive gate current. Once the NPN transistor is supplied with a base current, it draws a collector current amplified by the gain of

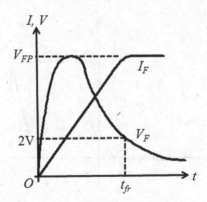

FIGURE 11.4   Turn-on characteristics of a power diode.

FIGURE 11.5   The SCR symbol.

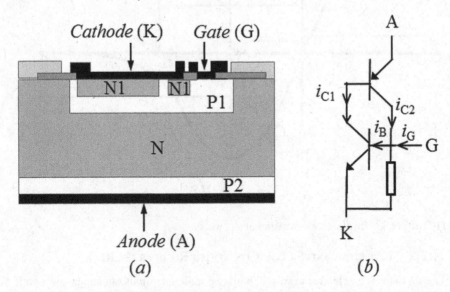

FIGURE 11.6   (a) A thyristor's silicon structure and (b) a simplified equivalent diagram.

the NPN transistor. The collector current is from the base of the PNP transistor. The PNP base current is amplified by the PNP current gain. Thus, even if the gate current is removed, there is a regenerative action which eventually leads to the thyristor remaining in an "ON" state, and a minimum anode current is supplied to trigger the thyristor.

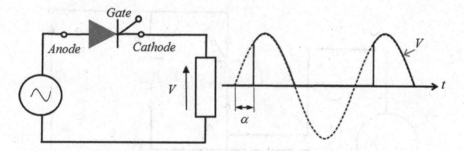

FIGURE 11.7   Single-pulse thyristor-controlled rectifier and its response.

The gate input of the thyristor is used to control the firing delay, which is defined as an angle α. Consider the single-pulse thyristor-controlled rectifier, with a resistive load as shown in Figure 11.7, and a firing angle delay α. The DC level is controlled by varying the firing delay angle, which in effect modulates the pulse width.

The thyristor has been used for motor speed control for several decades. A simple analog circuit, shown in Figure 11.8, can be implemented for controlling a DC motor without requiring any microcontroller. The voltage output from a thyristor is applied across the motor terminals so it can be controlled by varying the thyristor turn-on delay. When the delay is a minimum, the motor speed is a maximum and vice versa.

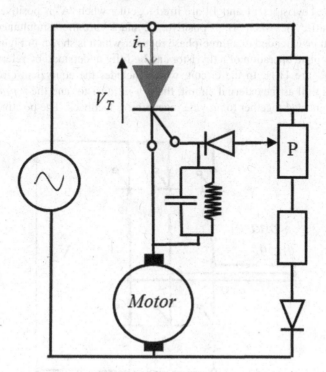

FIGURE 11.8   Motor speed controller.

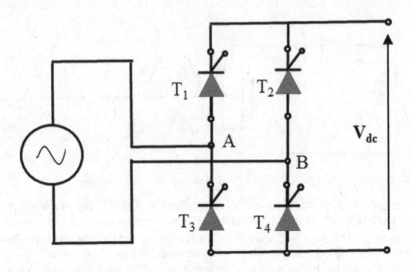

FIGURE 11.9   Fully controllable full-wave rectifier circuit.

The thyristor may also be used to provide fully controllable full wave rectification. A circuit using four thyristors is used just as in a standard full wave rectifier using four silicon diodes. Such a fully controllable full-wave rectifier circuit in shown in Figure 11.9. Thyristors T1 and T4 are fired together when "A" is positive, while on the other half-cycle, when "B" is positive, T2 and T3 are fired simultaneously. The concept can be extended to a three-phase rectifier which is shown in Figure 11.10.

The complete operation of a thyristor can be fully understood by referring to the circuit in Figure 11.11. In the circuit, which includes the equivalent circuit of the thyristor as well as the external circuit, the *p-n-p* transistor and the *n-p-n* transistor are interconnected together to provide regenerative feedback. The positive feedback

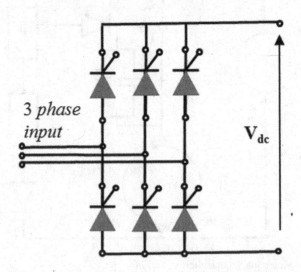

FIGURE 11.10   Three-phase, fully controllable full-wave rectifier circuit.

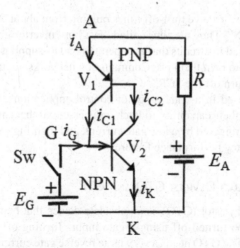

FIGURE 11.11   Equivalent circuit to explain the operation of the thyristor.

trigger serves to switch on the thyristor and also to maintain it in the "ON" state. However, it cannot be turned off by a control signal. Thus, it is only half-controllable. The complete static equations for the operation of the thyristor, in terms of the currents shown in Figure 11.11 and the additional internal collector-to-base currents are:

$$I_{c1} = \beta_1 I_A + I_{CB1}, \quad I_{c2} = \beta_2 I_K + I_{CB2}, \quad I_K = I_A + I_G \text{ and } I_A = I_{c1} + I_{c2}. \quad (11.1)$$

Hence it follows that,

$$I_{c2} = \beta_2 I_A + \beta_2 I_G + I_{CB2}. \quad (11.2)$$

Thus,

$$I_A = I_{c1} + I_{c2} = \beta_1 I_A + I_{CB1} + \beta_2 I_A + \beta_2 I_G + I_{CB2}. \quad (11.3)$$

Consequently,

$$I_A = \frac{\beta_2 I_G + I_{CB1} + I_{CB2}}{1 - (\beta_1 + \beta_2)}. \quad (11.4)$$

Equation (11.4) defines the regenerative action of the thyristor, as the positive feedback drives the thyristor currents to become unstable, when $(\beta_1 + \beta_2) = 1$. Consequently, the thyristor latches on to the "ON" state.

There are two main types of SCRs. Phase control thyristors are reverse biased, at least for a few milliseconds subsequent to a conduction period and are turned off by natural commutation. No fast-switching feature is either desired or possible of these devices. They are available at voltage ratings in excess of 5 KV, starting from about 50 V and current ratings of about 5 KA. The largest converters for high voltage DC transmission are built with series-parallel combination of these devices.

Inverter grade thyristors have turn-off times ranging from about 5 to 50 μs after they are switched to "ON". They are thus called "fast" or "inverter grade" SCRs. These SCRs are mainly used in circuits that are operated on DC supplies and no alternating voltage is available to turn them off. Commutation networks are used along with the basic converter to turn off the SCR'.

The SCRs are used in a wide range of control applications. They are installed in functions where their capability to latch on is essential, because of the ease with which they can be triggered by a low gate current, and their high voltage and current capability provide key performance features.

11.2.3 CONTROLLED DEVICES: GTO AND GTR

The gate turn-off thyristor (GTO), is an obsolete device that behaves like a normal thyristor, but can be turned off using a gate input. Turning off the GTO requires complex circuitry. The GTO needs a very large reverse gate current to turn it off. The gate drive design is complex due to this very large reverse gate current at turn-off.

The giant transistor (GTR) is a complex device with multiple breakdown and saturation characteristics, which essentially behaves like a bipolar junction transistor. It too is an obsolete device.

11.2.4 THE MOSFET

The MOSFET is generally a low voltage device that is turned on by applying a voltage in excess of a threshold voltage (15 V) and turned off by reducing this voltage to zero. The symbol of a typical MOSFET including the notation to denote the terminals is shown in Figure 11.12.

High voltage MOSFETs are also available up to a voltage of 600 V, but with a limited current. For higher current capability, several MOSFETs could be connected in parallel. The internal (dynamic) resistance between drain and source during the

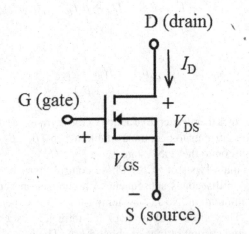

FIGURE 11.12 The symbol and terminal notation of a typical MOSFET.

"ON" state, RDS$_{ON}$, limits its power-handling capability. The resistive losses are high, at high voltages due to RDS$_{ON}$ and are dominant in HF application (>100 kHz). The main application of MOSFETs is in switched-mode power supplies.

The operation of the HEMT, which is shown in Table 11.1, is a little different to other types of FET and as a result, it is able to give a very much enhanced performance over the standard MOSFETs, particularly in microwave RF applications.

11.2.5 THE IGBT

The IGBT is like a MOSFET with an additional p-type semiconductor layer [1–3]. The equivalent circuit of an IGBT, shown in Figure 11.13, is constructed with two semiconductor devices, a MOSFET and a PNP transistor. The symbol of an IGBT and the typical manner in which its terminals are denoted are also shown in the same figure.

The typical ratings of an IGBT are, voltage: $V_{CE} < 4.5\,kV$, current: $I_C < 1.5\ kA$. They are used at switching frequencies up to 100 kHz. The typical application frequencies are, however, limited to 20–50 kHz. A typical set of characteristics, which are similar in shape to a combination of BJT or GTR and MOSFET characteristics, are shown in Figure 11.14.

The ON-state losses in an IGBT are much smaller than those of a power MOSFET, and are comparable with those of a GTR. The IGBT is easy to drive and in this respect, it is quite similar to a power MOSFET. As a device, the IGBT is faster than a GTR, but slower than power MOSFET. The behavior of the gate is similar to that of a MOSFET and it is easy to turn on and off. The losses in an IGBT, like a BJT, are low due to the low on-state collector-emitter voltage (2–3 V).

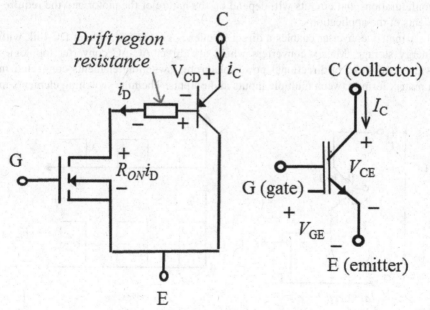

FIGURE 11.13 The equivalent circuit of an IGBT and its symbol.

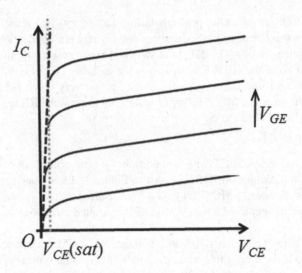

FIGURE 11.14   Typical voltage-current characteristics of an IGBT.

11.2.6 Applications

A typical application of an IGBT is to provide a controlled power source to an application. It is often used in a full-bridge, half-bridge and other related switched configurations of inverters and as controllers for high performance motors. However, various applications have different requirements. Typically, it is controlled by pulse width modulation (PWM) as shown in Figure 11.15 and in Figure 2.6. The actual configurations and circuits will depend on the nature of the motor and the requirements of the applications.

A matrix converter enables a direct frequency conversion without DC-link with energy storage. Matrix converters which are direct AC-AC converter topologies which consist of bi-directional power electronic switching elements connected in a matrix form between multiple inputs and outputs. The nine switching elements in

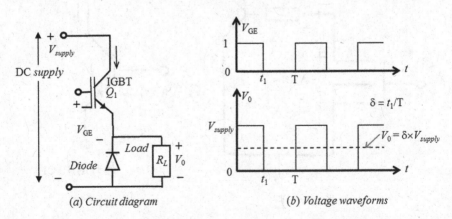

(a) Circuit diagram   (b) Voltage waveforms

FIGURE 11.15   PWM control of an IGBT.

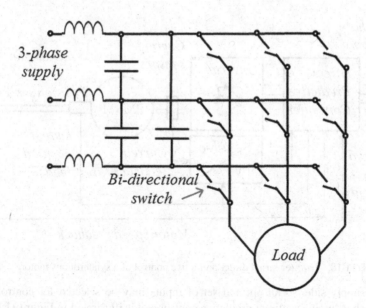

FIGURE 11.16 Typical functional diagram of a matrix converter.

a three-phase matrix converter, shown in Figure 11.16, facilitate the flow of electric power in both directions (from supply to load and vice versa) through the matrix converter, so variable amplitudes and frequencies at the output are delivered by switching the fixed input supply voltage with various modulation algorithms to variable speed drives. A typical application of a matrix converter for controlling a permanent magnet synchronous motor driving a linear actuator is shown in Figure 11.17. A matrix converter can be used in permanent magnet synchronous motor drives where the speed and position of the motor are not measured.

The availability of power electronic converters facilitates the development of new methods for control of high-performance electric drives. One approach involves the prediction of the future values of load currents for all voltages that may be available

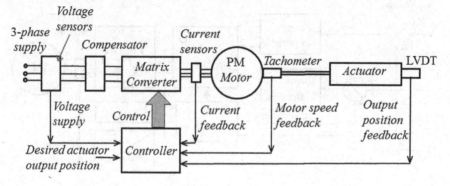

FIGURE 11.17 Control of a synchronous motor driving a linear actuator using a matrix converter.

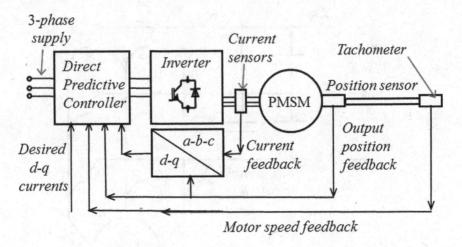

FIGURE 11.18   Inverter-based direct predictive control of a synchronous motor.

at the supply side, so an optimal set of inputs may be selected for control. This approach is known as direct predictive control and is illustrated in Figure 11.18.

A relatively recent development is of multilevel inverters, reviewed by Bana et al. [12], which are capable of generating a multi-level output through a suitable arrangement of power-switching semiconductor devices and voltage sources. They are useful for high-voltage applications because of their ability to synthesize output voltage waveforms with an improved harmonic spectrum and can attain higher voltages even with a limited maximum rating for the device.

The development of these multi-level inverters combined with space vector modulation has facilitated the development of fast and reliable direct torque controllers (DTC) for induction motors, as shown by Hakami, Alsofyani and Lee [13]. The DTC controller proposed by Hakami, Alsofyani and Lee [13] using an IGBT based three-level neutral-point clamped (NPC) inverter is illustrated in Figure 11.19. Figure 11.20 shows an open-end-winding induction motor, which is controlled by a

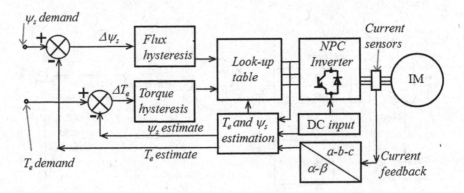

FIGURE 11.19   DTC control of an induction motor using a three-level neutral-point clamped inverter.

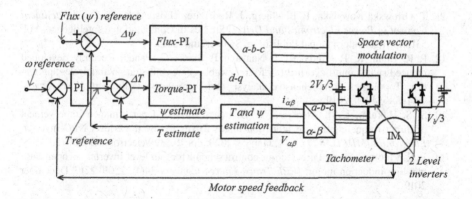

FIGURE 11.20   DTC control of an induction motor using a pair of two-level inverters combined space vector modulation.

pair of two-level inverters based on the methodology of DTC combined space vector modulation, as proposed by Vinod and Shiny [14].

CHAPTER SUMMARY

Semiconductors and power electronics are expected to play a key role in the evolution of all-electric aircraft. For this reason, an overview of semiconductors and power electronic components is presented in this chapter.

REFERENCES

1. B. J. Baliga, *Power Semiconductor Devices*, PWS Publishing Company, ISBN 0-534-94098-6, 1996.
2. B. J. Baliga, Trends in power semiconductor devices, *IEEE Trans. Electron Devices*, 43(10):1717–1731, October 1996.
3. V. K. Khanna. *Insulated Gate Bipolar Transistor (IGBT): Theory and Design*, IEEE Press, Wiley-Interscience, Hoboken, NJ, 2003.
4. P. R. Wilson, Advanced aircraft power electronics systems-the impact of simulation, standards and wide band-gap devices, *CES Trans. Elec. Mach. Syst.*, 1(1):72–82, March 2017.
5. X. Guo, Q. Xun, Z. Li, S. Du, Silicon carbide converters and MEMS devices for high-temperature power electronics: A critical review, *Micromachines (MDPI)* 10(406):1–26, 2019, doi:10.3390/mi10060406.
6. B. M. Wilamowski, J. D. Irwin (Eds.), *Power Electronics and Motor Drives*, 2nd Edition, CRC Press, Boca Raton, FL, 2011.
7. A. Emadi, *Energy-Efficient Electric Motors*, 3rd Edition, CRC Press, Boca Raton, FL, 2004.
8. S. Filizadeh, *Electric Machines and Drives: Principles, Control, Modeling, and Simulation*, CRC Press, Boca Raton, FL, 2017.
9. J. Kabziński (Ed.), Advanced control of electrical drives and power electronic converters, *Stud. Syst. Decis. Control*, Vol. 75, Springer International Publishing, Basel, Switzerland, 2017.
10. W. Piotr (Ed.), *Dynamics and Control of Electrical Drives*, Springer-Verlag, Berlin Heidelberg, 2011.

11. T. Orłowska-Kowalska, F. Blaabjerg, J. Rodríguez (Eds.), *Advanced and Intelligent Control in Power Electronics and Drives*, Studies in Computational Intelligence, Vol. 531, Springer International Publishing, Switzerland, 2014.

12. P. R. Bana, K. P. Panda, R. T. Naayagi, P. Siano, G. Panda, Recently developed reduced switch multilevel inverter for renewable energy integration and drives application: Topologies, comprehensive analysis and comparative evaluation, *IEEE Access*, 7:54888–54909, 2019.

13. S. S. Hakami, I. M. Alsofyani, K.-B. Lee, Low-speed performance improvement of direct torque control for induction motor drives fed by three-level NPC inverter, *Electronics (MDPI)*, 9(77):1–14, January 2020, doi:10.3390/electronics9010077.

14. B. R. Vinod, G. Shiny, Direct torque control scheme for four level-inverter fed open-end-winding induction motor, *IEEE Trans. Energy Conver.*, 34(4): 2209–2217, December 2019.

12 Flight Control and Autonomous Operations

12.1 INTRODUCTION TO FLIGHT CONTROL

The basic principles governing the design and operation of aircraft flight control systems are well known. Generally the term "flight control systems" for manned or autonomous flight includes not only the primary flying control systems, including digital fly-by-wire control systems, but also a host of secondary/slat and flap (high-lift) flight control systems, rudder and yaw controllers, stabilizer control systems, spoiler control electronics and the associated monitoring systems, autopilots and flight control units, actuator control electronics and any associated remote electronics units, collision avoidance systems, vehicle management systems/computers and systems for prognostic assessment and health management. From a hardware standpoint, it also includes the multiple redundant flight control and flight management computers, as well as pilot controls such as control sticks and sidesticks with their associated trim and feel control systems. The flight control computers compute and transmit all primary surface (rudder, elevators, ailerons, flaperons and stabilizers) actuator commands, pre-flight and other signals to control and maintain normal flight. Also referred to as high-lift or secondary flight controls, slat and flap controllers enable optimum take-off and landing speeds by increasing wing lift. The primary purpose of the slats and flaps is to increase the area of the wing, thus enabling the aircraft to reduce speed while generating the same lift. Thus, controlling the slats and flaps can be used to increase or decrease the area generating lift, which enables the pilot to control the speed. Slats and flaps are extended and retracted, and the spoilers are raised and lowered using suitable actuators. The actuator controller hardware transmits commands directly, or through remote electronic units, to surface actuators and provides surface commands in specialized flight and system modes. Finally, a flight control system also includes data interface capabilities and data buses defined in accordance with one of several standards such as ARINC, as well as data concentration and distribution facilities and flight data acquisition systems. In this chapter only the basic concepts and principles of flight control systems and optimal flight path synthesis relevant to electric aircraft will be examined. For a general overview of flight control systems, the reader is referred to Vepa [1].

12.1.1 RANGE AND ENDURANCE OF AN ELECTRIC AIRCRAFT

The range of an aircraft depends on various factors and is calculated as the total distance that can be traversed during the flight. The range is important to performance parameters, as it then determines the key conditions that must be met to maximize range. These conditions are generally maintained and tracked by the flight control system and set by the flight management system. It depends to a large extent on the total energy available on board the aircraft, the nature of the propulsion system, the mass of the aircraft and the overall aerodynamic properties of the aircraft.

Considering previous research relating to range and endurance estimation, Sachs [2, 3] has presented the flight performance characteristics of electric aircraft. Traub [4] and Hepperle [5] presented relations for the battery capacity in terms of the range and endurance. Donateo et al. [6] presented a new approach to calculating endurance for electric flight while differentiating between gross and net estimations. Traub [7] provided a technique for validating endurance of an electric aircraft. Avanzini and Giulietti [8], Avanzini, de Angelis and Giulietti [9] discussed the effect of introducing the Peukert effect, introduced by Peukert [10]. The Peukert effect, due to which the effective capacity of the battery decreases with increasing current taken from the battery, is discussed by Vepa [11].

The range for a classical, fuel-burning propeller driven aircraft R_f is given by the so-called Breguet range equation and is a function of the initial to final mass ratio m_0/m_f the lift to drag ratio C_L/C_D, the specific fuel consumption sfc, the local acceleration due to gravity g and the propulsive efficiency η_p. It is given by

$$R_f = \{\eta_p/(g \cdot sfc)\}(C_L/C_D)\log(m_0/m_f) \qquad (12.1)$$

In the case of battery-powered electric aircraft the mass of the aircraft does not change significantly and hence the range equation is different. Typically, if it is assumed that the aircraft climbs rapidly to the cruise height after take-off and descends rapidly prior to landing, the range is determined by the cruise velocity and the time of the cruise. Thus, the cruise range for an all-electric aircraft R_{ce} is given by

$$R_{ce} = \int_0^{t_c} v_c dt = v_c t_c. \qquad (12.2)$$

In Equation (12.2) it was assumed that the cruise velocity v_c is almost constant, and consequently the range is directly proportional to the time of cruise. In the case of a battery-powered electric aircraft, the time to cruise is equal to the time to drain the battery power available for cruise, which, under ideal conditions, could be expressed in terms of the mass of the battery m_{batt}, the specific energy available during cruise E_c and the power drawn from the batteries P_{batt}, and may be expressed as

$$t_c = m_{batt}E_c/P_{batt}. \qquad (12.3)$$

The power required by the aircraft may be expressed in terms of the power drawn from the batteries P_{batt} as

$$P_a = \eta_a P_{batt}. \qquad (12.4)$$

Furthermore, the specific energy available during cruise E_c may be expressed as a fraction of the total specific energy available to the aircraft E_a as

$$E_c = \eta_c E_a. \qquad (12.5)$$

The power required by the aircraft during cruise is that required to overcome the drag which could be related to the lift generated by the aircraft during cruise. However, from considerations of equilibrium flight, the lift is equal to the weight of the aircraft. Thus, it follows that

$$P_a = v_c D = v_c (D/L)L = v_c (C_D/C_L)m_0 g. \tag{12.6}$$

Thus Equation (12.3) may be expressed as

$$t_c = m_{\text{batt}} E_c/P_{\text{batt}} = \eta_c \eta_a m_{\text{batt}} E_a (C_L/C_D)/(v_c m_0 g). \tag{12.7}$$

Hence the cruise range of an electric aircraft R_{ce} is given by

$$R_{ce} = \{\eta_{ca}/(g \cdot m_{sp})\}(C_L/C_D)(m_{\text{batt}}/m_0), \tag{12.8}$$

where $\eta_{ca} = \eta_c \eta_a$ and $m_{sp} = 1/E_a$ is the specific mass per unit of energy delivered to the aircraft by the battery. Thus, the ratio of R_e to R_f is given by

$$\frac{R_{ce}}{R_f} = \frac{sfc}{m_{sp}} \frac{(m_{\text{batt}}/m_{e0})}{\log(m_{f0}/m_f)}. \tag{12.9}$$

Thus, to maintain the same range as a fossil fuel-powered aircraft, the mass of the battery required is

$$\frac{m_{\text{batt}}}{m_{e0}} = \frac{m_{sp}}{sfc} \log(m_{f0}/m_f). \tag{12.10}$$

But the initial mass of the electric aircraft includes the batteries and so, $m_{e0} = m_{ei} + m_{\text{batt}}$. Hence it follows that,

$$\frac{m_{\text{batt}}}{m_{ei}} = \left(\frac{sfc}{m_{sp}} \frac{1}{\log(m_{f0}/m_f)} - 1 \right)^{-1}. \tag{12.11}$$

Equation (12.11) determines the mass of the battery required. It could be very large unless,

$$(m_{sp}/sfc)\log(m_{f0}/m_f) \ll 1. \tag{12.12}$$

Based on the current battery technology, the above condition on m_{sp} is rather difficult to meet in general and could be met only for certain relatively small fuel-powered aircraft.

Returning to Equation (12.8) and noting that the drag coefficient may be expressed as the drag polar or in terms of the aspect ratio AR and Oswald's efficiency factor e, as

$$C_D = C_{D_0} + KC_L^2, \quad K = 1/(\pi AR \, e), \tag{12.13}$$

it can be shown that the lift to drag ratio C_L/C_D is a maximum when

$$C_{D_0} = KC_L^2. \tag{12.14}$$

Firstly, it is noted that, given the projected area of the wing planform A, the air density at the cruise altitude ρ, and assuming equilibrium flight, the cruise velocity can now be expressed as

$$v_c = \sqrt{v_c^2} = \sqrt{2L/\rho A C_L} = \sqrt{2m_0 g\sqrt{K}/\rho A C_{D_0}}. \tag{12.15}$$

Secondly, when the ratio C_L/C_D is a maximum,

$$C_L/C_D = \sqrt{C_{D_0}/K}\Big/2C_{D_0} = 1\Big/\left(2\sqrt{KC_{D_0}}\right). \tag{12.16}$$

The maximum cruise range of the all-electric aircraft $R_{ce,\max}$ is given by

$$R_{ce,\max} = \left\{\eta_c\eta_a E_a/g\right\}\left(1\Big/\left(2\sqrt{KC_{D_0}}\right)\right)\left(m_{\text{batt}}/m_0\right). \tag{12.17}$$

To determine the maximum endurance, one could first write an expression for the cruise endurance as

$$t_c = R_{ce}/v_c = \left\{\eta_{ca}E_a/g\right\}\left(C_L^{3/2}/C_D\right)\left(m_{\text{batt}}/m_0\right)\sqrt{\rho A/(2m_0 g)}. \tag{12.18}$$

The maximum of the ratio $C_L^{3/2}/C_D$ is obtained when $3C_{D_0} = KC_L^2$. Moreover

$$C_L^{3/2}/C_D = 3C_L^{3/2}/4C_L^2 = \left(27K\right)^{1/4}\Big/\left(4C_{D_0}^{1/4}\right) = \left(27K/(256C_{D_0})\right)^{1/4}. \tag{12.19}$$

Hence the maximum cruise endurance is

$$t_{c,\max} = \left\{\eta_{ca}E_a/g\right\}\left(27K/(256C_{D_0})\right)^{1/4}\left(m_{\text{batt}}/m_0\right)\sqrt{\rho A/(2m_0 g)}. \tag{12.20}$$

Only a few of the parameters in Equations (12.17) and (12.20) are operational parameters or have an impact on operational considerations. However, Equations (12.17) and (12.20) can be used to define some of the optimal design and operational requirement to achieve a maximum range or maximum endurance. Thus, assuming that the design requirements are met, one could define the requirements of the flight control and flight management systems to achieve the maximum possible range or endurance for a given aircraft. It is also possible to define the optimum path that the flight control system must track in order to minimize the direct operating cost or some measure of the flight cost.

12.1.2 Equivalent Air Speed, Gliding Speed and Minimum Power to Climb

Given that the lift is given in terms of the aerodynamic lift coefficient as,

$$L = \left(1/2\right)\rho U_0^2 C_L A, \tag{12.21}$$

one may also define an *equivalent sea level* air speed, based on the flow having the same kinetic energy at sea level i.e.

$$(1/2)\rho_0 U_E^2 = (1/2)\rho U_0^2. \tag{12.22}$$

Hence it follows that

$$U_E = U_0 \sqrt{\rho/\rho_0}. \tag{12.23}$$

The above expression determines the variation of speed with altitude when the aircraft is flying level and at constant speed.

Considering gliding flight, the vertically down "sink speed" may be found from

$$L = (1/2)\rho U_0^2 C_L A = mg\cos(-\gamma). \tag{12.24}$$

Hence it follows that the equivalent sea level gliding speed is

$$U_{Eg} \approx \sqrt{\frac{2mg\cos(-\gamma)}{\rho_0 \, C_L A}} \sin(-\gamma) = \sqrt{\frac{2mg}{\rho_0 \, C_L A}} \times \sqrt{\frac{C_L/C_D}{\left(1+\left(C_L/C_D\right)^2\right)^{3/2}}}. \tag{12.25}$$

Assuming $C_L/C_D \ll 1$,

$$U_{Eg} = \sqrt{\frac{2mg}{\rho_0 C_L A}} \sqrt{C_L/C_D} = \sqrt{\frac{2mg}{\rho_0 C_D A}}. \tag{12.26}$$

When $C_L/C_D \gg 1$, which is generally the case,

$$U_{Eg} = \sqrt{\frac{2mg}{\rho_0 C_L A}} \sqrt{\left(C_L/C_D\right)^{-5/2}}. \tag{12.27}$$

The above asymptote also provides an upper bound for the gliding speed of descent.

In the case of climbing flight, it may be assumed that the thrust is composed of two components, the direct engine thrust, T_e, and the thrust due to the power delivered to a propeller, P, if and when there is one. Thus, for climbing flight,

$$L = mg\cos\gamma, \tag{12.28a}$$

$$m\frac{dU}{dt} = T - D - mg\sin\gamma = T_e + \frac{\eta P}{U} - D - mg\sin\gamma, \tag{12.28b}$$

where η is the efficiency of the propeller in delivering useful power. The latter equation may be expressed as

$$m\frac{d}{dt}\left(\frac{U^2}{2}\right) + mgU\sin\gamma = U\left(T_e - D\right) + \eta P = m\frac{d}{dt}\left(\frac{U^2}{2}\right) + mg\frac{d}{dt}(h), \tag{12.29}$$

where the height rate $\dot{h}$ or rate of climb and the rate of forward travel are respectively given by

$$\dot{h} = U \sin \gamma, \quad \dot{d} = U \cos \gamma. \tag{12.30}$$

In general, the equation for the height rate or the rate of climb must be integrated to obtain the speed in terms of other parameters. In the case of climb with constant speed, introducing the rate of climb as $u_c = U \sin \gamma$, the steady rate of climb is given by

$$u_{c0} = \frac{\eta P + (T_e - D)U_0}{mg} = \frac{\eta P}{mg} + \frac{(T_e - D)U_0}{L} \cos \gamma$$

$$= \frac{\eta P}{mg} + \left(\frac{T_e}{mg} - \frac{D \cos \gamma}{L} \right) U_0. \tag{12.31}$$

Hence, eliminating U_0 and $\cos \gamma$,

$$u_{c0} = \frac{\eta P}{mg} + \left(\frac{T_e}{mg} - \frac{C_D}{C_L} \sqrt{1 - \left(\frac{u_{c0}}{U_0} \right)^2} \right) \sqrt{\frac{2mg}{\rho\, C_L A}} \left(1 - \left(\frac{u_{c0}}{U_0} \right)^2 \right)^{\frac{1}{4}}. \tag{12.32}$$

This is an implicit equation that could be solved numerically for u_{c0}.

The vertical velocity of an airplane depends on the flight speed and the inclination of the flight path or the climb angle. In fact, the rate of climb is the vertical component of the flight path velocity. The maximum achievable vertical velocity of an aircraft is dependent on the altitude at which the aircraft flies.

When $u_{c0} = 0$,

$$\frac{\eta P_0}{mg} + \frac{T_{e0}}{mg} \sqrt{\frac{2mg}{\rho\, C_L A}} = \frac{C_D}{C_L} \sqrt{\frac{2mg}{\rho\, C_L A}}, \tag{12.33}$$

which allows one to estimate the minimum power required to climb.

When the direct thrust provided by the engine is equal to zero, $T_{e0} = 0$,

$$P_0 = \frac{C_D}{C_L} \frac{mg}{\eta} \sqrt{\frac{2mg}{\rho\, C_L A}}. \tag{12.34}$$

Hence P_0 is a minimum when C_D is a minimum. Thus, the minimum power to climb is given by

$$P_0 = \frac{C_D}{C_L} \frac{mg}{\eta} \sqrt{\frac{2mg}{\rho\, C_L A}} = 2 \frac{Kmg}{\eta} \sqrt{\frac{2mg}{\rho A}} \sqrt{\frac{C_{D_0}}{K}}. \tag{12.35}$$

12.2   FLIGHT PATH OPTIMIZATION

A typical flight management system attempts to compute the climb, cruise and descent profiles that minimize the direct operating costs (DOC), expressed as:

$$J_{\text{DOC}} = C_i \int_0^{t_f} \left(i + C_{ti} \right) dt, \tag{12.36}$$

where, C_i=unit cost of battery usage (cost units/amps or cost units/units of current), i=current drawn from the battery (amps/hour or amps/time unit), C_{ti}=the Cost Index (amps/hour or amps/time unit) i.e. the ratio of the time cost $Ct = C_{ti}C_i$ (cost units/time unit) to battery usage cost, scaled to the units of current and t_f=the time of flight.

The concept of "energy height", h_e, is defined as the sum of the potential and kinetic energies of the aircraft per unit weight and is given by:

$$h_e = E/mg = \left(mgh + (1/2)mV^2 \right) \Big/ mg = h + (1/2)V^2/g. \tag{12.37}$$

where m is the aircraft mass, $V = V_{\text{TAS}}$ the true airspeed and h the altitude. The rate of change of the energy height is known as the "specific excess power", P_e, and is

$$P_e = \frac{dh_e}{dt} = \frac{dh}{dt} + \frac{V}{g}\frac{dV}{dt}. \tag{12.38}$$

Using the velocity V and the flight path angle γ, as natural coordinates and referring to Figure 12.1, the equations of motion may be written as

$$m\frac{dV}{dt} = T - D - mg\sin\gamma, \quad mV\frac{d\gamma}{dt} = L - mg\cos\gamma. \tag{12.39}$$

Thus, from the first of the equations of motion

$$\frac{V}{g}\frac{dV}{dt} + V\sin\gamma = V\left(\frac{T - D}{mg} \right). \tag{12.40}$$

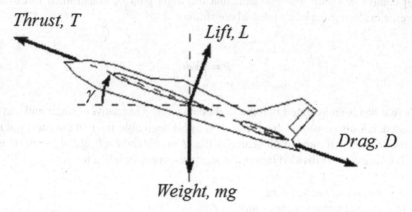

FIGURE 12.1   Forces acting on point mass model of aircraft.

But the rate of climb is given by $dh/dt = V \sin \gamma$. Hence, it follows that

$$P_e = \frac{dh_e}{dt} = \frac{dh}{dt} + \frac{V}{g}\frac{dV}{dt} = V\left(\frac{T-D}{mg}\right). \tag{12.41}$$

Thus, the rate of climb is given by

$$\frac{dh}{dt} = V\left(\frac{T-D}{mg} - \frac{1}{g}\frac{dV}{dt}\right) = V\left(n - \frac{1}{g}\frac{dV}{dt}\right), \tag{12.42}$$

where n is the "specific excess thrust" in gs.

The concepts of "energy height" and "specific excess power" are useful in determining how best to climb to a predetermined altitude and airspeed. For a minimum time solution, one approach is to adopt the maximum available thrust to maximize the rate of climb, while the other is to climb with a fixed, rate-limited maximum rate of climb.

Given also a model for the charge flow rate from the battery as,

$$dC_b/dt = -i, \tag{12.43}$$

the DOC performance index, J_{DOC}, may be expressed as

$$J_{DOC} = \int_0^{t_f}\left[\left(\frac{c_i}{c_i + c_t}\right)i + \left(1 - \frac{c_i}{c_i + c_t}\right)\right]dt \equiv \int_0^{t_f} P dt, \tag{12.44}$$

where c_f is the cost of fuel and c_t is the cost of time, $C_i = c_i/(c_i + c_t)$, and the cost index is the ratio, $C_{ti} = c_t/c_i$.

To complete the statement of the optimization problem, and to solve it as an optimal control problem, it is essential that the flight path be constrained to cover a specific range written as an integral constraint:

$$R = \int_0^{t_f} V dt, \tag{12.45}$$

where it has been assumed that the flight path angle γ is relatively small and can be ignored. It is also assumed that the wind speed is negligible and that the energy at the beginning and at the end of flight are $E = E_0$ at $t = t_0 = 0$ and $E = E_f$ at $t = t_f$, respectively.

The trajectory is divided into three segments, corresponding to

i)   Climb (increasing energy),
ii) Cruise (constant energy) and
iii) Descent (decreasing energy).

To facilitate an independent solution for each segment, we may split the cost functional into three contributing terms corresponding to climb, cruise and descent cost. Making the substitution, $dt = dE/\dot{E}$, the cost function may be expressed as,

$$J_{DOC} = \int_{E_0}^{E_{\max}} \left(P/|\dot{E}|\right)_{\dot{E}>0} dE + \gamma\left(E_{\max}\right) R_c + \int_{E_f}^{E_{\max}} \left(P/|\dot{E}|\right)_{\dot{E}<0} dE, \quad (12.46)$$

where $E_{\max}$ the cruise energy, R_c the cruise range, $\gamma(E_{\max})$ the cruise efficiency given by,

$$\gamma\left(E_{\max}\right) = \left(P/V_c\right)_{\dot{E}=0,\, E=E_{\max}}, \quad (12.47)$$

and V_c the cruise velocity.

The cruise range, R_c is related to the full range R by the relation

$$R_c = R - R_{up} - R_{dn}. \quad (12.48)$$

So,

$$R_c = \int_{E_0}^{E_f} V/\dot{E}\, dE - \int_{E_0}^{E_{\max}} \left(V/|\dot{E}|\right)_{\dot{E}<0} dE - \int_{E_f}^{E_{\max}} \left(V/|\dot{E}|\right)_{\dot{E}<0} dE, \quad (12.49)$$

where R_{up} and R_{dn} are respectively given by

$$R_{up} = \int_{E_0}^{E_{\max}} \left(V/|\dot{E}|\right)_{\dot{E}<0} dE, \quad R_{dn} = \int_{E_f}^{E_{\max}} \left(V/|\dot{E}|\right)_{\dot{E}<0} dE. \quad (12.50)$$

The cruise range must satisfy the inequality constraint

$$R_c = \int_{E_0}^{E_f} \frac{V}{\dot{E}}\, dE - \int_{E_0}^{E_{\max}} \left(V/|\dot{E}|\right)_{\dot{E}<0} dE - \int_{E_f}^{E_{\max}} \left(V/|\dot{E}|\right)_{\dot{E}<0} dE \geq 0. \quad (12.51)$$

Substituting for the cruise range in the cost function

$$J_{DOC} = \int_{E_0}^{E_{\max}} \left(\frac{P - \gamma\left(E_{\max}\right)V}{\dot{E}}\right)_{\dot{E}>0} dE + \gamma\left(E_{\max}\right) \int_{E_0}^{E_f} \frac{V}{\dot{E}}\, dE$$
$$+ \int_{E_f}^{E_{\max}} \left(\frac{P - \gamma\left(E_{\max}\right)V}{|\dot{E}|}\right)_{\dot{E}<0} dE. \quad (12.52)$$

Thus, the problem is to minimize the cost function, J_{DOC}, subject to the constraint that $R_c \geq 0$.

12.2.1 THE OPTIMAL CONTROL METHOD

Applications of optimal control to flight path optimization have been considered by Barufaldi, Morales and Silva [12], Kaptsov and Rodrigues [13], Falck et al. [14], Settele and Bittner [15, 16], Sachs, Lenz and Holzapfel [17] and Menon, Sweriduk and Bowers [18].

The cruise range may be expressed as

$$R_c = \int_{E_0}^{E_f} \mathcal{R}_c \, dE. \tag{12.53}$$

The cost integral may be expressed as

$$J_{\mathrm{DOC}} = \int_{E_0}^{E_f} P \, dE. \tag{12.54}$$

The Lagrangian to be minimized is given by,

$$L = P + \lambda \mathcal{R}_c, \tag{12.55}$$

where the Lagrange multiplier or co-state λ satisfies the requirement that, $\lambda \geq 0$. An augmented cost function is defined as

$$J_{\mathrm{augmented}} = \int_{E_0}^{E_f} L \, dE. \tag{12.56}$$

The minimization may be achieved by minimizing an augmented cost function independent for cruise, climb and descent.

12.2.2 CRUISE OPTIMIZATION: OPTIMAL CONTROL FORMULATION

In cruise, since the velocity may be considered to be equal to a constant, the rate of change of energy height is equal to the rate of change of altitude.

Thus, the equations of motion are

$$\dot{x} = V - V_w, \quad \dot{h} = V(T - D)/mg = u, \quad \dot{C}_b = -i, \tag{12.57}$$

where x is the lateral x-$axis$ position, V_w the headwind velocity and u the rate of climb.

The cost function is the only payoff at the terminal time and may be expressed in terms of the remaining charge in the battery at the end of the flight, $C_b(t_f)$, and the terminal time, t_f, as

$$J_{cost} = \int_0^{t_f} 0 \cdot dt, \quad \varphi(t_f) = c_i C_b(t_f) + c_t t_f, \quad J_t = J_{cost} + \varphi(t_f). \qquad (12.58)$$

Following optimal control theory, the Lagrangian and Hamiltonian are respectively defined as

$$L = 0 + \lambda_x(\dot{x} - V + V_w) + \lambda_h(\dot{h} - u) + \lambda_C(\dot{C}_b + i),$$

$$H = \lambda_x(V - V_w) + \lambda_h u - \lambda_C i. \qquad (12.59)$$

Hence, the Lagrangian may be expressed as

$$L = \lambda_x \dot{x} + \lambda_h \dot{h} + \lambda_C \dot{C}_b - H. \qquad (12.60)$$

Thus, we need to minimize

$$J_{augmented} = \int_0^{t_f} L\, dt + \varphi(t_f) = \int_0^{t_f} \left(\lambda_x \dot{x} + \lambda_h \dot{h} + \lambda_C \dot{C}_b - H\right) dt + \varphi(t_f). \qquad (12.61)$$

Integrating the integral term by parts

$$J_{augmented} = -\int_0^{t_f} \left(\dot{\lambda}_x x + \dot{\lambda}_h h + \dot{\lambda}_C C_b + H\right) dt$$

$$+ \left(\lambda_x x + \lambda_h h + \lambda_C C_b\right)\Big|_{t=0}^{t=t_f} + \varphi(t_f). \qquad (12.62)$$

Minimizing the integral part of the augmented performance index

$$J_{augmented, int} = -\int_0^{t_f} \left(\dot{\lambda}_x x + \dot{\lambda}_h h + \dot{\lambda}_C C_b + H\right) dt, \qquad (12.63)$$

with respect to the states and control input, the co-states, λ_x, λ_h and λ_C, can be shown satisfy the differential equations:

$$\dot{\lambda}_x = -\frac{\partial H}{\partial x}, \quad \dot{\lambda}_h = -\frac{\partial H}{\partial h}, \quad \dot{\lambda}_C = -\frac{\partial H}{\partial C_b},$$

and

$$\partial H/\partial u = 0 \quad \text{or} \quad \lambda_h = 0, \qquad (12.64)$$

where $H = \lambda_x \left(V - V_w \right) + \lambda_h u - \lambda_c i$. The boundary conditions at time, $t=0$ and at $t=t_f$ are $\left[x, h, C_b \right]_{t=0} = \left[x_0, h_0, C_{b0} \right]$ and $\left[x, h, C_b \right]_{t=t_f} = \left[x_f, h_f, C_{bf} \right]$. In addition, transversality at the time, $t=t_f$ requires that $H=c_t$ and $\lambda_c=c_i$.

The seven unknowns are the three states, x, h and m, the three co-states λ_x, λ_h and λ_C, and the final time, t_f. Further corner conditions require that H, λ_x, λ_m and the optimum trajectory are continuous at the junctions between climb and cruise and cruise and descent.

During cruise, the altitude is held constant and the equations reduce to

$$\dot{x} = V - V_w, \quad \frac{dC_b}{dt} = -i, \quad \dot{\lambda}_x = -\frac{\partial H}{\partial x}, \quad \dot{\lambda}_C = -\frac{\partial H}{\partial C_b}, \tag{12.65}$$

and the Hamiltonian is defined as,

$$H = \lambda_x \left(V - V_w \right) - \lambda_c i. \tag{12.66}$$

To solve the optimal control problem, we need to maximize H with respect to V at a given time and for this purpose, the co-states may be regarded as constants, since they are functions of time only.

12.2.3 OPTIMIZATION PROCEDURE: OPTIMUM CRUISE VELOCITY, OPTIMUM TRAJECTORY SYNTHESIS

The equation for the Hamiltonian, H, is the equation for a straight line in the (V, i) plane.

A suitable optimization procedure is to draw lines of constant H and select the one that is tangential to the $i(V)$ versus V curve. The point of tangency yields the optimum cruise velocity. In Figure 12.2, the slope of the optimum constant H line is $p = \lambda_x / \lambda_C$ and intercepts the V axis at

$$V = V_{\text{intercept}} = V_w + H/\lambda_x. \tag{12.67}$$

The optimization procedure therefore consists of minimizing the function $p(V)$ with respect to the airspeed.

The function $p(V)$ is:

$$p_{\min}(V) = \min_{V} \left[i(V)/V - \left(V_w + H/\lambda_x \right) \right]. \tag{12.68}$$

The co-state λ_C may then be calculated from $\lambda_C = \lambda_x / p_{\min}(V)$. The procedure may be extended to the climb/descent segments of the vertical flight path. The altitude is found by matching the flight profile at the corners to the climb/descent segments. The output of the trajectory synthesis is a flight plan table with a typical set of x values, altitudes, weights and relevant co-states (λ_C) corresponding to a set of time marker points.

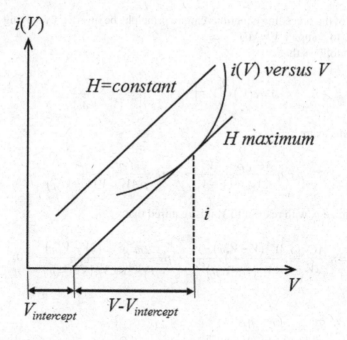

FIGURE 12.2 Cruise optimization.

Given that:

$$P = \eta TV = \eta DV = \eta \frac{1}{2}\rho V^3 \left(C_{Do} + KC_L^2\right)A = V_{\text{batt}}i,$$

$$C_L = 2L/\rho V^2 A = 2mg/\rho V^2 A, \tag{12.69}$$

where η is the product of propulsive efficiency and the electrical supply efficiency, C_{D0} is the profile drag coefficient, ρ is the air density, A is the aircraft's projected planform area and V_{batt} is the battery supply voltage,

$$i = \eta \frac{A}{2V_{\text{batt}}}\left(C_{Do}\rho V^3 + K \frac{4m^2 g^2}{\rho V A^2}\right) = \eta \frac{AC_{Do}\rho}{2V_{\text{batt}}} V^3 + \eta K \frac{2m^2 g^2}{\rho V_{\text{batt}} A} \frac{1}{V}. \tag{12.70}$$

The above equation for $i = i(V)$ and its derivative with V may be expressed, respectively, as,

$$i = \eta \frac{AC_{Do}\rho}{2V_{\text{batt}}}\left(V^3 + K \frac{4m^2 g^2}{\rho^2 C_{Do}} \frac{1}{V}\right),$$

$$\frac{di}{dV} = \eta \frac{AC_{Do}\rho}{2V_{\text{batt}}}\left(3V^2 - K \frac{4m^2 g^2}{\rho^2 A^2 C_{Do}} \frac{1}{V^2}\right). \tag{12.71}$$

The first of the preceding equations can, in principle, be inverted by solving a quartic equation, to compute $V = V(i)$.

It also follows that

$$p = i(V) \Big/ \left\{ V - \left(V_w + \frac{H}{\lambda_x} \right) \right\} = \frac{i(V)}{(V - V_{wH})}, \tag{12.72}$$

which reduces to

$$p = \eta \frac{A C_{Do} \rho}{2 V_{batt}} \frac{V^3}{(V - V_{wH})} + \eta K \frac{2 m^2 g^2}{\rho A V_{batt}} \frac{1}{V (V - V_{wH})}. \tag{12.73}$$

To minimize p, with respect to V, it is required that

$$\frac{dp}{dV} = \eta \frac{A C_{Do} \rho}{2 V_{batt}} \frac{3 V^4 (V - V_{wH}) - V^5}{V^2 (V - V_{wH})^2} - \eta K \frac{2 m^2 g^2}{\rho A^2 V_{batt}} \frac{(2V - V_{wH})}{V^2 (V - V_{wH})^2} = 0, \tag{12.74}$$

or

$$\frac{dp}{dV} = \frac{\eta A}{V_{batt} V^2 (V - V_{wH})^2}$$

$$\times \left\{ \frac{C_{Do} \rho}{2} V^4 (2V - 3V_{wH}) - K \frac{2 m^2 g^2}{\rho A^2} (2V - V_{wH}) \right\} = 0. \tag{12.75}$$

Thus, the quintic equation

$$V^4 (2V - 3V_{wH}) - K \frac{4 m^2 g^2}{\rho^2 A^2 C_{Do}} (2V - V_{wH}) = 0, \tag{12.76}$$

is numerically solved for V, with the additional requirement that the above quantic equation for V must have just one real root for V at the point of tangency and within the flight envelope of the aircraft.

Alternatively, the point of tangency is obtained by solving

$$p = \frac{di(V)}{dV} = \frac{i(V)}{V - V_{wH}}. \tag{12.77}$$

Hence, V_{wH} is obtained from

$$\frac{di}{dV} = \eta \frac{A C_{Do} \rho}{2 V_{batt}} \left(3 V^2 - K \frac{4 m^2 g^2}{\rho^2 A^2 C_{Do}} \frac{1}{V^2} \right)$$

$$= \eta \frac{A C_{Do} \rho}{2 V_{batt}} \left(\frac{V^3}{(V - V_{wH})} + K \frac{4 m^2 g^2}{\rho^2 A^2 C_{Do}} \frac{1}{V (V - V_{wH})} \right). \tag{12.78}$$

Consequently,

$$V_{wH} = V - \frac{\left(V^5 + K \dfrac{4m^2g^2}{\rho^2 A^2 C_{Do}} V \right)}{\left(3V^4 - K \dfrac{4m^2g^2}{\rho^2 A^2 C_{Do}} \right)} = \frac{\left(2V^5 - 2K \dfrac{4m^2g^2}{\rho^2 A^2 C_{Do}} V \right)}{\left(3V^4 - K \dfrac{4m^2g^2}{\rho^2 A^2 C_{Do}} \right)}. \tag{12.79}$$

Thus,

$$V^4 \left(2V - 3V_{wH} \right) - K \frac{4m^2g^2}{\rho^2 A^2 C_{Do}} \left(2V - V_{wH} \right) = 0. \tag{12.80}$$

12.2.4 Modeling with the Peukert Effect

The impact of the Peukert effect on the capacity of a lithium battery is further accentuated due to the sensitivity of the battery to temperature and low temperatures severely attenuating the rated battery capacity. The discharge curves of a lithium battery, including the Peukert effect, may be obtained by the method outlined by Traub [19]. The Peukert effect results in the effective capacity of the battery decreases with increasing current taken from the battery. Therefore, the effective battery discharge current i_{dch} increases with rising battery current in accordance with the law:

$$i_{dch} = i(V) \left(i/i_{nom} \right)^{\chi - 1}. \tag{12.81}$$

Hence,

$$i(V) = \left(i_{drain} i_{nom}^{\chi - 1} \right)^{1/\chi}. \tag{12.82}$$

Thus, eliminating $i(V)$ from Equation (12.71), it reduces to

$$V^3 - \frac{2V_{batt} \left(i_{drain} i_{nom}^{\chi - 1} \right)^{1/\chi}}{\eta A C_{Do} \rho} V + K \frac{4m^2 g^2}{C_{Do} \rho^2 A^2} = 0. \tag{12.83}$$

A plot of the optimal flight velocity $V = V_{opt}$ with V_{wH} is shown in Figure 12.3.

When the parameter

$$K \frac{4m^2 g^2}{\rho^2 A^2 C_{Do}} \gg V^4, \tag{12.84}$$

is very large in comparison with the fourth power of V, Equation (12.80) is reduced to, $2V - V_{wH} = 0$. Thus, the relationship between $V = V_{opt}$ and V_{wH} is mostly linear when $V < 50$ as observed in Figure 12.3. This plot is for an aircraft with a mass of 2,000 kg flying at a cruise altitude of 6,100 m and based on a profile drag coefficient, $C_{Do} = K = 0.02$. (In a real aircraft the drag coefficient could be up to 20 to 40 times

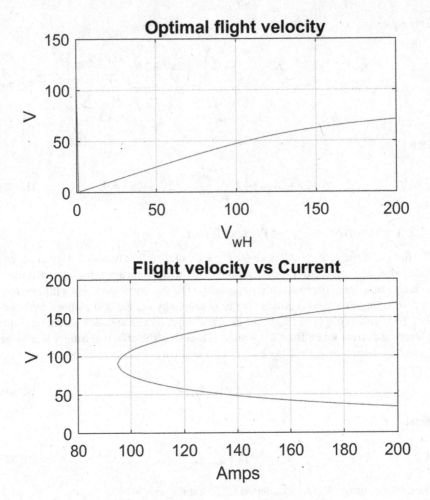

FIGURE 12.3 Typical plots of the optimal flight velocity $V = V_{opt}$ with V_{wH}, and of the flight velocity with current drawn from the battery.

higher.) The value of χ in Equation (12.82) is set to 1.05 and $i_{nom} = 20$ Amps. The value chosen for K which determines the induced drag coefficient, implies a very high aspect ratio wing. Also shown in Figure 12.3 is the plot of the velocity versus the total current drawn from the battery. The latter plot is only valid for $i > 94$ Amps, as below this current the solution for the velocity is not a real number.

As for the optimum solutions for the velocity, there are only three real solutions for the optimum velocity, one of which is negative. In fact, for a given current $i > 94$ Amps, there are only two real solutions for the flight velocity, and both the lower and higher values are shown in Figure 12.3. From the graph it can be seen that a large current of about 135 Amps is necessary to achieve a flight velocity of 50 m/s, corresponding to 180 kph! From the upper plot in Figure 12.3, it is noted that only the lower branch in the lower plot is a feasible optimum. From the lower plot in Figure 12.3, it may also be observed that the minimum current that must be drawn is about 94 Amps corresponding to a single cruise velocity of 80 m/s or 288 kph. A

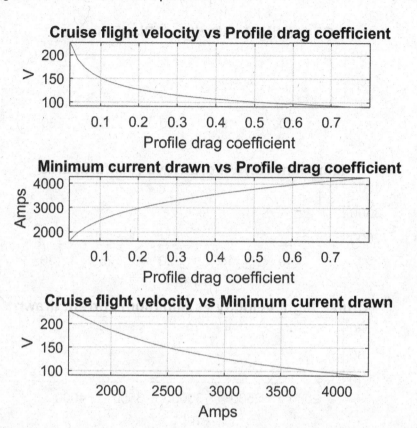

FIGURE 12.4 Cruise flight velocity and minimum current drawn versus C_{Do}.

distance of 400 km is covered in 1.4 h, and the minimum battery capacity required is 132 Amp hrs. When traveling at the lower speed of 180 kph, the battery capacity required is 70% more. Thus, one is able to estimate the minimum battery capacity required for an aircraft cruising at an altitude of 6,100 m.

Figure 12.4 shows the plotted cruise flight velocity and minimum current drawn versus C_{Do}, which is varied from 0.02 to 0.8 for a typical, small, general aviation aircraft with $i_{nom} = 80$ Amps in Equation (12.82). They are also plotted against each other. The minimum battery capacity required rapidly goes up as C_{Do} and K are increased. In Figure 12.5, the minimum current and the cruise flight velocity versus the wing planform area are plotted. Yet the benefits of optimizing the performance are also apparent. It is important that at the design stage it is ensured that the optimum flight speeds are within the flight envelope of the aircraft, so the aircraft can benefit from using the optimal solution. The minimum current goes down as the planform area is increased, although there is a corresponding decrease in the cruise flight velocity.

12.3   INTEGRATED FLIGHT AND PROPULSION CONTROL

The use of a set of distributed thrust generating propellers, and/or fans designed to provide primary propulsion for an electric aircraft, as a flight control system to

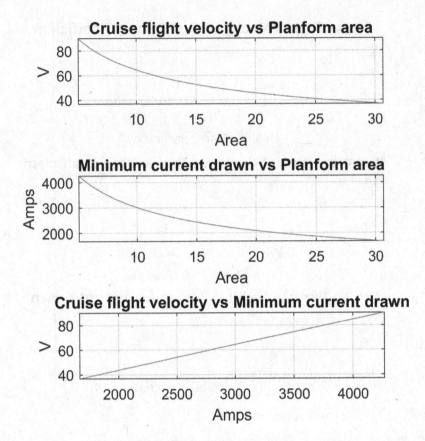

FIGURE 12.5 Cruise flight velocity and minimum current drawn versus wing planform area.

control the direction of flight as well as the orientation of aircraft can lead to significant reductions in weight. Moreover, the optimization of the integrated electric propulsion and flight control system can offer significant performance improvements: increased thrusts for propulsion and flight control, reduced current consumption, increased motor life and general improved airplane performance, with an increase in the range and endurance of the aircraft. The design, software and hardware developments, such as the use of HTS motors and associated testing requirements have also been shown to be feasible. However, the design of flight control laws associated with such distributed propulsion control system will require redesign from first principles.

The thrust generated propeller is generally a function of the atmospheric density at the altitude of flight and the flight velocity, in addition to the primary control variable which is the current input to the motor. While the atmospheric density is altitude-dependent, as far the longitudinal motion is concerned, the flight velocity has two components, the forward and the vertical velocity components. The primary purpose of all the thrusters on board the aircraft is to provide thrust in the forward direction, and for this reason the thruster dynamics would be coupled with longitudinal dynamics of the aircraft. Thus, when the thrusters are used to control the lateral motions of the aircraft, they tend to enhance the coupling between the lateral and longitudinal dynamics which must be taken into consideration in the design of lateral stability augmentation systems and lateral autopilots. In the next section, we will illustrate the development of a linear model of the aircraft dynamics which can be used to derive the controls for the stability augmentation and autopilot design problems.

12.3.1 MODEL-BASED DESIGN OF CONTROL LAWS FOR DISTRIBUTED PROPULSION-BASED FLIGHT CONTROL

Consider first a hypothetical aircraft endowed with multiple thrusters. The longitudinal perturbation dynamics in terms of the non-dimensional forward perturbation velocity Δu_s, the perturbation angle of attack $\Delta \alpha_s$, the perturbation pitch rate Δq_s, the perturbation pitch angle $\Delta \theta_s$ and the perturbation altitude Δh_s [1], given the equilibrium level, horizontal, forward flight velocity of the aircraft U_s^e, in terms of the stability derivatives in the stability axes, may be shown to be of the form

$$
\begin{bmatrix} \Delta \dot{u}_s \\ \Delta \dot{\alpha}_s \\ \Delta \dot{q}_s \\ \Delta \dot{\theta}_s \\ \Delta \dot{h}_s \end{bmatrix} = \begin{bmatrix} x_u & U_s^e x_w & 0 & -g & 0 \\ z_u/U_s^e & z_w & 1+z_q/U_s^e & 0 & 0 \\ m_u & U_s^e m_w & m_q & 0 & 0 \\ 0 & 0 & 1 & 0 & 0 \\ 0 & -U_s^e & 0 & U_s^e & 0 \end{bmatrix} \begin{bmatrix} \Delta u_s \\ \Delta \alpha_s \\ \Delta q_s \\ \Delta \theta_s \\ \Delta h_s \end{bmatrix} + \sum_{j=1}^{M} \begin{bmatrix} x_{\tau j} \\ z_{\tau j}/U_s^e \\ m_{\tau j} \\ 0 \\ 0 \end{bmatrix} \Delta \tau_j. \tag{12.85}
$$

In the above, it is assumed that there are M thrusters and the thrust lever angle for each of the thrusters is assumed to be defined by $\Delta \tau_j$. Generally, $\Delta \tau_j$ is a function of the non-dimensional forward perturbation velocity Δu_s, the perturbation angle of attack $\Delta \alpha_s$, the perturbation altitude Δh_s and the perturbation current to the electric motor, Δi_j. Thus, $\Delta \tau_j$ may be expressed by the linear relation

$$
\Delta \tau_j = t_u \Delta u_s + t_\alpha \Delta \alpha_s + t_h \Delta h_s + t_i \Delta i_j. \tag{12.86}
$$

The motor dynamics may be expressed in terms of the first order model

$$
\tau_{mj} \Delta \dot{i}_j + \Delta i_j = \Delta i_{jd}, \tag{12.87}
$$

where τ_{mj} is the motor time constant and Δi_{jd} is the commanded current input. Thus, one can eliminate $\Delta \tau_j$ and write the augmented state equations of longitudinal motion in the form

$$
\begin{bmatrix} \Delta \dot{u}_s \\ \Delta \dot{\alpha}_s \\ \Delta \dot{q}_s \\ \Delta \dot{\theta}_s \\ \Delta \dot{h}_s \\ \Delta \dot{i}_i \end{bmatrix} = \mathbf{A}_{elg} \begin{bmatrix} \Delta u_s \\ \Delta \alpha_s \\ \Delta q_s \\ \Delta \theta_s \\ \Delta h_s \\ \Delta i_j \end{bmatrix} + \sum_{j=1}^{M} \mathbf{B}_{elg\,j} \Delta i_{jd}. \tag{12.88}
$$

The above model may be used in the synthesis of longitudinal stability augmentation and longitudinal autopilot control laws.

In the case of the lateral perturbation dynamics, which are written in terms of the non-dimensional side perturbation velocity Δv_s, the perturbation roll rate Δp_s, the perturbation yaw rate Δr_s and the perturbation bank angle $\Delta\phi_s$ and the lateral stability derivatives in the stability axes as

$$
\begin{bmatrix} \Delta\dot{v}_s \\ \Delta\dot{p}_s \\ \Delta\dot{r}_s \\ \Delta\dot{\varphi}_s \end{bmatrix} = \begin{bmatrix} y_v & y_p & y_r-1 & g \\ l_v & l_p & l_r & 0 \\ n_v & n_p & n_r & 0 \\ 0 & 1 & 0 & 0 \end{bmatrix} \begin{bmatrix} \Delta v_s \\ \Delta p_s \\ \Delta r_s \\ \Delta\varphi_s \end{bmatrix} + \sum_{j=1}^{M} \begin{bmatrix} y_{\tau_j} \\ l_{\tau_j} \\ n_{\tau_j} \\ 0 \end{bmatrix} \Delta\tau_j, \qquad (12.89)
$$

where $\Delta\tau_j$ is still expressed as

$$
\Delta\tau_j = t_u\Delta u_s + t_\alpha\Delta\alpha_s + t_h\Delta h_s + t_i\Delta i_j. \qquad (12.90)
$$

Thus, one can eliminate $\Delta\tau_j$ and write the complete lateral equations of motion in the form

$$
\begin{bmatrix} \Delta\dot{v}_s \\ \Delta\dot{p}_s \\ \Delta\dot{r}_s \\ \Delta\dot{\varphi}_s \end{bmatrix} = \mathbf{A}_{\text{elt}} \begin{bmatrix} \Delta v_s \\ \Delta p_s \\ \Delta r_s \\ \Delta\varphi_s \end{bmatrix} + \mathbf{B}_{\text{elt}} \begin{bmatrix} \Delta u_s \\ \Delta\alpha_s \\ \Delta q_s \\ \Delta\theta_s \\ \Delta h_s \\ \Delta i_j \end{bmatrix}. \qquad (12.91)
$$

Thus, the equations of lateral motion are coupled with equations of longitudinal motion, which must also be considered for the synthesis of lateral stability augmentation and lateral autopilot control laws. For the synthesis techniques, the reader is referred to Vepa [1].

12.4   FLIGHT MANAGEMENT FOR AUTONOMOUS OPERATION

To understand the key principles of the design of autonomous electric aircraft, it is important to start with an appreciation of autonomous behavior. "Autonomous behavior" begins with the establishment of a plan to accomplish some desired goal or set of goals subject to some given set of resources and a set of constraints. The focus is more on the need for "autonomy" rather than on "unmanned" operations. Autonomous systems arise from a synergy of various interacting subsystems. The principle of synergy, simply stated, is that the sum is greater than the parts. When all the subsystems of an electric aircraft are assembled so they work together, one has a functioning electric aircraft. Thus, synergy involves a very special kind of integration. One is naturally interested in defining all the subsystems that make up an autonomous system and integrating them to deliver synergy. Before considering the key elements of an autonomous control systems, which is responsible for providing functional autonomy to a vehicle or a machine, it is important to understand the features of an automatic control system and the features of an autonomous control system—the principal differences as well as the common features between them. In an automatic control system, the decision-making is done outside the controller. The controller is usually under the supervision of an operator or pilot, who is responsible for establishing the set-points and plans. There is no need for the controller to understand the environment in which it is functioning, as that is part of the pilot's responsibility. Thus, there is no need for situational or spatial awareness of the working environment. Consequently, the controller has limited or low authority. The automatic control system is responsible for tracking a given set-point or a plan which can be considered to be an ordered collection of set-points. The control system must continuous and smoothly track a desired set-point or a commanded trajectory. Most automatic control systems, such as an aircraft's flight control systems, must necessarily be sufficiently stable and possess an appropriate degree of robustness.

12.4.1 Autonomous Control Systems

In an autonomous control system, the decision-making is within the remit of the controller which is endowed with considerable independence. The controller is not under the supervision of an operator or pilot. The controller itself is responsible for defining set-points and plans. Moreover, an autonomous control system is capable of spatial awareness and/or situation assessment, which in turn implies that it has the responsibility to monitor the working or immediate environment in which the vehicle or machine is functioning. Thus an autonomous controller has greater authority where the additional authority and the responsibility is to manage its own "flight" plan, manage all external communications and maintain surveillance, monitor all internal systems for faults and synthesize its own "flight" path or trajectory, re-configure/re-plan if necessary, manage contingencies and react when there is an obstacle (or conflict) to change its plan, course or set-points.

The additional responsibilities fall into four categories which can stated as: mission-related issues, spatial and/or situational assessment, including environmental monitoring, "flight" (trajectory) management and optimization issues and issues related to system level faults such as fault monitoring and diagnosis. Both automatic control systems and autonomous control systems are expected to deal with uncertainties. They are both robust in their design. Feedback architecture is common to both, with adaptive or self-tuning control loops. Fault tolerance is also a feature of both. However, it is only autonomous control systems that are expected to deal with changes in mission (mission re-planning), obstacles or conflicts (route re-planning, conflict resolution), reconfigure to deal with faults both internal and "external", (systems re-configuration) and deal with contingencies and newer constraints (motion re-planning). To sum it up, autonomous control systems are also endowed with the capacity for re-planning (including fast real-time re-planning) and re-configuration in addition to the ability to synthesize plans, including mission-related plans, route or motion planning, all of which may be essential. Monitoring tasks may need to be integrated for optimum use of resources. For example, when navigating the vehicle, it may be necessary to perform simultaneous localization and mapping (SLAM) where localization refers to position-fixing and mapping involves building a map of the immediate or working environment.

12.4.2 ROUTE PLANNING

Route planning generally involves a solution to a problem similar to the travelling salesman problem (TSP), where the systems establishes an optimum route that it has to traverse based on some considerations of cost. There has been a renewed interest in drone-assisted TSPs. The classical solution is based on dynamic programming [20], [21] and on heuristic or genetic algorithms which tend to mimic biological systems in some sense. One geometric heuristic method involves first constructing a convex hull that contains all the cities that must be visited by the travelling salesman followed by sequentially inserting each of the enclosed cities based on the largest angle subtended by each of them with two consecutive cities on the tour. The main aim of these algorithms is to reduce the total "cost" or "cost index".

12.4.3 MISSION PLANNING FOR AUTONOMOUS OPERATIONS

Motion planning involves the synthesis of optimal trajectories from waypoint to waypoint. An optimal trajectory between two waypoints is usually constructed in real time by solving an optimal control problem which involves the minimization of a performance metric, subject to constraints. The optimal real-time solution is obtained by solving a two-point boundary value problem (TP-BVP). Examples of optimal trajectory synthesis include kinematic motion planning without reference to the dynamics and optimal steering. The synthesis of optimal trajectories requires the minimization of a performance index subject to motion constraints as differential equations. It is customary to use Lagrange multipliers, or co-states, to incorporate the differential constraints, and the solution naturally leads to a TP-BVP, which is then solved to obtain generic solutions for each phase. A classical solution to this problem is Dubin's solution [22] which places no cost on fuel (which may be high!). The methodology is used to synthesize reference trajectories in real time based on the use of a point mass, kinematic model of aircraft and assuming that the cost index, which is the ratio between the unit costs of time and of fuel, is small, as illustrated by McGee and Hedrick [23]. There are indeed several other approaches to optimal trajectory synthesis, based on the use of interpolating polynomials, notions of decision trees (graph-based algorithms such as Dijkstra's algorithm, A*, D*, ...), using waypoints A (start) to G (finish), using free domains (configuration space) and by linking a waypoint A to a waypoint G. Dijkstra's algorithm targets finding the shortest path in a graph where edges'/arcs' weights are already known. Dijkstra's algorithm is a special case of dynamic programming. It finds the shortest path which depends purely on local path cost. The A*, D* algorithms are extensions of Dijkstra's algorithm. There are also several configuration space-based algorithms for trajectory synthesis, sampling-based algorithms, as discussed in Karaman and Frazzoli [24], such as the probabilistic road map (PRM), rapidly exploring random tree (RRT) algorithms and combinatorial algorithms. While the former are faster and require less computing power, the latter are more complete. Yan, Liu and Xiao [25] have applied the A* and PRM algorithms to a typical UAV planning problem.

12.4.4 SYSTEMS AND CONTROL FOR AUTONOMY

Clearly autonomous control involves new tasks such as mission, route or motion planning, new navigation-related tasks (localization, mapping, SLAM) and new approaches to vehicle guidance. These newer tasks are facilitated to a large extent due to the availability of newer suites of sensors and actuators. One can therefore break down the study of autonomous systems into a study of the systems for autonomy as well as a study of the required control for autonomy and apply to a number of prototype electric aircraft. This would facilitate capturing a range of new algorithms, the study of new or arising methods, the establishment of generic principles and possibly scaling-up to apply the methods to large electric aircraft. The systems for autonomy involve mechanical components (drive, power and chassis), power systems, electronics and computations, sensors for perception, proprioceptive sensors and other external sensing, newer mission payloads, communication and control systems and "smart systems" and interfaces.

Control for autonomy would involve the dynamics and control from an operational perspective, the study of machines/mechanisms for mobility and manipulation, understanding of machine-level control, the development of perception algorithms, reasoning and decision-making methods, reactive control, adaptation and learning and the study of human–machine interfaces.

12.5 FLIGHT PATH PLANNING

A typical path-following system is expected to precisely control the vehicle's position in three-dimensional space; in particular, with precise path-following for straight courses defined *a priori*, irrespective of wind velocity and turbulence. The maximum errors of altitude and horizontal path following are both fixed by predefined limits. Further airspeed control and regulation is also simultaneously required to track a particular command speed and to avoid stall or overloading due to speeding. The process of path-following requires a path planned *a priori*. Terrain-following path planning and control systems have been designed to meet the above requirements [26]. The principle of flight path planning will be illustrated by considering a few cases.

12.5.1 PATH PLANNING IN THREE DIMENSIONS USING A PARTICLE MODEL

One of the simplest approaches to path planning in three dimensions is to use a particle model. Thus, the body equations of motion may be simplified and expressed in terms of the velocity vector, $\mathbf{v}_b$ and the position vector $\mathbf{r}_b$ by setting the body angular velocity vector ω to zero, and including the drag forces as

$$m\dot{\mathbf{v}}_b + m\omega \times \mathbf{v}_b = T\mathbf{u} - K_d \text{diag}(\mathbf{v}_b)\mathbf{v}_b, \quad \dot{\mathbf{r}}_b + \omega \times \mathbf{r}_b = \mathbf{v}_b, \quad (12.92)$$

$$\omega = 0. \quad (12.93)$$

While the equations of motion could be modified to include the variations in the drag force in three directions, as well as the weight and given by

$$\frac{d}{dt}\begin{bmatrix} x \\ y \\ z \end{bmatrix} = \begin{bmatrix} v_x \\ v_y \\ v_z \end{bmatrix}, \quad (12.94a)$$

$$m\frac{d}{dt}\begin{bmatrix} v_x \\ v_y \\ v_z \end{bmatrix} = T\begin{bmatrix} \cos\psi \sin\delta \\ \sin\psi \sin\delta \\ \cos\delta \end{bmatrix} - \begin{bmatrix} K_D & 0 & 0 \\ 0 & K_N & 0 \\ 0 & 0 & K_L \end{bmatrix}\begin{bmatrix} v_x^2 \\ v_y^2 \\ v_z^2 - w_0^2 \end{bmatrix}. \quad (12.94b)$$

with

$$K_L w_0^2 = mg, \quad \mathbf{u} = \begin{bmatrix} \cos\psi \sin\delta & \sin\psi \sin\delta & \cos\delta \end{bmatrix}^T. \quad (12.95)$$

However, in the interest of simplicity, it will be assumed that $K_d = K_D = K_N = K_L$ and that $K_L w_0^2 = 0$. Thus,

$$\frac{d}{dt}\mathbf{r}_b = \mathbf{v}_b, \quad m\frac{d}{dt}\mathbf{v}_b = T\mathbf{u} - K_d \text{diag}(\mathbf{v}_b)\mathbf{v}_b. \quad (12.96)$$

The vector $\mathbf{u}$ defines the direction of the thrust vector. When one is interested in the problem of finding the steering control,

$$\mathbf{u} = \mathbf{u}(t), \quad t_0 \le t \le t_f, \quad (12.97)$$

the torque direction time history is sought, such that it minimizes the functional cost:

$$J = 0.5|\mathbf{r}(t) - \mathbf{r}_d|^2\Big|_{t=t_f} = \Phi\{\mathbf{r}(t)\}\Big|_{t=t_f}, \quad (12.98)$$

subject to Equations (12.94) and (12.95), with, $\mathbf{r} = \mathbf{r}_b$, $\mathbf{v} = \mathbf{v}_b$.

Introducing the single state vector, $\mathbf{x} = \begin{bmatrix} \mathbf{r}^T & \mathbf{v}^T \end{bmatrix}^T$, so the Equations (12.94a and 12.94b) are expressed as

$$\frac{d}{dt}\mathbf{x}^T = \frac{d}{dt}\begin{bmatrix} \mathbf{r}^T & \mathbf{v}^T \end{bmatrix} = f^T, \quad f(t) = \begin{bmatrix} \mathbf{v} \\ T\mathbf{u} - K_d \text{diag}(\mathbf{v})\mathbf{v} \end{bmatrix}. \quad (12.99)$$

To solve the optimization problem, six Lagrangian multipliers or co-states are introduced, given by the two column vectors $\lambda_r(t)$ and $\lambda_v(t)$, denoted by a single column vector, $\lambda(t)$. Following Bryson and Ho [27], a Hamiltonian function is defined as,

$$H = \lambda^T(t) f(t). \tag{12.100}$$

Hence,

$$H = \lambda_r^T \mathbf{v} - \lambda_v^T \left(K_d \text{diag}(\mathbf{v}) \mathbf{v} \right) + \lambda_v^T T \boldsymbol{u}. \tag{12.101}$$

The necessary conditions [27] for the first variation of J to be zero include the co-state differential equations,

$$\frac{d}{dt} \lambda(t) = -\frac{\partial H}{\partial \mathbf{x}} = \begin{bmatrix} \mathbf{0} \\ \lambda_r - 2K_d \text{diag}(\lambda_v) \mathbf{v} \end{bmatrix}. \tag{12.102}$$

The optimality conditions are

$$\partial H / \partial \psi = \lambda_v^T \cdot T \begin{bmatrix} -\sin\psi \sin\delta & \cos\psi \sin\delta & 0 \end{bmatrix}^T = 0, \tag{12.103}$$

and

$$\partial H / \partial \delta = \lambda_v^T \cdot T \begin{bmatrix} \cos\psi \cos\delta & \sin\psi \cos\delta & -\sin\delta \end{bmatrix}^T = 0. \tag{12.104}$$

Hence,

$$\lambda_v = |\lambda_v| \begin{bmatrix} \cos\psi \sin\delta & \sin\psi \sin\delta & \cos\delta \end{bmatrix}^T = -|\lambda_v| \boldsymbol{u}. \tag{12.105}$$

Thus, the two-parameter control vector $\boldsymbol{u}$ can be expressed as

$$\boldsymbol{u} = -\lambda_v / |\lambda_v|. \tag{12.106}$$

The choice of the sign in Equation (12.106) will depend on the direction of the desired torque, whether forward or reverse torque. Thus, the closed-loop equations of motion are

$$d\mathbf{x}/dt = f, \quad \boldsymbol{u} = -\lambda_v / |\lambda_v|. \tag{12.107}$$

To complete the definition of the optimal solution, the boundary conditions at $t=t_f$ for the co-state system are found by applying the transversality conditions. The transversality conditions ensure that the initial and final states are selected optimally within the feasible regions of the states. For the transversality conditions, one may write

$$\lambda_r(t_f) = \partial \Phi \{\mathbf{r}(t)\} / \partial \mathbf{r} \Big|_{t=t_f} = (\mathbf{r}(t_f) - \mathbf{r}_d), \tag{12.108}$$

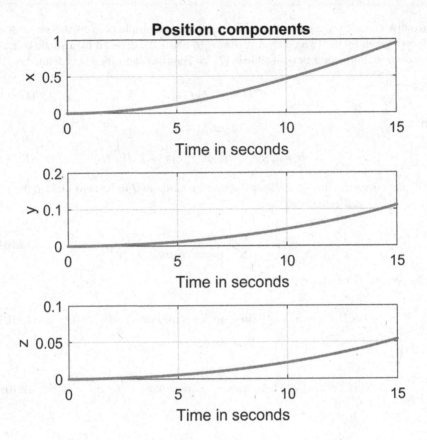

FIGURE 12.6 Position components optimal trajectory in three dimensions.

$$\lambda_v(t_f) = \partial\Phi\{\mathbf{r}(t)\}/\partial\mathbf{v}\big|_{t=t_f} = 0. \tag{12.109}$$

The solution is obtained by solving a TP-BVP. A typical optimal trajectory in three dimensions, as well as the time history of the velocities, using non-dimensional position units, are shown in Figures 12.6 and 12.7 respectively.

The problem is much easier to solve if decomposed into two separate problems: defining the path in the horizontal plane, and considering the lateral path planning problem independently.

12.5.2 PATH PLANNING IN THE HORIZONTAL PLANE

Following McGee, Spry and Hedrick [28] the equations of motion, based on a kinematic model, and the desired trajectory are respectively defined in the horizontal plane as

$$\frac{d}{dt}\begin{bmatrix} x \\ y \\ \psi \end{bmatrix} = \begin{bmatrix} \cos\psi \\ \sin\psi \\ u \end{bmatrix}, \quad \mathbf{x}_d(t) = \begin{bmatrix} x \\ y \\ \psi \end{bmatrix} = \begin{bmatrix} x_f - \beta t \\ y_f \\ \psi_f \end{bmatrix}. \tag{12.110}$$

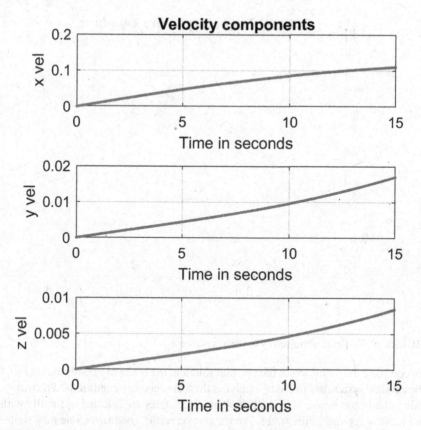

FIGURE 12.7   Velocity components optimal trajectory in three dimensions.

The cost function to be minimized is

$$J = \int_0^{t_f} L\,dt + 0.5\left|\mathbf{x}(t) - \mathbf{x}_d\right|^2 \Big|_{t=t_f} = \frac{1}{2}\int_0^{t_f} u^2\,dt + \Phi\{\mathbf{x}(t)\}\Big|_{t=t_f}, \qquad (12.111)$$

$$H = L + \lambda^T \mathbf{f} = \frac{1}{2}u^2 + \lambda_1 \cos\psi + \lambda_2 \sin\psi + \lambda_3 u. \qquad (12.112)$$

The optimality conditions are

$$\frac{d}{dt}\lambda(t) = -\frac{\partial H}{\partial \mathbf{x}} = -\begin{bmatrix} 0 & 0 & 0 \\ 0 & 0 & 0 \\ -\sin\psi & \cos\psi & 0 \end{bmatrix}\lambda(t), \quad \lambda(t) = \begin{bmatrix} \lambda_1 \\ \lambda_2 \\ \lambda_3 \end{bmatrix}; \qquad (12.113)$$

$$\frac{\partial H}{\partial u} = u + \lambda_3 = 0, \quad \Rightarrow \quad u = -\lambda_3. \qquad (12.114)$$

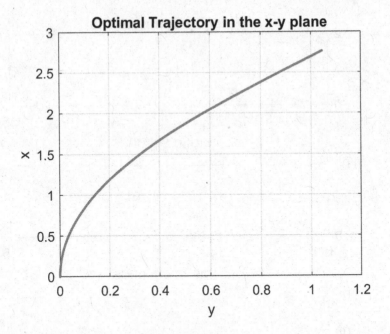

FIGURE 12.8 Optimal trajectory in the *x-y* plane.

To complete the definition of the optimal solution, the boundary conditions at $t=t_f$ for the co-state system are found by applying the transversality conditions. The transversality conditions ensure that the initial and final states are selected optimally within the feasible regions of the states. For the transversality conditions, one may write

$$\lambda\left(t_f\right)=\partial\Phi\left\{\mathbf{x}\left(t\right)\right\}/\partial\mathbf{x}\Big|_{t=t_f}=\left(\mathbf{x}\left(t_f\right)-\mathbf{x}_d\right). \tag{12.115}$$

The solution is obtained by solving a two-point boundary value problem. A typical optimal trajectory in the *x-y* plane, as well as the yaw angle time history, are shown in Figures 12.8 and 12.9 respectively.

12.5.3 PATH-FOLLOWING CONTROL

Once the reference paths are defined it is essential that path-following control laws are defined so the autonomous aircraft is able to track the reference trajectories. The problem has been discussed in Vepa [1] and the reader is referred to this reference for further details.

CHAPTER SUMMARY

There is clearly a need to re-think optimal flight plans and flight controllers that are expected to track these flight plans designed specifically for electric aircraft. In this chapter these requirements are briefly reviewed.

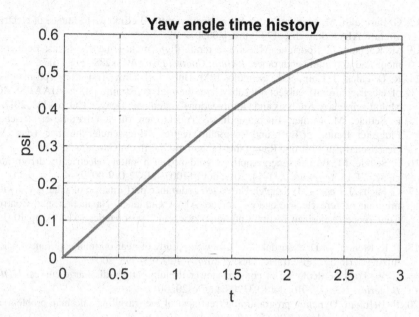

FIGURE 12.9 The yaw angle time history.

REFERENCES

1. R. Vepa, *Flight Dynamics Simulation and Control of Aircraft: Rigid and Flexible*, CRC Press, Boca Raton, FL, August 18, 2014, 660 pages.
2. G. Sachs, Flight performance issues of electric aircraft. AIAA Atmospheric Flight Mechanics Conference, American Institute of Aeronautics and Astronautics (AIAA), Minneapolis, MN, 2012.
3. G. Sachs, Unique range performance properties of electric aircraft, AIAA Atmospheric Flight Mechanics (AFM) Conference, American Institute of Aeronautics and Astronautics, Boston, MA, 2013.
4. L. W. Traub, Range and endurance estimates for battery-powered aircraft, *J. Aircr.*, 48(2):703–707, 2011.
5. M. Hepperle, Electric flight – Potential and limitations, 2012, http://www.elib.dlr.de, Accessed August 16, 2017.
6. T. Donateo, A. Ficarella, L. Spedicato, A. Arista, M. Ferraro, A new approach to calculating endurance in electric flight and comparing fuel cells and batteries, *Appl. Energy*, 187:807–819, 2017.
7. L. W. Traub, Validation of endurance estimates for battery powered UAVs, *Aeronaut. J.*, 117(1197):1155–1166, 2013.
8. G. Avanzini, F. Giulietti, Maximum range for battery-powered aircraft, *J. Aircr.*, 50(1):304–307, 2013.
9. G. Avanzini, E. L. de Angelis, F. Giulietti, Optimal performance and sizing of a battery-powered aircraft, *Aerosp. Sci. Technol.*, 59:132–144, 2016.
10. W. Peukert, Über die Abhängigkeit der Kapazität von der Entladestromstärke bei Bleiakkumulatoren, *Elektrotech. Z.*, 20:20–21, 1897.
11. R. Vepa (Ed.), Batteries: Modelling and state of charge estimation, Chapter 7, pp 323–347. In *Dynamic Modelling, Simulation and Control of Energy Generation*, Lecture Notes on Energy Series, Vol. 20, Springer, London, UK, 2013, ISBN 978-1-4471-5399-3.

12. G. Barufaldi, M. Morales, R. G. Silva, Energy optimal climb performance of electric aircraft. AIAA Scitech 2019 Forum, p 0830, 2019.

13. M. Kaptsov, L. Rodrigues, Electric aircraft flight management systems: Economy mode and maximum endurance, *J. Guid. Control Dyn.*, 41(1):288–293, 2017.

14. R. D. Falck, J. Chin, S. L. Schnulo, J. M. Burt, J. S. Gray, Trajectory optimization of electric aircraft subject to subsystem thermal constraints, 18th AIAA/ISSMO Multidisciplinary Analysis and Optimization Conference, Denver, CO, p 4002, 2017.

15. F. Settele, M. Bittner, Energieoptimale Trajektorien für ein batterie-elektrisches Flugzeug. Deutscher Luft- und Raumfahrtkongress; Hochschule München, Deutsche Gesellschaft für Luft-und Raumfahrt-Lilienthal-Oberth eV, Garching, 2017.

16. F. Settele, M. Bittner, Energy-optimal guidance of a battery-electrically driven airplane, *CEAS Aeronaut. J.*, 1–14, 2019, doi:10.1007/s13272-019-00398-x.

17. G. Sachs, J. Lenz, F. Holzapfel, Unlimited endurance performance of solar UAVs with minimal or zero electric energy storage. AIAA Guidance, Navigation, and Control Conference, American Institute of Aeronautics and Astronautics (AIAA), pp 10–13, 2009.

18. P. K. Menon, G. D. Sweriduk, A. H. Bowers, Study of near-optimal endurance-maximizing periodic cruise trajectories, *J. Aircr.*, 44(2):393–398, 2007.

19. L. W. Traub, Calculation of constant power lithium battery discharge curves, *MDPI Batteries*, 2(2):17, 2016, doi:10.3390/batteries2020017.

20. R. Bellman, Dynamic programming treatment of the travelling salesman problem, *J. Assoc. Comput. Mach.*, 9(1):61–63, 1962.

21. M. Held, R. M. Karp, A dynamic programming approach to sequencing problems, *J. Soc. Ind. Appl. Math.*, 1(1):10, 1962.

22. L. E. Dubins, On curves of minimal length with a constraint on average curvature, and with prescribed initial and terminal positions and tangents, *Am. J. Math.*, 79(3):497–516, 1957.

23. T. G. McGee, J. K. Hedrick, Optimal path planning with a kinematic airplane model, *J. Guid. Control Dyn.*, 30(2):629–632, 2007.

24. S. Karaman, E. Frazzoli, Sampling-based algorithms for optimal motion planning, *Computing Research Repository (CoRR)*, arXiv:abs/1105.1186, 2011.

25. F. Yan, Y.-S. Liu, J.-Z. Xiao, Path planning in complex 3D environments using a probabilistic roadmap method, *Int. J. Autom. Comput.*, 10(6):525–533, 2013.

26. P. Williams, Three-dimensional aircraft terrain-following via real-time optimal control, *J. Guid. Control Dyn.*, 30(4):1201–1205, 2007.

27. A. E. Bryson, Jr., Y. C. Ho. *Applied Optimal Control*, Ginn and Company, Waltham, MA, 1969.

28. T. G. McGee, S. Spry, J. K. Hedrick, Optimal path planning in a constant wind with a bounded turning rate, AIAA Paper, AIAA Guidance, Navigation, and Control Conference and Exhibit, pp 1–11, Reston, VA, 2005.

Index

AC motors, *see* Alternating-current motors
Acoustic analogy, 169
Active flow control, control laws, 234–236
  hybrid active laminar flow control, 244–245
  integral equations, boundary layer, 236–238
  inverse boundary layer method, uniform
    solutions, 238–239
  typical airfoil, 245–253
  uniform and prescribed shape factor, 239–241
  velocity profiles, 243–244
  vorticity–velocity formulation, control flow
    inputs, 241–243
Actuator disk/disc, 121–125, 137
Advance ratio, 118, 123, 129, 136
  ducted propeller, 140
  vs. propulsive efficiency, 123, 124
Advective derivative, 175, 188
Aeolian tones, 179, 181
Aeroacoustics and low noise design, 6
  aeroacoustic analogies, 161–173
  Farassat's formulation, FW–H equation,
    190–191
  far-field noise formulation, rotating source,
    191–195
  Lighthill, Ffowcs Williams and Hawkins
    and Kirchhoff integral methods,
    173–177
  Lilley's analogy and application to ducts,
    195–202
  monopoles, dipoles and quadrupoles, 177–179
    tonal characterization, aeroacoustically
      generated noise, 179–180
  propeller noise fields, theoretical modeling,
    186–190
  propellers and motor application, 180–181
    airfoil and propeller noise sources,
      181–184
    Hamilton-standard procedure, propeller
      aerodynamic loading, 184–186
Aeroacoustic tones, 179
Aerodynamic balance, 62
Aerodynamic drag, 62–66
Aerodynamic lift coefficient, 296
Airbus E-Fan prototype, 4
Aircraft propulsion, HTS design for, 165
Airfoil/aerofoil
  flow over an, 61–62, 114
  pitching moment, 61
  principles of, 61
  section

geometry, 117, 118
  lift and drag, 118–119
  sharp edge, 105
Airframe design, 61
Alignment torque, *see* Reluctance torque
Alkaline cell, 38–39
All-electric propulsion, 1, 3–4, 158
  all-electric X-57 plane, 6
  battery technology, 4
  future work, 7
  issues, 6–7
  low noise design, 6
  motors for, 49
  technology limitations, 7
Alternating-current (AC) motors
  characteristics, 20
  controlling, 22–24
  induction motors, 21–24
  synchronous motor, 18–20
Approximation coefficients, 228
Aqueous electrolyte-based Li-air battery, 45
Armature
  current–voltage relationship, 15
  torque, 11–13, 15
Asymmetric half-bridge converter, 26
Automatic control systems, 313–314
Autonomous behavior, 312
Autonomous flight, 293
Autonomous systems, 312, 315

Back-EMF, in synchronous motors, 20
Band gap energy, 263
Bardeen, Cooper and Schrieffer (BCS)
  theory, 155
Barrier potential, 275, 276
Barrier-slot rotor, 25
Base-collector junction, 276
Base-emitter junction, 276
Base flow velocity, 217
Base region, 276
Batteries
  anode, 29
  applications, 46
  capacity, 32
  capacity fading, 33
  cathode, 29
  charge capacity, 32, 33
  Daniell cell, principle of, 29, 30
  definition, 29
  discharge rates, 32

dry battery cell, 30, 31
dynamic models
    cell level model, 34–35
    circuit models, 34, 35
    empirical model, 33–34
    hybrid models, 34
    physical models, 33
    SOC estimation, 35–37
for electric aircraft, 46
electrodes, 29, 30
history, 29
rechargeable, 31–32
specifications, 30–31
structure, 30, 31
temperature effects, 32–33
types and characteristics, 37–45
Battery efficiency model, 33
BCS theory, *see* Bardeen, Cooper and Schrieffer
        theory
BEM theory, *see* Blade element momentum
        theory
B-H curve, ferromagnetic material, 51, 52
Biot–Savart law, 140–142, 144–145, 147
Bipolar junction transistor (BJT), 277
Blackbody radiation, 269
Blade element momentum (BEM) theory,
        128–137
    ducted propellers, application to, 137–140
Blade element theory, 123–126, 128–129
Blade passing frequency (BPF), 186
Blasius solution, 77–80
    displacement thickness, 86
    dissipation integral, 87
    energy thickness, 87
    Head's shape factor, 88
    momentum thickness, 86
    skin-friction-related stress, 87
Boundary layer
    analysis, 71–74
    Blasius solution, 77–80, 86–88
    blowing, suction or porosity, 99–100
    control, 234
    definition, 66–67
    displacement thickness, 80–81
    fence, 211
    energy thickness, 81–82
    equations, 74–75
        reduction, 100–102
    Holstein and Bohlen method, 90
    integral equations, 236–238
    inverse, 238–239
    laminar, 97–98, 104, 208, 216
    momentum, 209
    momentum thickness, 81
    Navier–Stokes equations, 66–69
        non-dimensionalizing and linearizing,
            70–71

Pohlhausen's method, 88–89
refined velocity profiles, 91–96
shape factor, 245–249, 253
theory, 67
Thwaites correlation technique, 102–104
transition and separation, 104–109, 216
turbulent (*see* Turbulent boundary layer)
two-dimensional boundary layer equations,
        76–77
velocity profile, 207, 235, 250
viscous energy dissipation, 69–70
von Karman momentum integral equation,
        82–86, 88–90, 112
vorticity and stress, 75–76
BPF, *see* Blade passing frequency
Breguet range equation, 294
Broadband noise, 182, 184
Brushless DC motors, 11, 26–27
    advantage, 27
    dynamic modeling, 27
    electronic commutation, 27
    switching and commutation, 27–28
    three-phase, 27, 28

Cavities, 213
Clark transformation, 22, 24
Closure equations, 101
Coercivity, 52–53
Continuous suction, 208
Continuous wavelet transform (CWT), 225
Control laws, model-based design of, 311–312
Convective derivative, 175, 188
Cooper pair, 155
Cost integral, 302
Covalent bond, 262
Cruise efficiency, 301
Cruise flight velocity and minimum current
    vs. C_{Do}, 309
    vs. wing planform area, 310
Cruise optimization, 304, 305
Cruise range, 294, 295, 301, 302
Cruise velocity, 296
Cryostats, HTS motors, 161–162
Curie temperature, 51
    of NdFeB, 53
Current *vs.* voltage characteristics, silicon photo-
        diode, 261
CWT, *see* Continuous wavelet transform

D'Alembert's paradox, 66
Daniell cell, principle of, 29, 30
Dark current, 260, 266–267
DBD, *see* Dielectric-barrier-discharge plasma
        actuator
DC-AC converter, 279
DC-AC voltage source converter, 279
"D cell," 29

DC motors, *see* Direct current motors
De-centralization, 6
DEP, *see* Distributed electric propulsion
Depletion region, 275, 276
Depletion zone, 263
Depth of discharge (DOD), 31
Detailed coefficients, 228
Diamagnetism, 50
Dielectric-barrier-discharge (DBD) plasma
             actuator, 207, 218, 219
    schematic diagram, 208
Dijkstra's algorithm, 314, 315
Dimples, 210
Dipoles, 170, 177–179
Direct current (DC) motors
    characteristics, 13, 14
    classification, 14, 15
    control, 16–18
    dynamic modeling, 14–16
    principles, 11–13
Direct operating costs (DOC), 299
Direct torque controllers (DTC)
    using three-level neutral-point clamped
             inverter, 290
    using two-level inverters combined space
             vector modulation, 291
Discrete wavelet transform (DWT), 226, 228
Displacement thickness, 80–81, 86, 236
Distributed electric propulsion (DEP), 152
Distributed propulsion, 152
DOC, *see* Direct operating costs
DOD, *see* Depth of discharge
Dopants, 261, 262
Doped semiconductor, 275
    layer, 280
Drag coefficient, 63, 66
    blade element momentum (BEM) theory, 133,
             137, 139
Drag forces, 61, 316
Drag polar, 64
Drag reduction, 114–115
Drela's method, 113–114
DTC, *see* Direct torque controllers
Dubin's solution, 314
Ducted propellers, 137–140
DWT, *see* Discrete wavelet transform
Dynamic equilibrium, 14

Efficiency, solar cell, 267
Electrical energy conversion efficiency, 137
Electrolyte, 29, 32, 35, 38–40
    aqueous, 43–45
    gel polymer, 42–43
    liquid, 41–42
    non-aqueous, 43–45
    qualities, 41
    solid-state, 42

Electromagnetic effect, 50
Electromotive force (EMF)
    back-EMF, 12–13, 20
    induced, 12, 22
Electron–hole pair generation, 263, 264
EMF, *see* Electromotive force
Endurance, 293–296
Energy height, 299, 300
Energy thickness, 81–82, 87
Equivalent circuit
    of IGBT, 287
    multi-junction solar cell, 271
    silicon photo-diode, 260
    of thyristor, 284, 285
    of silicon photo-diode, 260
Equivalent sea level, 297
Extrinsic semiconductors, 275, 276

Faraday's disc, principle of operation, 163, 164
Faraday's law, 34
Farassat's formulation, F–H equation, 190–191
Far-field noise, 201
    formulation based on rotating source,
             191–195
Fast Fourier Transform (FFT), 225
Feedback control, 218
Fences, 210, 211
Ferrimagnetism, 52
Ferrite magnets, 53
Ferromagnetism, 51, 52
FET, *see* Field effect transistors
FF, *see* Fill factor
Ffowcs Williams-Hawkings (FW-H) equation,
             171, 172
FFT, *see* Fast Fourier Transform
Fick's laws, 34
Field effect transistors (FET), 277
Field oriented control (FOC), 23
Fill factor (FF), 265, 267
First-generation cells, 270
Flat type rotor, 25
Flight control and autonomous operations
    electric aircraft, range and endurance of,
             293–296
    equivalent air speed, 296–298
    flight management, 312–313
        automatic control system, 313–314
        mission planning, 314–315
        route planning, 314
        systems and control for, 315
    flight path optimization, 299–301
        optimal control formulation, 302–304
        optimal control method, 302
        optimum cruise velocity, 304–307
        optimum trajectory synthesis, 304–307
        Peukert effect, 307–309
    flight path planning

in horizontal plane, 318–320
in three dimensions using particle model,
         315–318
gliding speed, 296–298
integrated flight and propulsion control,
         309–311
    control laws, model-based design of,
         311–312
minimum power required to climb, 298
overview of, 293
Flight management, autonomous operation,
         312–313
    automatic control system, 313–314
    mission planning, 314–315
    route planning, 314
    systems and control for, 315
Flight path optimization, 299–301
    optimal control formulation, 302–304
    optimal control method, 302
    optimum cruise velocity, 304–307
    optimum trajectory synthesis, 304–307
    Peukert effect, 307–309
Flight path planning
    in horizontal plane, 318–320
    in three dimensions using particle model,
         315–318
Flow control
    active (see Active flow control, control laws)
    definition, 207
    passive methods (see Passive methods)
Flow separation noise, 181–182
Flux pinned superconductors, 157
Flux pinning, 156
Flux pumping, 157
FOC, see Field oriented control
Formulation 1A, 190
Forward bias, 276
Fully controllable full-wave rectifier circuit, 284

Galerkin method, 201
Gallium arsenide (GaAs) cells, 265, 279
Gallium nitride devices, 279
Gate turn-off thyristor (GTO), 277, 286
Gel polymer electrolytes, 42
Giant transistor (GTR), 286
Gibbs energy, 44
Glauert wall jet, 219
Gliding speed, 296–298
Goldstein singularity, 102
Green's function approach, 170, 174
GTO, see Gate turn-off thyristor
GTR, see Giant transistor
Gurney flap, 213–214

Halbach array-based electric motor, 140
Halbach arrays
    axial field, 54, 55

axial field Halbach disc, 54, 56
    complex, 55–57
    linear, 54
    magnetic field modeling, 57–59
    magnetization directions, 54, 55
    radial, 57, 58
    ring type structures, 57, 58
Half-cell, 29, 30, 38–40, 42
Hall-effect sensors, 27
Hamiltonian function, 303, 304, 317
Hamilton-standard procedure, propeller
         aerodynamic loading, 184–186
Harmonic noise, 184
Head's method, 111–113
Head's shape factor, 88
Heavy duty zinc-chloride cells, 38
Heavy rare earth element (HRE), 53
"High-lift" propellers, 5
High temperature coercivity, 53
High-temperature superconductors (HTS), 310
    critical surface characteristics, 156
    features, 156–157
    Meissner effect, 156
    Meissner state, 156
    superconductivity, 155–156
High torque permanent magnet motors, 49
Hole diffusion current, 259
Holstein and Bohlen method, 90
Homopolar motors, 14
    DC motor, 163
    modern drum type motor structure, 163–164
    principle of operation, 163, 164
    superconducting, 165
Hot-electron extraction, 270
HRE, see Heavy rare earth element
HTS, see High-temperature superconductors
HTS motors, 140
    aircraft propulsion
         benefit for, 157–158
         design for, 165
    all-electric propulsion system, 158
    cryostats, 161–162
    DC motors, 159
    field winding, 157–159
    flux pump, 159
    homopolar motors, 163–165
    liquid nitrogen-cooled, 158
    reluctance motors, 158
    synchronous and induction motors, 160–161
    3-phase HTS PMSM control, 162–163
Hybrid electric propulsion, 1, 3–4
Hybrid electrolyte-based Li-air battery, 45
Hyfish, unmanned jet, 3

Ideal efficiency, 263, 264
IGBTs, see Insulated gate bipolar transistors
Impurities, 275

Induction motors, 49, 290, 291
  controlling, 22–24
  HTS, 160–161
  rotor winding types, 21
  squirrel-cage rotor, 21–22
Inflow velocity ratio, 129–130
Ingard–Myers boundary condition, 201
Initial mass, electric aircraft, 295
Insulated gate bipolar transistors (IGBTs), 22–23, 277, 279, 287–288
Integrated flight and propulsion control, 309–311
  control laws, model-based design of, 311–312
Intrinsic coercivity, 53
Intrinsic semiconductors, 275
Inverter-based direct predictive control, 290
I-V characteristics, power diode, 280, 281

Junction field effect transistor (JFET), 277

Kalman filter approach, 36–37
Kalman filtering, 216, 219
Kirchhoff's equation, 173
Klebanoff patterns, 105
Kutta–Joukowski condition, 61

Lagrange multiplier, 144, 302, 314
Lagrangian function, 303
Lamb vector, 195
Laminar boundary layer vortex shedding noise, 182
Laminar flow control, 234
Lateral autopilot control laws, 312
Lateral perturbation dynamics, 312
Lateral stability augmentation, 311, 312
Lead-acid batteries, 39–40
Lenz's law, 21, 58
Lift coefficient, 63, 66
  blade element momentum (BEM) theory, 133, 137, 139
  characteristic, 62, 119
Lift force, 61
Lifting line theory, 140–147
Lighthill's analogy, 171
Lighthill stress tensor, 175
Lilley equation, 199
Lilley's analogy, 195–202
Linear Halbach array, 54
Lithium-air (Li-air) batteries, 43–45
Lithium-ion (Li-ion) batteries, 30, 40–42
Lithium–sulfur (Li–S) batteries, 42–43
Load current, 267
Loading noise, 191
Longitudinal autopilot control laws, 312
Longitudinal motion, state equations of, 311
Longitudinal stability augmentation, 312
Lorentz force, 57, 163, 164
Lorentz's law, 13

Ludwig Prandtl's analysis, 71–72
Ludwig-Tillman correlation, 113

Magnetic field intensity, 50
Magnetic flux, 157
Magnetic flux density, 12, 58
Magnetic materials, 49–53
Magnetization, 50
Magneto-striction, 52
Mallat's algorithm, 226
Material derivative, 175
Matrix converter, 289
Maximum endurance, 296
Maximum energy product (MEP), 52
Maximum power point (MPP), 268–269
Maximum power point tracking (MPPT), 268–269
Maxwell's laws, of electromagnetism, 58
Maxwell X-57's Mod II vehicle, 5
Meissner effect, 156
Meissner state, 156
MEP, see Maximum energy product
Metal-air battery, 43–45
Metal-oxygen battery, see Metal-air battery
Micro-VGs, 210–211
Mission planning, autonomous operations, 314–315
Momentum equation, 73–74
  integral, 82–86, 88–90, 100, 112
Momentum theory, 120–121
Momentum thickness, 81, 86, 236
Monopoles, 177–179
More-electric propulsion, 1
Morphing, 207
MOSFET, 277, 286–287
Mother wavelet, 225, 226
Motion planning, 314, 315
Motor area coefficient, 12
Motor dynamics, 311
Motor speed controller, 283
MPP, see Maximum power point
MPPT, see Maximum power point tracking
Multi-junction silicon PV cells, 270–271
  power output of, 271–273

Natural laminar flow, 234
Navier–Stokes (NS) equation, 66–69, 217
  ducted propeller, 137
  non-dimensionalizing and linearizing, 70–71
Neodymium magnets, 49
Neutral-point clamped (NPC) inverter, 290
Newton's second law of motion, 67
N-factor, 216
Nickel-cadmium batteries, 29, 40
Non-aqueous electrolyte-based Li-air battery, 45
Notch type rotor, 25
NPC, see Neutral-point clamped inverter

NPN transistor, 281, 282
NS, *see* Navier–Stokes equation
n-type
   impurities, 275
   layer, 263
   semiconductor, 276

Optimal control
   formulation, 302–304
   method, 302
Optimal flight velocity, 308
Optimal real-time solution, 314
Optimal trajectory, 320
Optimization procedure, 304
Optimum cruise velocity, 304–307
Optimum path, 296
Optimum trajectory synthesis, 304–307
Oswald's efficiency factor, 64
Overall reaction, 38–43

Parallel output impedance, 272
Paramagnetism, 50–51
Park transformation, 22, 24
Passive flow control, 208, 214
Passive flow separation control, 209
Passive methods
   coupled with plasma actuators, 214–215
   of flow control, 208–209
      cavities, 213
      dimples, 210
      fences, 210, 211
      Gurney flap, 213–214
      micro-VGs, 210–211
      riblets, 209–210
      vortex generators (VGs), 210–212
      vortilons, 211–213
      winglets, 212–213
Path-following, 315
Path-following control, 320
Permanent magnet motors, high torque, 49
Permanent magnet(s), 11–12, 14, 19, 24, 26, 27
   Halbach array of, 54–57
   with reduced rare earth elements, 53
Permanent-magnet synchronous motor (PMSM),
      279, 289
   3-phase HTS PMSM control, 162–163
Perturbation stream function, 217, 218, 241, 242
Peukert effect, 294, 307–309
Peukert's law, 33
Photo conductive cells, 261–262
Photo conductivity, 259
Photoconductor, 261, 262
Photocurrent, 259, 260, 271
Photo-diode, 259–262, 265
Photoelectric effect, 259
Photon, 259, 263, 266, 270
Photovoltaic cells

maximum power point tracking (MPPT),
      268–269
multi-junction silicon PV cells, 270–271
   power output of, 271–273
photoconductive cells, 261–262
photoelectric effect, history of, 259
photovoltaic (PV) effect, 262–263
Shockley–Queisser limit, 269–270
silicon photo diodes, 259
solar cell, 263–265
   characteristics, 265–266
   power output of, 266–268
Photovoltaic (PV) effect, 262–263
Pipe tones, 180
Pitching moment, 61
Planck's law, 263
Plasma, defined, 207
Plasma actuators, 207
   control laws, active flow control, 234–236
      hybrid active laminar flow control, 244–245
      integral equations, boundary layer,
         236–238
      inverse boundary layer method, uniform
         solutions, 238–239
      typical airfoil, 245–253
      uniform and prescribed shape factor,
         239–241
      velocity profiles, 243–244
      vorticity–velocity formulation, control
         flow inputs, 241–243
   flow control, 207–208
   passive methods, flow control, 208–209
      cavities, 213
      dimples, 210
      fences, 210, 211
      Gurney flap, 213–214
      micro-VGs, 210–211
      riblets, 209–210
      vortex generators (VGs), 210–212
      vortilons, 211–212, 231
      winglets, 212–213
   passive methods coupled with, 214–215
   plasma actuation, 207–208
   skin-friction drag reduction, 215–216
      feedback control of transition, 216–219
      flow modeling, DBD plasma actuators,
         219–224
      laminar flow regulation over an airfoil,
         229–234
      simulated flow features decomposition,
         224–225
      wavelet decomposition and de-noising,
         225–226
      wavelet decomposition based on wavelet
         transform, 226–229
PMSM, *see* Permanent-magnet synchronous
   motor

p-n junction type
   silicon photo diode, 259, 260
   typical silicon photovoltaic cell, 262, 263
Pohlhausen's method, 88–89
Point of separation, 106
Point of tangency, 306
Point source model, 191
Poisson equation, 59
Position components optimal trajectory, three
     dimensions, 319
Potential barrier, 276, 280
Powell's analogy, 196
Power coefficients, 118, 140
   shaft, 130–132, 137
   thrust, 130, 131
Power diodes, 280–281
   current-voltage characteristics of, 281
   turn-off characteristics of, 281
   turn-on characteristics of, 282
Power electronic devices, 277–280
   application, 288–291
   gate turn-off thyristor (GTO), 277
   giant transistor (GTR), 286
   insulated gate bipolar transistor (IGBT),
     277, 279
   MOSFET, 277, 286–287
   power diodes, 280–281
   silicon-controlled rectifier (SCR), 277,
     281–286
   thyristor, 281–286
Prandtl's tip flow correction factor, 134
Pridmore-Brown equation, 200, 201
Primary batteries, 37–39
Primary flying control systems, 293
PRM, *see* Probabilistic road map
Probabilistic road map (PRM), 315
Propeller induced noise, 203
Propeller noise fields, theoretical modeling,
     186–190
Propellers
   actuator disk/disc, 121–125
   advance ratio, 118, 129
   airfoil section, 117, 118
   BEM theory (*see* Blade element momentum
     theory)
   blade circulation distribution, 148–151
   blade element theory, 123–126, 128–129
   blade section characteristic angles, 117, 118
   for distributed propulsion, 152
   inflow dynamics and modeling, 126–127
   lifting line theory, 140–147
   momentum theory, 120–121
   and motor application, 180–181
     airfoil and propeller noise sources,
       181–184
     Hamilton-standard procedure, propeller
       aerodynamic loading, 184–186

standard features and design considerations,
     151–152
   thrust and torque, 127–128
Propulsion motors, 5–6
Propulsive efficiency, 118, 128, 136
   vs. advance ratio, 123, 124
   vs. shaft horse power (SHP), 123, 124
Proton exchange membrane (PEM) fuel cell
     power plant, 3–4
Pseudo-piston model, 184
p-type
   impurities, 275
   layer, 263
   semiconductor, 276
Pulse width modulation (PWM), 23, 27, 288
   space-vector pulse width modulation (SV
     PWM), 162
PV, *see* Photovoltaic effect
PWM, *see* Pulse width modulation
Pylon, 191

Quadrupole, 170, 177–179
Quantum-dot phenomenon, 270
Quintic equation, 306

Radial Halbach array, 57, 58
Randles-Warburg model, 35
Range and endurance, 4
Rapidly exploring random tree (RRT), 315
Rare earth elements, 49
Rechargeable batteries, 31–32
Recombination, 269
Rectifier, 280
Reduction-oxidation reaction, 30, 32
Reluctance motors, 11
   construction types, 24–25
   reluctance/alignment torque, 25
   rotor types, 25
   switched, 26
Reluctance torque, 25
Remanence, 52–53
Remnant magnetic moment, 51
Reverse bias, 276
Reverse saturation current, 272
Reynolds number, 73, 104, 107–109, 118–119
Riblets, 209–210
Room-temperature superconductor, 155
Rossiter modes, 184
Rotational tonal noise, 186–187
Route planning, 314
RRT, *see* Rapidly exploring random tree

Samarium–cobalt (SmCo5) magnets, 49, 53
Scalable Convergent Electric Propulsion
     Technology Operations Research
     (SCEPTOR), 5, 6
Scaling function, 228

SCEPTOR, *see* Scalable Convergent Electric
        Propulsion Technology Operations
        Research
Schottky diodes, 280
SCR, *see* Silicon-controlled rectifier
Screech tones, 180
SE, *see* Systems engineering
Secondary alkaline batteries, 40
Secondary batteries, 37, 39–40
Second-generation cells, 270
Selenide cells, 266
Selenium film, 262
Semiconductors, 275–276
Shaft power efficiency, 144
Shape factor, 247
Shape parameter, 237
Shockley–Queisser limit, 269–270
Short circuit current, 266, 267
Short circuit current density, 266
Short-Time-Fourier-Transform (STFT), 225
Silicon carbide (SiC), 279
Silicon-controlled rectifier (SCR), 277, 281–286
        symbol, 282
Silicon photo diodes, 259–261
        current *vs.* voltage characteristics of, 261
        equivalent circuit of, 260
        p-n junction type, 259, 260
Silver oxide battery, 39
Simultaneous localization and mapping
        (SLAM), 314
Single-pulse thyristor-controlled rectifier, 283
Single state vector, 316
Sink speed, 297
SIT, *see* Static induction transistor
Skin-friction coefficient, 100, 237
Skin-friction drag, 215–216
        feedback control of transition, 216–219
        flow modeling, DBD plasma actuators,
            219–224
        laminar flow regulation over an airfoil,
            229–234
        simulated flow features decomposition,
            224–225
        wavelet decomposition and de-noising,
            225–226
        wavelet decomposition based on wavelet
            transform, 226–229
Skin-friction-related stress, 87
SLAM, *see* Simultaneous localization and
        mapping
Sliding mode control, 269
Slip speed, 22
SOC, *see* State of charge
Solar cell, 263–265
        characteristics, 265–266
        power output of, 266–268

Solar Impulse aircraft, 4
Sound pressure level, 173, 177
Space-vector pulse width modulation
        (SV PWM), 162
Spanwise fences, 210
Spatial growth rates, 229, 235, 250
Specific energy, 294
Specific excess power, 299, 300
Specific excess thrust, 300
Spectrum losses, 269
Squirrel-cage rotor, 21–22
Stacked cells, 271
Stall torque, 13
State of charge (SOC), 34
        estimation
            Coulombic measurements, 35–36
            electrolyte specific gravity
                measurement, 35
            fully loaded battery voltage method, 35
            open circuit battery voltage method, 36
            stabilized float current measurement, 35
        evolution, 37, 38
        Kalman filter approach, 36–37
State of health, 31
Static induction transistor (SIT), 277
Stepper motors, 11, 26
STFT, *see* Short-Time-Fourier-Transform
Stokes derivative, 67
Substantial derivative, 197
Substantive derivative, 67
Subsystems, 2–3
Suction coefficient, 237
Superconducting materials, 155–156
        features, 156–157
Superconductivity, 155–156
SV PWM, *see* Space-vector pulse width
        modulation
Switched reluctance motor, 26
Synchronous motors, 18–19
        driving control, matrix converter, 289
        HTS, 160–161
        loading and back-EMF, 20
        synchronous speed, 19
        three-phase motors, 19–20
        torque-speed characteristic, 20
        types, 19
Systems engineering (SE), 2–3
Systems-thinking, 2

Thickness noise, 190
Thin airfoil theory, 141
Third-generation cells, 270
Three-phase, fully controllable full-wave rectifier
        circuit, 284
Thrust coefficient, 118
        of ducted propeller, 138, 140

Thwaites correlation technique, 102–104
Thyristor, 281–286
Thyristor-controlled rectifier, 283
Tip vortex formation noise, 182
Tollmien-Schlichting (T-S) waves, 104, 216
Total derivative, 67
TP-BVP, *see* Two-point boundary value problem
Trailing-edge bluntness vortex shedding
        noise, 182
Trajectory synthesis, 304, 315
Transfer function, 17
    closed loop system, 18
Transistors, 276–278
Transitional boundary layer, 108–109
Transport by diffusion, 275
Transport by drift, 275
Trapped field superconductors, 157
Travelling salesman problem (TSP), 314
T-S, *see* Tollmien-Schlichting waves
TSP, *see* Travelling salesman problem
Turbo-generator motor drive, 158
Turbulence stress tensor, 170
Turbulent boundary layer, 209, 216
    Drela's method, 113–114
    entrainment equation, 111–113
    kinematic momentum and displacement
        thicknesses, 110
    prediction, 111
    Reynolds number, 109, 110
    shape factor, 110
    turbulent fluid flows, 110
Turbulent boundary layer trailing edge
        noise, 181
Turbulent freestream flow noise, 182
Turn-off characteristics, power diode, 280, 281
Two-point boundary value problem (TP-BVP),
        314, 318
Type II superconductors, 156
Type I superconductors, 156

UAV, *see* Unmanned aerial vehicle
Universal motors, 14
Unmanned aerial vehicle (UAV), 1–2

Valence electrons, 262
Velocity components optimal trajectory, three
        dimensions, 319
Vertical velocity, 298
VGs, *see* Vortex generators
Voltage-current characteristics, of IGBT,
        287, 288
Voltage-source inverter (VSI), 22–23
    three-phase VSI, 162, 163
Voltaic pile, 29
Von Karman momentum integral equation,
        82–86, 88–90, 112
Vortex generators (VGs), 210–212
Vortex state, 156
Vorticity transport equations, 68–69
Vortilons, 211–213
VSI, *see* Voltage-source inverter

Wall-jet, 221
Walz–Thwaites' criterion, transition/separation,
        106–108
Wavelet decomposition, 216, 226, 229
    algorithm, 228
Wavelet packet, 227
    decomposition, 229
Wavelet transforms (WTs), 226
Weibull fit model, 34
Wentzel–Kramers–Brillouin (WKB)
        method, 202
Wing fences, 210
Winglets, 212–213
Wingtip-mounted cruise propellers, 5
WKB, WKB Wentzel–Kramers–Brillouin
        method
WTs, *see* Wavelet transforms

Yaw angle time history, 321
Yttrium barium copper oxide (YBCO), 157, 158

Zener diode, 276
Zener voltage, 276
Zinc-air battery, 39
Zinc-carbon battery, 37–38

Printed in the United States
by Baker & Taylor Publisher Services

Printed in the United States
by Baker & Taylor Publisher Services